# Intelligent Control and Learning Systems

Volume 2

**Series Editor**

Dong Shen, School of Mathematics, Renmin University of China, Beijing, Beijing, China

The Springer book series Intelligent Control and Learning Systems addresses the emerging advances in intelligent control and learning systems from both mathematical theory and engineering application perspectives. It is a series of monographs and contributed volumes focusing on the in-depth exploration of learning theory in control such as iterative learning, machine learning, deep learning, and others sharing the learning concept, and their corresponding intelligent system frameworks in engineering applications. This series is featured by the comprehensive understanding and practical application of learning mechanisms. This book series involves applications in industrial engineering, control engineering, and material engineering, etc.

The Intelligent Control and Learning System book series promotes the exchange of emerging theory and technology of intelligent control and learning systems between academia and industry. It aims to provide a timely reflection of the advances in intelligent control and learning systems. This book series is distinguished by the combination of the system theory and emerging topics such as machine learning, artificial intelligence, and big data. As a collection, this book series provides valuable resources to a wide audience in academia, the engineering research community, industry and anyone else looking to expand their knowledge in intelligent control and learning systems.

Ronghu Chi · Yu Hui · Zhongsheng Hou

# Data-Driven Iterative Learning Control for Discrete-Time Systems

Springer

Ronghu Chi
College of Automation and Electronic Engineering
Qingdao University of Science and Technology
Qingdao, Shandong, China

Yu Hui
College of Automation and Electronic Engineering
Qingdao University of Science and Technology
Qingdao, Shandong, China

Zhongsheng Hou
School of Automation
Qingdao University
Qingdao, Shandong, China

ISSN 2662-5458 ISSN 2662-5466 (electronic)
Intelligent Control and Learning Systems
ISBN 978-981-19-5949-3 ISBN 978-981-19-5950-9 (eBook)
https://doi.org/10.1007/978-981-19-5950-9



This Springer imprint is published by the registered company Springer Nature Singapore Pte Ltd.
The registered company address is: 152 Beach Road, #21-01/04 Gateway East, Singapore 189721, Singapore

# Preface

This monograph studies the design and analysis of data-driven iterative learning control. “Data-driven” means that the controller is designed and analyzed by using online/offline I/O data of the plants merely, instead of depending on the first-principle model information of the plants. “Iterative” indicates that the dynamic process is repeatable over a finite time interval. “Learning” in control means to change the controller itself to improve control performance through a prior understanding of the controlled plant and external environment attained through its own learning ability. This monograph mainly presents a series of “Data-Driven Iterative Learning Control” (DDILC) methods to overcome the disadvantages of the traditional iterative learning control methods.

The main distinct merits of the DDILC presented in this monograph are summarized as: (i) their learning gains can be iteratively tunable by an adaptive updating law using real time I/O data, thus improving the robustness against nonrepetitive uncertainties; (ii) they have an iteratively monotonic convergence that is most expected in practical applications to achieve a good transient response; (iii) their stability, convergence, and robustness can be guaranteed theoretically with rigorous analysis by using the contraction mapping principle and iterative dynamic linearization as the basic mathematical tools. So, this monograph is a significant breakthrough for all the traditional PID-type ILC whose fixed learning gains make it weak against uncertainties, the adaptive ILC that is limited to linearly parametric uncertain systems, and the optimization-based ILC that requires the exact linear model of the systems.

As the basic mathematic tools, four different Iterative Dynamic Linearization (IDL) methods are presented to build a linear I/O dynamic relationship of the controlled system between two consecutive iterations at all finite time instants. Then, this monograph proposes four DDILC schemes consisting of both an optimal iterative learning law and an iterative adaptive parameter updating law based on the linear data models derived from the IDL methods. Furthermore, this monograph explores the critical issues of incomplete data information, specified point tracking, higher order law, system constraint, nonrepetitive uncertainty, and event-triggered strategy to facilitate real applications.

From this monograph, the readers will learn the new concepts of Iterative Pseudo Partial Derivative (IPPD), Iterative Pseudo Gradient (IPG), the new generalized Lipschitz condition, the innovative IDL methods, and the novel analysis method using contraction mapping principle and fixed-point theorem, all of which constitute the theoretical basis of DDILC. Additionally, the extensions of the DDILC to the key challenges mentioned above further enhance its wide applications in practice. As a result, the readers can generate breakthrough ideas and insights from the creative DDILC methods to deal with complex nonlinear systems in real industries.

The monograph, which has 12 chapters, is organized as follows. Chapter 1 is an introduction to the general topics and problems treated in this book. Chapter 2 presents a compact form iterative dynamic linearization-based DDILC (CFIDL-DDILC) for a class of nonlinear repetitive systems. In the sequence, Chaps. 3 and 4 provide the significant extensions of Chap. 2 to the partial form iterative dynamic linearization-based DDILC (PFIDL-DDILC), full form iterative dynamic linearization-based DDILC (FFIDL-DDILC), and state transition-based iterative dynamic linearization-based DDILC (STIDL-DDILC), respectively. All these IDL methods build a linear I/O dynamic relationship of the nonlinear systems between two iterations which is called an iterative Linear Data Model (iLDM) in this monograph and becomes the basis of the subsequent controller design and analysis. In Chaps. 5–7, the DDILC design and analysis are re-explored by considering the incomplete information of package dropouts, varying trial lengths, and data-quantization, respectively. Chapter 8 discusses the synthetic design of DDILC for specified point tracking tasks. Chapter 9 develops a higher order DDILC to enhance the control performance. Chapter 10 takes the I/O constraints into account for the DDILC design and analysis. Chapter 11 presents an ESO-based DDILC method to deal with nonrepetitive uncertainties. Finally, Chap. 12 introduces an iterative event-triggered mechanism into the DDILC method.

The research reported in this monograph has benefited greatly from collaborations with a number of Ph.D. students and input from many friends and collaborators worldwide and it would simply be impossible to ensure that we included them all by name here. We are, however, grateful to our co-authors and colleagues, Biao Huang, University of Alberta, Shangtai Jin, Beijing Jiaotong University, Chiang-Ju Chien, Huafan University, Xuhui Bu, Henan Polytechnic University, Na Lin, Ruikun Zhang, and Huimin Zhang, Qingdao University of Science and Technology, for their contributions to this monograph. Meantime, the first author would like to express the greatest conceivable gratitude that is owed to his wife and son, Puzhao Li and Yongqing Chi, for their sacrifice, patience, and understanding throughout his study and research. Without their support and love, he could not have completed this monograph. Financial support from the National Science Foundation of China under Grant 61873139 and the Taishan Scholar Program of Shandong Province of China is gratefully acknowledged. Finally, we must thank the able team at Springer-Verlag for the great work that contributed to the successful completion.

Qingdao, China<br>June 2022

Ronghu Chi

# Contents

# Chapter 1
# Introduction

## 1.1 Learning in Control

Artificial intelligence (AI) has experienced a great resurgence as an essential part of the advanced technology in the industry. In computer science, AI means the learning of intelligent agents. Learning involves psychology, physiology, biology, brain science, and all life-related disciplines (Xu and Tan, 2003). In general, learning refers to the action of a system to adapt and change its behavior based on input/output (I/O) observations (Moore, 2012).

Learning in the control can be classified as single objective learning, pattern-based multiple objective learning, and quantitative biological learning (Xu and Hou, 2005). Learning control means to change itself to improve control performance through a prior understanding of the controlled plant and external environment attained through its own learning ability and it is the basis of intelligent autonomous control. Different from machine learning, deep learning, neural network, expert system, and other methods commonly used in data mining processing such as data clustering analysis and feature extraction, the learning algorithm used in process control in the field of automation has clear requirements for real-time, certainty, security, and so on, to make the precise adjustment and tracking of the control process be realized (Wang and Chen, 2020).

Many control systems have this learning ability to respond to changes in its environment, including feedback control systems, adaptive control systems, and any type of artificial neural network equipped with a weight update algorithm (Moore, 2012). For example, adaptive control can use the output errors as feedback to adjust the input while the parametric uncertainties can be learned according to the input and output data. However, in a broader sense, one can regard the control methods based on the error-feedback principle as a kind of passive learning strategy by nature where only when the output errors have been produced, can the control action be tuned accordingly (Chi et al., 2019). Therefore, the feedback-based control systems are always accompanied with a delay in response to the control performance.

 
R. Chi et al., *Data-Driven Iterative Learning Control for Discrete-Time Systems*, Intelligent Control and Learning Systems 2,
https://doi.org/10.1007/978-981-19-5950-9_1

## 1.2 Iterative Learning Control

In this monograph, the primary goal is centered on iterative learning control (ILC) (Arimoto et al., 1984). The term "iterative" indicates a kind of action that requires the dynamic process to be repeatable, i.e., the dynamic system is deterministic and the tracking task is repeatable over a finite tracking interval (Xu and Tan, 2003). This kind of control problems are frequently encountered in many industrial processes, such as wafer manufacturing process, batch reactor process, various assembly lines, etc.

The motivation for iterative learning control comes from a deeper recognition that knowledge can be learned from experience. In other words, when a control task is performed repeatedly, we gain extra information from a new source: past input and tracking error profiles, which can be viewed as a kind of "experience". The "experience" serves as a new source of knowledge related to the dynamic process model, and accordingly reduces the need for the process model information. The new knowledge learned from "experience" provides the possibility of improving the tracking control performance.

Iterative learning control (Arimoto et al., 1984) was motivated by observing the behavior of systems that operate repetitively. It learns from the past experiences by straightforwardly using the control inputs and tracking errors in the previous iterations and thus can improve control performance gradually with repeated operations. In other words, ILC is an approach to improve the transient response performance of systems that operate repetitively, and finally realizes that the output of the system completely tracks the desired trajectory over the entire time interval. From the perspectives of AI, ILC is a supervised learning method in the machine learning category (Wang and Chen 2020; Chi et al. 2019). In addition, the purpose of ILC is to directly learn the desired control input on the basis of the previous input action through an iterative analytical learning law which enhances its learning capacity.

In general, ILC uses an active feedforward learning mechanism to learn from the experience of the previous operations. The idea is straightforward: use the control information of the previous trials to improve the control performance of the present trial. A typical ILC is realized through memory-based learning and its schematic diagram is shown in Fig. 1.1 (Xu and Tan, 2003), where $k$ denotes the $k$th trial, $y_k$ denotes the system output in the $k$th trial, $u_k$ denotes the control input in the $k$th trial, and $y_d$ denotes the desired output of the system.

Figure 1.1 illustrates the basic idea of a typical ILC. Each time the system operates, its input and output signals ($u_k(t)$ and $y_k(t)$, respectively) are stored in memory. The learning control algorithm then evaluates the performance error $e_k(t) = y_d(t) - y_k(t)$. Based on the error signal, the learning controller then computes a new input signal $u_{k+1}(t)$, which is stored for use on the next trial. The next input command is chosen in such a way as to guarantee that the performance error will be reduced on the next trial. One can find that all of the control inputs of the current iteration can be determined before the start of this iteration. So, the typical ILC adopts a pure

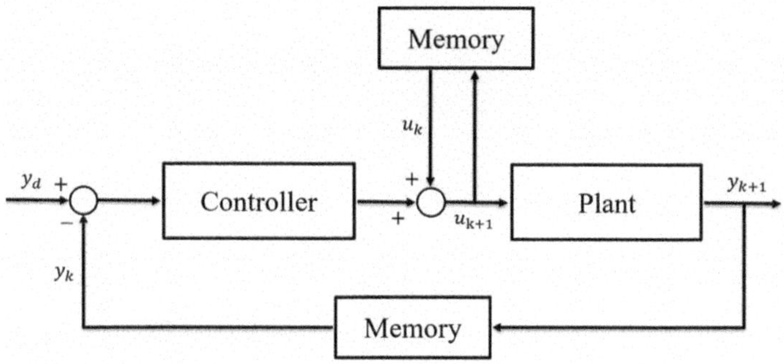


**Fig. 1.1** Schematic of feedforward ILC

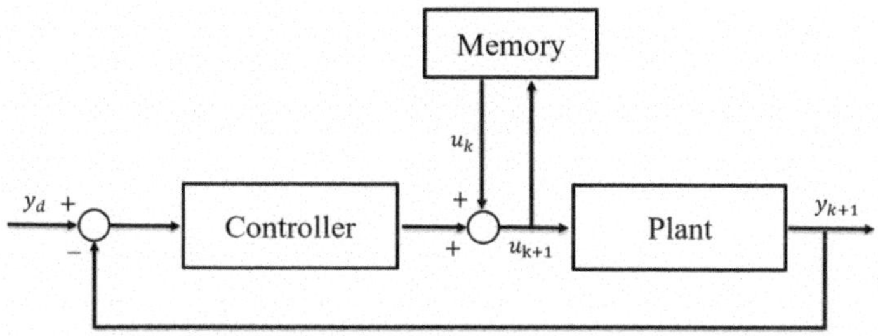


**Fig. 1.2** Schematic of feedback ILC

feedforward mechanism by removing the feedback signals in the time domain from the current control loop.

As an alternative to the feedforward ILC scheme, a feedback ILC strategy is also used frequently, as shown in Fig. 1.2. One can see that, in addition to the standard feedback loop, a set of memory components is used to record the control signals of the preceding trial, $u_k(t)$, which is incorporated into the present control, $u_{k+1}(t)$, in a pointwise manner. The feedback ILC only needs to store the control input information of previous batches, which reduces the amount of storage. Different from the feedforward ILC, this control strategy requires online calculations.

## 1.3 PID-type ILC

Arimoto and his collaborators first proposed D-type, PD-type, and PID-type ILC algorithms, introduced λ-norm, analyzed, and obtained sufficient conditions to ensure the convergence of the ILC algorithm, where the contraction mapping principle has become one of the basic methods of convergence analysis.

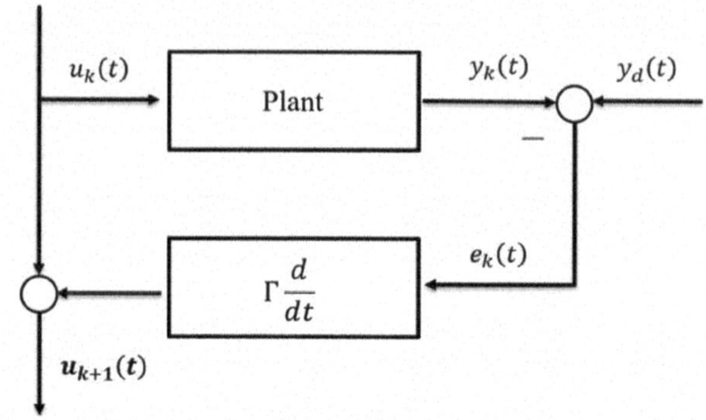


**Fig. 1.3** D-type ILC

The characteristic of the D-type ILC is that its correction term only uses the derivative signal of the output error, i.e., $\dot{e}_k(t)$. The simplest D-type iterative learning law is as follows:

$$u_{k+1}(t) = u_k(t) + \Gamma\dot{e}_k(t) \tag{1.1}$$

where $\Gamma$ is the gain matrix. The schematic of the D-type ILC is shown in Fig. 1.3.

Assume that the plant is a linear time-invariant (LTI) system, with signals defined over the interval $\left(0, t_f\right)$, and with a state-space description $(A, B, C, D)$. Also, assume that $CB$ has full rank. Arimoto showed that if the induced operator norm $\|I - CB\Gamma\|_i$ satisfies

$$\|I - CB\Gamma\|_i < 1 \tag{1.2}$$

and some initial condition requirements are met, then

$$\lim_{k\to\infty} y_k(t) = y_d(t). \tag{1.3}$$

Convergence of the error is in the sense of the λ norm, which is defined as

$$\|x(t)\|_\lambda = \sup_{0\le t\le t_f}\left\{e^{-\lambda t}\max_{1\le l\le r}\left|x_{l(t)}\right|\right\} \tag{1.4}$$

where $x(t)$ is an $r$-dimensional vector. This norm ensures pointwise convergence over the interval of interest. Arimoto et al. also gave convergence conditions for this type of learning control algorithm when it was applied to time-varying plants and certain nonlinear models encountered in robotics. In the sequence, Arimoto and his co-workers have proposed a variety of different learning control algorithms. The primary difference between the various approaches is in how the error is utilized in the learning control algorithm. The most general linear algorithm presented is expressed as follows:

$$u_{k+1}(t) = u_k(t) + \underbrace{L(t)e_k(t)}_{P} + \underbrace{\Psi(t)\int_0^t e_k(\tau)d\tau}_{I} + \underbrace{\Gamma(t)\dot{e}_k(t)}_{D}. \tag{1.5}$$

This algorithm essentially forms a PID-like system for processing the error while maintaining a linear effect on the past input signal. It is open-loop in the time domain, so there is no warranty of the stability of the dynamic system along the time axis. As far as the operation interval is finite, and the dynamics is either linear or global Lipschitz continuous, finite escape time phenomenon does not occur, hence the system state is always bounded. One can find that the traditional PID-type ILC is simple in the structure, the amount of calculation is small, and thus it is easy to implement. Further, it is worth pointing out that the PID-type ILC is almost a model-free method since neither model knowledge is required nor system state is included in the algorithm.

In more recent works, many scholars have considered how to use the knowledge gained from learning to improve the convergence rate and error performance, as well as to deal with iteration-related uncertainties. In Lee et al. (1994), Lee et al. (1996), and Liu and Gao (2010), the authors proposed a feedback-based learning control strategy and considered filtering techniques to enhance robustness against modeling errors and random disturbances. Based on the 2D model of the batch process, Liu and Gao (2010) proposed a robust closed-loop ILC method, which established sufficient conditions to ensure the robust stability of the closed-loop ILC system under the constraints of linear matrix inequalities. In Wang et al. (2012), Wang et al. (2010), and Liu et al. (2014), the authors proposed a PID control strategy based on set-point iterative learning, and obtained sufficient conditions for robust asymptotic convergence. In Park (1999), the authors studied the robustness of a PID-type ILC to deal with the initial state error. Márquez-Vera et al. (2014) proposed a PD-type ILC for bioreactors, and adopted a fuzzy method to calculate the control gain.

## 1.4 Norm Optimization-Based ILC

A key problem in the traditional PID-type ILC is how to select an appropriate learning gain in practical applications if little model information is available. Considering learning control for discrete-time linear systems, Togai and Yamano (1985) proposed the learning control algorithm

$$u_{k+1}(t) = u_k(t) + Ge_k(t+1) \tag{1.6}$$

where $t$ is an integer variable denoting discrete-time instant. For this update algorithm, the gain $G$ is optimized using gradient methods to minimize the quadratic cost of the error

$$J = \frac{1}{2} e_k^T(t+1) Q e_k(t+1) \tag{1.7}$$

between successive trials. The authors consider several techniques for choosing $G$, specifically using the steepest-descent, Gauss-Newton, and Newton-Raphson methods to find the optimal learning control law.

In Amann et al. (1996), Amann et al. re-describe the cost function in super vector, so that the optimal technology can be easily introduced to ILC tasks, and the norm optimization-based ILC (NOILC) method is proposed. Specifically, consider the following linear discrete-time system operating repeatedly on a finite time interval

$$\begin{cases} \boldsymbol{x}(t+1) = \boldsymbol{A}\boldsymbol{x}(t) + \boldsymbol{B}\boldsymbol{u}(t) & \boldsymbol{x}(0) = \boldsymbol{x}_0 \quad 0 \le t \le N \\ \boldsymbol{y}(t) = \boldsymbol{C}\boldsymbol{x}(t) & \boldsymbol{x} \in R^n \quad \boldsymbol{u} \in R^m \quad \boldsymbol{y} \in R^p. \end{cases} \tag{1.8}$$

For simplicity, the state-space matrices $\boldsymbol{A}$, $\boldsymbol{B}$, and $\boldsymbol{C}$ are assumed to be time-invariant. By defining the following two supervectors $\boldsymbol{y} = \left[\, \boldsymbol{y}^T(1)\ \boldsymbol{y}^T(2) \cdots \boldsymbol{y}^T(N) \,\right]$ and $\boldsymbol{u} = \left[\, \boldsymbol{u}^T(0)\ \boldsymbol{u}^T(1) \cdots \boldsymbol{u}^T(N-1) \,\right]^T$, the state-space Eq. (1.8) can be written as follows:

$$\boldsymbol{y} = \boldsymbol{y}_0 + \boldsymbol{G}\boldsymbol{u} \tag{1.9}$$

where

$$\boldsymbol{G} = \begin{bmatrix} \boldsymbol{CB} & & & \\ \boldsymbol{CAB} & \boldsymbol{CB} & & \\ \vdots & \vdots & \ddots & \\ \boldsymbol{CA}^{N-1}\boldsymbol{B} & \boldsymbol{CA}^{N-2}\boldsymbol{B} & \cdots & \boldsymbol{CB} \end{bmatrix}$$

is a block lower triangular matrix.

$$\boldsymbol{y}_0 = \left[\, (\boldsymbol{CA})^T\ \left(\boldsymbol{CA}^2\right)^T \ldots \left(\boldsymbol{CA}^N\right)^T \,\right]^T \boldsymbol{x}_0$$

is the initial condition response vector. It is usually assumed as $\boldsymbol{y}_0 = 0$ or $\boldsymbol{x}_0 = 0$.

On this basis, Amann, Owens, and Rogers propose the following objective function:

$$\min_{\boldsymbol{u}_k} \mid J_k = \left\{ \|\boldsymbol{e}_k\|_Q^2 + \|\boldsymbol{u}_k - \boldsymbol{u}_{k-1}\|_R^2 \right\}. \tag{1.10}$$

In (1.10), a penalty term for input changes is added to make the algorithm have an integral behavior along the batch axis. Without other constraints and certain conditions, Amann et al. derive a non-causal input update law

$$\boldsymbol{u}_k = \boldsymbol{u}_{k-1} + \boldsymbol{R}^{-1}\boldsymbol{G}^T\boldsymbol{Q}\boldsymbol{e}_k. \tag{1.11}$$

In 1996, Lee et al. (1996) rewrite the non-causal control law (1.11) into the following causal form according to the idea of solving the quadratic optimal tracking problem in finite time interval

$$\boldsymbol{u}_k = \boldsymbol{u}_{k-1} + \left(\boldsymbol{G}^T \boldsymbol{Q}\boldsymbol{G} + \boldsymbol{R}\right)^{-1} \boldsymbol{G}^T \boldsymbol{Q}\boldsymbol{e}_{k-1}. \tag{1.12}$$

Moreover, Barton and Alleyne (2010) study the application of the above NOILC method to multi-axis robots such that the tracking performance and robustness of the motion trajectory can be improved. Van de Wijdeven et al. (2009) propose a robust monotonic convergence analysis method for the linear quadratic NOILC in a finite time. Chu and Owens (2009) propose a successive projection method to improve the tracking performance and convergence speed of the NOILC algorithm for non-minimum phase systems. Lee and Lee (1998) show that the above quadratic ILC can be realized as output feedback to enhance the robustness of the system. According to this idea, Lee and Lee (1998) and Lee et al. (1994) explore the combination of model predictive control (MPC) and ILC, which has become one of the main directions in batch process research and attracted many attentions of scholars (Chen et al. 2014; Mishra et al. 2010; Amann et al. 1998; Lee et al. 1999; Wang et al. 2013; Jin et al. 2013; Chin et al. 2004; Oh and Lee 2016).

When the disturbances and noises of a batch process become large, an observer-based NOILC is proposed in Lee et al. (2000),

$$\boldsymbol{u}_{k+1} = \boldsymbol{u}_k + \boldsymbol{H}^Q \hat{\boldsymbol{e}}_{k|k} \tag{1.13}$$

where $\hat{e}_{k|k}$ is the estimation of the repetitive error existing in the measured error $\boldsymbol{e}_k$, and it is obtained by using the following observer

$$\hat{\boldsymbol{e}}_{k|k-1} = \hat{\boldsymbol{e}}_{k-1|k-1} - \boldsymbol{G}\Delta\boldsymbol{u}_k \tag{1.14}$$

$$\hat{\boldsymbol{e}}_{k|k} = \hat{\boldsymbol{e}}_{k|k-1} + \boldsymbol{K}\left(\boldsymbol{e}_k - \hat{\boldsymbol{e}}_{k|k-1}\right). \tag{1.15}$$

As stated in Lee et al. (2000), under such a research framework of NOILC, various practical problems, such as system constraints, disturbances, measurement noises, and unmodeled dynamics, can be easily considered as optimal problems. In many industrial processes, it is always necessary to impose some constraints on process variables to ensure the safety and smooth operation. Generally, the constraints on input and output are common, including constraints on the bounds of input and output values, and constraints on the change rates of the input and output. These constraints can be easily integrated into the NOILC design and analysis in a rigorous and systematic manner. In addition, NOILC has strict monotonic convergence along the iterative direction, which is one of the most desirable properties in practical engineering applications. Therefore, NOILC has been widely used in repeated operation systems, and its theoretical research has experienced a development process from linear to nonlinear, from model-based to data-driven.

In Cao et al. (2016), a discrete-time iterative learning Kalman filtering method is proposed by combining a filter to estimate the system state. In Moore and Verwoerd (2008), an $L_1$ optimal robust ILC is proposed to deal with bounded disturbances of the system. Xu and Tan (2002) took time optimization and robustness as their comprehensive goals, and proposed a new robust ILC optimal design and convergence analysis method with theoretical analysis and evaluation of the convergence rate, and the system dynamics effect on the learning performance is also studied wherein. A hierarchical ILC method is proposed in Sanzida and Nagy (2013) based on a linear time-varying perturbation model for the crystal cooling process. When the control variables are unmeasurable, Axelsson et al. (2014) propose an OILC method based on recursive Bayesian estimation, and extends it to the nonlinear discrete-time system based on linearization. Aiming at the problem of system constraints, a linear matrix inequality method (Nguyen and Banjerdpongchai, 2009) is proposed for robust NOILC based on quadratic performance indicators. In Lim and Barton (2014), a Pareto ILC method is proposed by optimizing multiple performance objectives. In Chu and Owens (2009), two acceleration algorithms of NOILC are proposed with convergence analysis. A multi-objective design method is proposed in Tousain et al. (2001) to attain the control gain of NOILC based on internal model principles. Liu and Wu (2002) propose an improved iterative learning algorithm based on optimization indicators to ensure that the control signal converges to the linear quadratic optimal control solution of the system. In Oh and Lee (2016), an iterative learning model predictive control method is proposed for batch processes with multivariable constraints, which comprehensively considers the constraints, time-varying problems, external disturbances, and random characteristics of the system. In Liu and Wang (2012), a constrained ILC method is proposed on the basis of the two-dimensional piecewise linearization description of batch processes.

## 1.5 Data-Driven Design and Analysis of ILC

It is worth pointing out that ILC (Arimoto et al., 1984) is originally proposed for nonlinear uncertain systems directly using I/O data for the controller design without requiring the exact knowledge of the system model and thus is classified as data-driven control (Hou and Jin, 2013). However, the above NOILC methods can only be categorized into "model-based control" because the knowledge of an accurate linear model of the system is required. When the model is inaccurate, the monotonic convergence of NOILC will be no longer guaranteed, and learning transients with large, rapid growth of the error or even instability may occur.

Practically, it is difficult to gain an explicit process model, especially in large scale and complex industrial processes such as oil refinery plants, traffic and communication networks, power grids, aeronautics and astronautics, etc. Even if a mathematical model may be obtained by first principles or identification techniques with a significant amount of effort, some other theoretical and practical difficulties also exist. For example: (i) the unmodeled dynamics and the poor robustness are inevitable problems; (ii) the structure of the plant is often difficult to determine; (iii) the more accurate the model is, the more complex the control law might be, which may lead to poorer robustness and lower reliability of the controlled system, and difficulties would be brought into practical implementation and application of the designed control system. Thus, model-based control strategies may not be always able to produce a satisfactory performance. It is desirable to have a control method that is less dependent on an explicit model. This motivates us to study a data-driven or data-based control method, which means that the controller's design merely uses the input and output measurement data and the controller itself does not contain any explicit model information about the controlled plant (Hou and Jin 2013; Hou 2014; Hou and Wang 2013).

Therefore, data-driven ILC strategies have attracted much attentions. In Janssens et al. (2011) and Janssens et al. (2012), the authors propose a data-driven ILC for linear time-invariant systems with the help of finite impulse response filtering technology where the impulse response model of the system can be estimated from the input and output data obtained in previous iterations, however, the control performance of the system mainly depends on the estimation accuracy of the impulse response model. Rădac et al. (2012), Radac and Precup (2016) propose a model-free data-driven iterative reference input tuning method for a linear time-invariant system based on experimental random search algorithms. In Volckaert et al. (2013), a general NOILC is proposed for constrained nonlinear systems by solving a nonlinear programming problem. Based on the fast fourier transform (FFT), a model-free ILC algorithm (Zhou et al., 2012) is proposed for an unknown or time-varying secondary path signal noise active suppression system. In Radac and Precup (2015), a data-driven ILC is proposed using the virtual reference feedback tuning method.

No matter what kind of learning control, model-based or data-based, the key point in the control system design is to establish a linear relationship between the system input and output. On the other hand, however, industrial processes are becoming more complex and larger in scale. Some so-called "hard nonlinear characteristics" including friction nonlinearity, saturation nonlinearity, dead zone nonlinearity, gap nonlinearity, time-delay nonlinearity, etc., exist in a large number of actual control systems and often cause undesired behavior of the control system (Slotine and Li, 1991). In dealing with nonlinear control systems, linearization of nonlinear model is one of the most popular methods such that the powerful mathematical tools and existing skills in linear systems can be applied to nonlinear systems. However, most traditional linearization methods, such as Taylor's linearization (Chen and Narendra, 2004), piecewise linearization (Xi et al., 1996), feedback linearization (Deng et al., 2008), input-output linearization (Kravaris and Chung, 1987), orthogonal function-approximation-based or polynomial functions approximation linearization (Dumont and Fu, 1993), etc., only provide an approximation to the original nonlinear systems, and some of them require an exactly known mathematical model of the system, or they have too many parameters. Therefore, the corresponding linearized models and control strategies may be questionable when applied in practice due to the model mismatch and complexity.

More recently, an iterative dynamic linearization (IDL) method has been developed for repetitive nonlinear discrete-time systems (Chi and Hou 2007; Chi et al. 2012, 2015a), which builds a linear I/O relationship between two consecutive iterations at all the time instants over a finite running interval. The important merits of this IDL method are that: (a) it is a data-driven method proposed directly for an unknown nonlinear system, where no model information of the original nonlinear plant is needed except for the I/O data; (b) the resulting linearization form, or called linear data model (LDM) in the literature, has a simple structure with few parameters. Therefore, the IDL method (Chi and Hou 2007; Chi et al. 2012, 2015a) is controller-oriented, and can directly draw on the tools and methods of linear systems for the controller design and analysis. Now, the IDL method has become one of the most important tools for data-driven ILC design and analysis. It provides a new method of studying closed-loop and open-loop data of the system. Moreover, it can be considered as a modeling of the data relationship in the open and closed loops of the system, which is essential for the control theory of complex systems with rich information.

In Chi and Hou (2007), Chi et al. (2012), and Chi et al. (2015a), by virtue of the IDL-based linear data models, several data-drive ILC (DDILC) methods have been proposed for the tasks of entire trajectory tracking, point-to-point tracking (multiple intermittent points tracking), and terminal point tracking. In Chi et al. (2015b), a terminal DDILC is proposed for nonrepetitive reference points, but the identical initial condition is required for the convergence analysis. Considering that the identical initial conditions are difficult to meet in practice due to positioning accuracy, measurement accuracy, state disturbances, etc., Refs. Liu et al. (2015b) and Chi et al. (2017a), respectively, use neural networks and high-order internal models to approximate the influence of the randomly varying initial conditions on the output of the system, and such approximations are further used in the DDILC design as compensations. In Chi et al. (2016), the iteration-varying initial states are considered as an unknown parameter which is then estimated in the DDILC design and analysis. For a terminal tracking task where the system state and output are unmeasurable at points other than the terminal point, a new DDILC method (Chi et al., 2015c) is proposed to use additional control information to improve control performance. In Chi et al. (2022), an indirect DDILC scheme is proposed for both linear and nonlinear systems to enhance the P-type controller by learning from set points. To effectively utilize the limited network channel to improve the quality and rate of signal transmission, an event-triggered DDILC is proposed in Lin et al. (2020). A high-order DDILC has been proposed in Chi et al. (2015c) and Chi et al. (2015d) to improve control performance. Also, the problem of computational complexity and I/O constraints have been discussed in Chi et al. (2018) and Chi et al. (2017b). The DDILC methods have been verified for their effectiveness by experiments and simulations with practical application models such as chemical batch processes (Chi et al. 2012, 2015a, b; Liu et al. 2015a, b; Chi et al. 2015c, d, e, 2016, 2017a, b, 2018), freeway traffic systems (Chi and Hou 2007; Chi et al. 2008; Jin et al. 2012), car suspension system (Chi et al., 2022), automatic train stop system (Jin et al., 2015), and permanent linear motor (Hui et al., 2019), etc. In view of the increasingly complex and larger-scale nonlinear practical systems, DDILC will become the main method of repetitive industrial control systems with a wide range of theoretical research prospects and potential applications.

## 1.6 Structure of This Monograph

This monograph belongs to the subject of control and systems theory. It studies a novel data-driven framework for the design and analysis of iterative learning control for nonlinear nonaffine discrete-time systems. A series of IDL methods are discussed first to build a linear data mapping with respect to the system's output and input between two consecutive iterations. On this basis, this monograph presents a series of data-driven ILC (DDILC) approaches with rigorous analysis. After that, this monograph also conducts significant extensions to the challenge problems of data dropout, varying trial lengths, data quantization, specified points tracking, higher order learning algorithms, input and output constraints, nonrepetitive uncertainties, event-triggered mechanisms, etc. The readers can learn recent progress on DDILC for complex systems in practical applications.

The chapters of this monograph are organized as follows. Chapter 1 is the introduction of this monograph. Four DDILC methods with different IDL techniques are presented in Chaps. 2, 3, and 4, respectively. Chapters 5, 6, and 7 make a preliminary discussion on the problems of incomplete information in the DDILC design and analysis, including package dropouts, varying trial lengths, and data quantization. Chapter 8 explores the DDILC design and analysis for specified points tracking tasks. In Chap. 9, the higher order learning algorithms are proposed for the DDILC with rigorous analysis. The problem of I/O constraints is discussed in Chap. 10 for the DDILC design and analysis. Chapter 11 presents an ESO-based DDILC method to deal with nonrepetitive uncertainties. The event-triggered mechanism is introduced in the DDILC design in Chap. 12.

## References

Amann N, Owens DH, Rogers E (1996) Iterative learning control for discrete-time systems with exponential rate of convergence. IEE Proc-Control Theory Appl 143(2):217–224

Amann N, Owens DH, Rogers E (1998) Predictive optimal iterative learning control. Int J Control 69(2):203–226

Arimoto S, Kawamura S, Miyazaki F (1984) Bettering operation of robots by learning. J Robot Syst 1(2):123–140

Axelsson P, Karlsson R, Norrlöf M (2014) Estimation-based norm-optimal iterative learning control. Syst Control Lett 73:76–80

Barton KL, Alleyne AG (2010) A norm optimal approach to time-varying ILC with application to a multi-axis robotic testbed. IEEE Trans Control Syst Technol 19(1):166–180

Cao Z, Lu J, Zhang R, Gao F (2016) Iterative learning Kalman filter for repetitive processes. J Process Control 46:92–104

Chen L, Narendra KS (2004) Identification and control of a nonlinear discrete-time system based on its linearization: a unified framework. IEEE Trans Neural Netw 15(3):663–673

Chen C, Xiong Z, Zhong Y (2014) Design and analysis of integrated predictive iterative learning control for batch process based on two-dimensional system theory. Chinese J Chem Eng 22(7):762–768

Chi R, Hou Z (2007) Dual-stage optimal iterative learning control for nonlinear non-affine discrete-time systems. Acta Automatica Sinica 33(10):1061–1065

Chi R, Hou Z, Sui S (2008) Non-parameter adaptive iterative learning control for the freeway traffic ramp metering. Control Theory Appl 25(6):1011–1015

Chi R, Wang D, Hou Z, Jin S (2012) Data-driven optimal terminal iterative learning control. J Process Control 22(10):2026–2037

Chi R, Hou Z, Huang B, Jin S (2015) A unified data-driven design framework of optimality-based generalized iterative learning control. Comput Chem Eng 77:10–23

Chi R, Wang D, Lewis FL, Hou Z, Jin S (2015) Adaptive terminal ILC for iteration-varying target points. Asian J Control 17(3):952–962

Chi R, Liu Y, Hou Z, Jin S (2015) Data-driven terminal iterative learning control with high-order learning law for a class of non-linear discrete-time multiple-input-multiple output systems. IET Control Theory A 9(7):1075–1082

Chi R, Liu Y, Hou Z, Jin S (2015) High-order data-driven optimal TILC approach for fed-batch processes. Can J Chem Eng 93(8):1455–1461

Chi R, Hou Z, Jin S, Wang D, Chien CJ (2015) Enhanced data-driven optimal terminal ILC using current iteration control knowledge. IEEE Trans Neural Netw Learn Syst 26(11):2939–2948

Chi R, Huang B, Wang D, Zhang R, Feng Y (2016) Data-driven optimal terminal iterative learning control with initial value dynamic compensation. IET Control Theory A 10(12):1357–1364

Chi R, Lin N, Zhang R, Huang B, Feng Y (2017) Stochastic high-order internal model-based adaptive TILC with random uncertainties in initial states and desired reference points. Int J Adapt Control Signal Process 31(5):726–741

Chi R, Liu X, Zhang R, Hou Z, Huang B (2017) Constrained data-driven optimal iterative learning control. J Process Control 55:10–29

Chi R, Hou Z, Jin S, Huang B (2018) Computationally efficient data-driven higher order optimal iterative learning control. IEEE Trans Neural Netw Learn Syst 29(12):5971–5980

Chi R, Hui Y, Huang B, Hou Z (2019) Adjacent-agent dynamic linearization-based iterative learning formation control. IEEE Trans Cybern 50(10):4358–4369

Chi R, Li H, Shen D, Hou Z, Huang B (2022) Enhanced P-type control: indirect adaptive learning from set-point updates. IEEE Trans Automat Contr. https://doi.org/10.1109/TAC.2022.3154347

Chin I, Qin SJ, Lee KS, Cho M (2004) A two-stage iterative learning control technique combined with real-time feedback for independent disturbance rejection. Automatica 40(11):1913–1922

Chu B, Owens DH (2009) Accelerated norm-optimal iterative learning control algorithms using successive projection. Int J Control 82(8):1469–1484

Deng H, Li H, Wu Y (2008) Feedback-linearization-based neural adaptive control for unknown nonaffine nonlinear discrete-time systems. IEEE Trans Neural Netw 19(9):1615–1625

Dumont GA, Fu Y (1993) Non-linear adaptive control via laguerre expansion of volterra kernels. Int J Adapt Control Signal Process 7(5):367–382

Hou Z (2014) Highlight and perspective on model free adaptive control. J Syst Sci Math Sci 34(10):1182

Hou Z, Wang Z (2013) From model-based control to data-driven control: survey, classification and perspective. Inf Sci 235:3–35

Hou Z, Jin S (2013) Model free adaptive control: theory and applications. Springer, Berlin

Hui Y, Chi R, Huang B (2019) Extended state observer-based data-driven iterative learning control for permanent magnet linear motor with initial shifts and disturbances. IEEE Trans Syst Man Cybern Syst 51(3):1881–1891

Janssens P, Pipeleers G, Swevers J (2012) A data-driven constrained norm-optimal iterative learning control framework for LTI systems. IEEE Trans Control Syst Technol 21(2):546–551

Janssens P, Pipeleers G, Swevers J (2011) Model-free iterative learning control for LTI systems and experimental validation on a linear motor test setup. In: 2011 American control conference. IEEE. https://ieeexplore.ieee.org/document/5990798

Jin S, Hou Z, Chi R, Liu X (2012) Data-driven model-free adaptive iterative learning control for a class of discrete-time nonlinear systems. Control Theory Appl 29(8):1001–1009

Jin S, Hou Z, Chi R (2015) Optimal terminal iterative learning control for the automatic train stop system. Asian J Control 17(5):1992–1999

Jin S, Hou Z, Chi R (2013) Data-driven predictive terminal iterative learning control for a class of discrete-time nonlinear systems. In: 32nd Chinese control conference. IEEE. https://ieeexplore.ieee.org/document/6639933

Kravaris C, Chung CB (1987) Nonlinear state feedback synthesis by global input/output linearization. AIChE J 33(4):592–603

Lee KS, Lee JH (1998) Model-based predictive control combined with iterative learning for batch or repetitive processes. Iterative Learning Control, Springer, Boston

Lee JH, Morari M, Garcia CE (1994) State-space interpretation of model predictive control. Automatica 30(4):707–717

Lee KS, Bang SH, Chang KS (1994) Feedback-assisted iterative learning control based on an inverse process model. J Process Control 4(2):77–89

Lee KS, Kim WC, Lee JH (1996) Model-based iterative learning control with quadratic criterion for linear batch processes. J Inst Control Robot Syst 2(3):148–157

Lee KS, Bang SH, Yi S, Son JS, Yoon SC (1996) Iterative learning control of heat-up phase for a batch polymerization reactor. J Process Control 6(4):255–262

Lee KS, Chin IS, Lee HJ, Lee JH (1999) Model predictive control technique combined with iterative learning for batch processes. AIChE J 45(10):2175–2187

Lee JH, Lee KS, Kim WC (2000) Model-based iterative learning control with a quadratic criterion for time-varying linear systems. Automatica 36(5):641–657

Lim I, Barton KL (2014) Pareto iterative learning control: optimized control for multiple performance objectives. Control Eng Pract 26:125–135

Lin N, Chi R, Huang B, Hou Z (2020) Event-triggered nonlinear iterative learning control. IEEE Trans Neural Netw Learn Syst 32(11):5118–5128

Liu T, Gao F (2010) Robust two-dimensional iterative learning control for batch processes with state delay and time-varying uncertainties. Chem Eng Sci 65(23):6134–6144

Liu T, Wang Y (2012) A synthetic approach for robust constrained iterative learning control of piecewise affine batch processes. Automatica 48(11):2762–2775

Liu T, Wang X, Chen J (2014) Robust PID based indirect-type iterative learning control for batch processes with time-varying uncertainties. J Process Control 24(12):95–106

Liu T, Wang D, Chi R (2015) Neural network based terminal iterative learning control for uncertain nonlinear non-affine systems. Int J Adapt Control Signal Process 29(10):1274–1286

Liu Y, Chi R, Hou Z (2015) Neural network state learning based adaptive terminal ILC for tracking iteration-varying target points. Int J Autom Comput 12(3):266–272

Liu S, Wu T (2002) Iterative learning control based on optimal index. In: The 4th world congress on intelligent control and automation

Márquez-Vera MA, Ramos-Velasco LE, Suárez-Cansino J, Márquez-Vera CA (2014) Fuzzy iterative learning control applied in a biological reactor using a reduced number of measures. Appl Math Comput 246:608–618

Mishra S, Topcu U, Tomizuka M (2010) Optimization-based constrained iterative learning control. IEEE Trans Control Syst Technol 19(6):1613–1621

Moore KL (2012) Iterative learning control for deterministic systems. Springer, Berlin

Moore KL, Verwoerd MH (2008) L1-optimal robust iterative learning controller design. In: 2008 American control conference. IEEE. https://ieeexplore.ieee.org/document/4587099

Nguyen DH, Banjerdpongchai D (2009) An LMI approach for robust iterative learning control with quadratic performance criterion. J Process Control 19(6):1054–1060

Oh SK, Lee JM (2016) Iterative learning model predictive control for constrained multivariable control of batch processes. Comput Chem Eng 93:284–292
Park KH (1999) A study on the robustness of a PID-type iterative learning controller against initial state error. Int J Syst Sci 30(1):49–59
Radac MB, Precup RE (2015) Optimal behaviour prediction using a primitive-based data-driven model-free iterative learning control approach. Comput Ind 74:95–109
Radac MB, Precup RE (2016) Model-free constrained data-driven iterative reference input tuning algorithm with experimental validation. Int J Gen Syst 45(4):455–476
Rădac MB, Precup RE, Petriu EM, Preitl S, Dragos CA (2012) Data-driven reference trajectory tracking algorithm and experimental validation. IEEE Trans Ind Inform 9(4):2327–2336
Sanzida N, Nagy ZK (2013) Iterative learning control for the systematic design of supersaturation controlled batch cooling crystallisation processes. Comput Chem Eng 59:111–121
Slotine JJE, Li W (1991) Applied nonlinear control. Prentice Hall, Englewood Cliffs
Togai M, Yamano O (1985) Analysis and design of an optimal learning control scheme for industrial robots: a discrete system approach. In: 1985 IEEE conference on decision and control. IEEE. https://ieeexplore.ieee.org/document/4048541
Tousain R, Van der Meché E, Bosgra O (2001) Design strategy for iterative learning control based on optimal control. In: the 40th IEEE conference on decision and control. IEEE. https://ieeexplore.ieee.org/document/980905
Van de Wijdeven J, Donkers T, Bosgra O (2009) Iterative learning control for uncertain systems: robust monotonic convergence analysis. Automatica 45(10):2383–2391
Volckaert M, Diehl M, Swevers J (2013) Generalization of norm optimal ILC for nonlinear systems with constraints. Mech Syst Signal Process 39(1–2):280–296
Wang F, Chen J (2020) Intelligent Control: Method and Application. China Science and Technology Press, Beijing
Wang Y, Zisser H, Dassau E, Jovanovič L, Doyle FJ III (2010) Model predictive control with learning-type set-point: application to artificial pancreatic $\beta$-cell. AIChE J 56(6):1510–1518
Wang Y, Liu T, Zhao Z (2012) Advanced PI control with simple learning set-point design: Application on batch processes and robust stability analysis. Chem Eng Sci 71:153–165
Wang L, Freeman CT, Chai S, Rogers E (2013) Predictive-repetitive control with constraints: from design to implementation. J Process Control 23(7):956–967
Xi YG, Wang F, Wu GH (1996) Nonlinear multi-model predictive control. IFAC Proc Vol 29(1):2359–2364
Xu J, Tan Y (2003) Linear and nonlinear iterative learning control. Springer, Heidelberg
Xu J, Hou Z (2005) Learning control: the state of the art and perspective. Acta Automatica Sinica 31(6):943–955
Xu J, Tan Y (2002) Robust optimal design and convergence properties analysis of iterative learning control approaches. Automatica 38(11):1867–1880
Zhou Y, Yin Y, Zhang Q, Gan W (2012) Model-free iterative learning control for repetitive impulsive noise using FFT. In: International symposium on neural networks. Springer, Berlin

# Chapter 2
# Compact Form Iterative Dynamic Linearation Based DDILC

## 2.1 Introduction

Iterative learning control (ILC) (Arimoto et al. 1984) was originally proposed as an intelligent learning mechanism for robot manipulators. Many ILC methods have been developed and widely applied to achieve perfect tracking performance by repeating the tracking task with learning. In general, the ILC methods can be classified into three categories according to the design and analysis methods, i.e., contraction mapping (CM)-based ILC (CM-ILC), composite energy function (CEF)-based ILC (CEF-ILC), and normal optimization-based ILC (NOILC).

CM-ILC methods refer to designing linear iterative learning algorithms by the use of contraction mapping and fixed point theory. Most of PID-type ILC schemes (Arimoto et al. 1984; Bristow et al. 2006; Ahn et al. 2007; Saab 1994; Park et al. 1999; Sun and Wang 2002) belong to such a category and the pointwise convergence property of tracking error over finite time interval is achieved in the sense of lambda norm with the requirements on global Lipschitz and identical initialization conditions. The problems of CM-ILC are that (i) the fixed learning gains are not robust to the uncertainties; (ii) it is hard to deal with locally Lipschitz continuous systems and there may not exist a finite $\lambda$ to suppress the diversifying dynamic impact; (iii) it is hard to choose a proper learning gain if no model information is available.

CEF-ILC is complementary to the CM-ILC and can deal with state tracking problem in the presence of locally Lipschitz continuous nonlinearities, whereas CM-ILC solves the output tracking problem with globally Lipschitz continuous nonlinearities. Many adaptive ILC methods (Tayebi 2004; Xu and Xu 2004; Rotariu et al. 2008; Chi et al. 2008) have been proposed by virtue of the CEF design and analysis. The learning gain of CEF-ILC can be updated iteratively using the I/O data that makes the CEF-ILC good at addressing system uncertainties. However, most of the CEF-ILC schemes are limited to parametric uncertain systems. In other words, the design and analysis of CEF-ILC depend on the model information of the systems, i.e., the parametric structures known as a priori.



R. Chi et al., *Data-Driven Iterative Learning Control for Discrete-Time Systems*, Intelligent Control and Learning Systems 2,
https://doi.org/10.1007/978-981-19-5950-9_2

NOILC is proposed for known linear systems (Hwang et al. 1991; Amann et al. 1996; Lee et al. 2000; Gunnarsson and Norrlof 2001; Sun and Alleyne 2014). The explicit optimal cost function is given and minimized to design an optimal ILC algorithm, and monotonic tracking error convergence along the iteration axis is guaranteed. The main advantage of NOILC is that it overcomes the limitation of poor transient performance along the iteration axis. However, NOILC requires an accurate known linear model of the system for designing a learning algorithm. In addition, the learning gain keeps unchanged which makes NOILC not robust to the uncertainties.

In summary, model dependence is a critical drawback that commonly exists in the above three types of ILC methods and prevents them from extensive applications. The reason is that realistic systems, such as the urban traffic system and petrochemical system, are becoming more complex with strong nonlinearities, high orders, and large uncertainties, all of which make it difficult or even impossible to attain their mechanistic models. Even if the physical models are established by significant efforts (Hou and Jin 2013), they are commonly too complicated to be suitable for controller design and analysis. On the other hand, a huge amount of process data is stored for many industrial processes and contains valuable state information. Hence, how to utilize these process data directly for controller design, monitoring, prediction, diagnosis, and assessment of the industrial systems is of significant importance when accurate mechanistic models are unavailable. Consequently, data-driven control (Hou and Jin 2013; Hou et al. 2016; Yin et al. 2014; Chi and Hou 2007; Chi et al. 2015; Janssens et al. 2013; Radac et al. 2013; Volckaert et al. 2013) becomes an interesting and attractive topic in recent years, where no explicit model is required for the controller design and analysis.

Establishing a linear dynamic relationship between the system output and input is the fundamental basis for controller design whether it is data-driven or model-based (Hou et al. 2016). Therefore, this chapter develops a compact form iterative dynamic linearization (CFIDL) method and builds an iterative linear relationship of the system output increment at the current time instant between two consecutive iterations with respect to the corresponding system input increment at the previous one time instant between the same two consecutive iterations. The iterative linear I/O relationship is attained without depending on any model information of the controlled plant and it only exists virtually in the computers for the single purpose of the ILC design and analysis in a data-driven framework. Then, the CFIDL-based data-driven ILC (CFIDL-DDILC) is proposed for nonlinear nonaffine systems with no model information available.

The main contributions of this chapter are summarized as: (a) it overcomes the identical initial condition, which is a critical assumption in traditional ILC analysis; (b) it is similar to norm-optimal ILC in that the CFIDL-DDILC is also designed according to a quadratic index function and achieves a monotonic convergence along iteration direction. However, the proposed CFIDL-DDILC method is data-driven and does not depend on any explicit model information of the plant; (c) although the CFIDL-DDILC is similar to the traditional PID-type ILC method in both the controller form and the contraction mapping analysis tool, the learning gain of the control law can be updated iteratively using the measured I/O data, thus making it more robust against uncertainties.

This chapter is organized as follows. Section 2.2 shows the problem formulation. The CFIDL-DDILC is proposed in Sect. 2.3 with convergence analysis in Sect. 2.4. Simulation study is shown in Sect. 2.5. Section 2.6 gives some conclusions.

## 2.2 Problem Formulation

Consider a class of nonlinear nonaffine discrete-time systems,

$$\begin{aligned} y_k(t+1) = f(&y_k(t), y_k(t-1), \ldots, y_k(t-n_y), \\ &u_k(t), u_k(t-1), \ldots, u_k(t-n_u)) \end{aligned} \tag{2.1}$$

where $y_k(t)$ and $u_k(t)$ are system output and input, respectively, and $y_k(t) = 0$ and $u_k(t) = 0$, $\forall t < 0$; $n_y$ and $n_u$ are two unknown orders; $t \in \{0, 1, \ldots, N\}$ is time instant where $N \in Z^+$ represents the endpoint of the finite operation interval; $k \in \{0, 1, \ldots\}$ indicates the iteration number; $f(\cdots) : R^{n_u+n_y+2} \mapsto R$ is an unknown nonlinear scalar function, and $f(0, 0) = 0$.

Many nonlinear system models, such as the Hammerstein model (Narendra and Gallman 1966), the bilinear model (Billings 2013), etc., are the special cases of model (2.1).

Some assumptions are made on system (2.1).

**Assumption 2.1** The partial derivative of $f(\cdots)$ with respect to the $(n_y + 2)$th variable is continuous for all $t$ and $k$.

**Assumption 2.2** System (2.1) satisfies the generalized *Lipschitz* condition for all $k$, that is

$$|y_{k_1}(t+1) - y_{k_2}(t+1)| \le b|u_{k_1}(t) - u_{k_2}(t)|, \ \forall t \in \{0, 1, \ldots, N\} \tag{2.2}$$

for $u_{k_1}(t) \neq u_{k_2}(t)$ and any $k_1 \neq k_2, k_1, k_2 \ge 0$, where $y_{k_i}(t+1) = f(y_{k_i}(t), y_{k_i}(t-1), \ldots, y_{k_i}(t-n_y), u_{k_i}(t), u_{k_i}(t-1), \ldots, u_{k_i}(t-n_u))$, $i = 1, 2$, and $b$ is a positive constant.

**Remark 2.1** From a practical point of view, the two assumptions are reasonable and acceptable. Assumption 2.1 is a typical condition for general nonlinear systems in the field of control system design. Assumption 2.2 imposes an upper bound on the change rate of the system output driven by the change of the control input. From an energy viewpoint, it is reasonable that the energy change inside a system cannot go to infinity if the energy change of the control input is at a finite level.

To facilitate the presentation of the following theorem, define $\Delta y_k(t+1) = y_k(t+1) - y_{k-1}(t+1)$ as the output change and $\Delta u_k(t) = u_k(t) - u_{k-1}(t)$ as the input change between two consecutive iterations.

**Theorem 2.1** *Consider nonlinear system (2.1) satisfying assumptions 2.1 and 2.2. If $|\Delta u_k(t)| \neq 0$, then there exists a time-iteration-varying parameter $\theta_k(t) \in R$, called iteration-related pseudo partial derivative (IPPD), such that system (2.1) can be transformed into the following compact form iterative dynamic linearization (CFIDL) data model:*

$$\Delta y_k(t+1) = \theta_k(t)\Delta u_k(t) \tag{2.3}$$

*and $|\theta_k(t)| \leq b$.*

***Proof*** From the definition of $\Delta y_k(t+1)$ and Equation (2.1), we have

$$\begin{aligned}\Delta y_k(t+1) &= f(y_k(t), y_k(t-1), \ldots, y_k(t-n_y), u_k(t), u_k(t-1), \ldots, u_k(t-n_u)) \\ &\quad - f(y_{k-1}(t), y_{k-1}(t-1), \ldots, y_{k-1}(t-n_y), u_{k-1}(t), u_{k-1}(t-1), \\ &\quad \ldots, u_{k-1}(t-n_u)) \\ &= f(y_k(t), y_k(t-1), \ldots, y_k(t-n_y), u_k(t), u_k(t-1), \ldots, u_k(t-n_u)) \\ &\quad - f(y_k(t), y_k(t-1), \ldots, y_k(t-n_y), u_{k-1}(t), u_k(t-1), \ldots, u_k(t-n_u)) \\ &\quad + f(y_k(t), y_k(t-1), \ldots, y_k(t-n_y), u_{k-1}(t), u_k(t-1), \ldots, u_k(t-n_u)) \\ &\quad - f(y_{k-1}(t), y_{k-1}(t-1), \ldots, y_{k-1}(t-n_y), u_{k-1}(t), u_{k-1}(t-1), \\ &\quad \ldots, u_{k-1}(t-n_u))\end{aligned} \tag{2.4}$$

Using Assumption 2.1 and the differential mean value theorem, from (2.4), we have

$$\Delta y_k(t+1) = \frac{\partial f}{\partial u_k(t)}(u_k(t) - u_{k-1}(t)) + \psi_k(t) \tag{2.5}$$

where $\partial f/\partial u_k(t)$ is a proper partial derivative value of $f(\cdots)$ with respect to $u_k(t)$ at some point in the interval $[u_k(t), u_{k-1}(t)]$, and

$$\begin{aligned}\psi_k(t) &= f(y_k(t), y_k(t-1), \ldots, y_k(t-n_y), u_{k-1}(t), u_k(t-1), \ldots, u_k(t-n_u)) \\ &\quad - f(y_{k-1}(t), y_{k-1}(t-1), \ldots, y_{k-1}(t-n_y), u_{k-1}(t), u_{k-1}(t-1), \\ &\quad \ldots, u_{k-1}(t-n_u)).\end{aligned}$$

Consider the following equation with a variable $\varsigma_k(t)$:

$$\psi_k(t) = \varsigma_k(t)\Delta u_k(t). \tag{2.6}$$

It is clear that (2.6) must have a solution $\varsigma_k(t)$ since $|\Delta u_k(t)| \neq 0$ for all $t \in \{0, 1, \ldots, N\}$ and $k \in \{0, 1, \ldots\}$.

Let $\theta_k(t) = \partial f/\partial u_k(t) + \varsigma_k(t)$. So (2.3) can be obtained from (2.5). Apparently, $|\theta_k(t)| \leq b$ according to Assumption 2.2. □

**Remark 2.2** From the proof of Theorem 2.1, one sees that the iterative dynamic linearization does not omit any information about the original plant. So, the obtained linearization data model (2.3) is completely equivalent to the original nonlinear system. However, the traditional linearization method is at best an approximation of the nonlinear systems. For example, Taylor's linearization ignores high-order terms and thus leads to an approximate model.

**Remark 2.3** The CFIDL builds a relationship between the system output change and input change for a repetitive dynamic plant along the iteration direction rather than the time direction. The time-dynamics of the repetitive process seem to be neglected, however, the system runs with time proceeding and the time-dynamics can be reflected by the system's real-time states.

**Remark 2.4** The development of linear data model (2.3) only depends on the I/O measurements of the closed-loop control plant. Neither structural nor parametric information about the plant is used explicitly or implicitly. Therefore, the CFIDL developed in this chapter is a data-driven method, and thus it can be used for most practical plants to deal with stronger nonlinearities.

**Remark 2.5** The CFIDL-based linear data model (2.3) virtually exists in the computer and only serves for the controller design and analysis. Further, it has a simple pseudo-linear structure; therefore, many techniques and tools of model-based control (MBC) design and analysis can be borrowed to develop variant ILC methods based on such a CFIDL data model.

**Remark 2.6** The CFIDL is one of the most important tools in studying the data-driven ILC theory and methods since it develops a way to discover the incidence relation of the online and off-line data of the processes for the purpose of controller design and analysis. The CFIDL method can be regarded as a modeling of the data relationship of the systems, which might find many applications in future control theory researches, especially for complex information-rich repetitive systems.

## 2.3 Controller Design

Given a desired trajectory $y_d(t)$, $t \in \{0, 1, \ldots, N\}$, the control target is to find a sequence of appropriate control inputs $u_k(t)$ such that the tracking error, $e_k(t+1) = y_d(t+1) - y_k(t+1)$, converges to zero as the iteration number $k$ approaches infinity.

Rewrite (2.3) as

$$y_k(t+1) = y_{k-1}(t+1) + \theta_k(t)\Delta u_k(t). \tag{2.7}$$

Define an index function of control input as

$$J(u_k(t)) = |e_k(t+1)|^2 + \lambda|\Delta u_k(t)|^2 \tag{2.8}$$

where $\lambda$ is a positive weighting factor.

According to (2.7) and the definition of $e_k(t+1)$, one can rewrite $J(u_k(t))$ as

$$\begin{aligned} J(u_k(t)) = &|y_d(t+1) - y_{k-1}(t+1) - \theta_k(t)(u_k(t) - u_{k-1}(t))|^2 \\ &+ \lambda|u_k(t) - u_{k-1}(t)|^2. \end{aligned} \tag{2.9}$$

By virtue of the optimal condition $\partial J/\partial u_k(t) = 0$, the learning control law of the control input is presented as

$$u_k(t) = u_{k-1}(t) + \frac{\rho\theta_k(t)e_{k-1}(t+1)}{\lambda + |\theta_k(t)|^2} \tag{2.10}$$

where $\rho > 0$ is a step-size constant series, which is added to make the generality of algorithm (2.10) and it is also used in the subsequent stability analysis.

Since IPPD, $\theta_k(t)$, is not available, here we use its estimation in the learning control law (2.10). Therefore, we rewrite (2.10) as follows:

$$u_k(t) = u_{k-1}(t) + \frac{\rho\hat{\theta}_k(t)e_{k-1}(t+1)}{\lambda + |\hat{\theta}_k(t)|^2} \tag{2.11}$$

where $\hat{\theta}_k(t)$ is the estimate of the IPPD parameter $\theta_k(t)$, and it is updated iteratively in terms of the optimal solution of the following criterion index function,

$$J(\hat{\theta}_k(t)) = |\Delta y_{k-1}(t+1) - \hat{\theta}_k(t)\Delta u_{k-1}(t)|^2 + \mu|\hat{\theta}_k(t) - \hat{\theta}_{k-1}(t)|^2 \tag{2.12}$$

where $\mu > 0$ is a positive weighting factor.

Rewriting (2.12),

$$\begin{aligned} J(\hat{\theta}_k(t)) = &|y_{k-1}(t+1) - (y_{k-2}(t+1) + \hat{\theta}_{k-1}(t)\Delta u_{k-1}(t)) - \\ &(\hat{\theta}_k(t) - \hat{\theta}_{k-1}(t))\Delta u_{k-1}(t)|^2 + \mu|\hat{\theta}_k(t) - \hat{\theta}_{k-1}(t)|^2. \end{aligned} \tag{2.13}$$

Using the optimal condition $\partial J/\partial\hat{\theta}_k(t) = 0$, one can obtain the parameter iterative updating law as follows:

$$\hat{\theta}_k(t) = \hat{\theta}_{k-1}(t) + \frac{\eta\Delta u_{k-1}(t)(\Delta y_{k-1}(t+1) - \hat{\theta}_{k-1}(t)\Delta u_{k-1}(t))}{\mu + |\Delta u_{k-1}(t)|^2} \tag{2.14}$$

where $0 < \eta < 2$ is a step-size constant added to make the generality of algorithm (2.14).

To make the condition $|\Delta u_k(t)| \neq 0$ satisfied for all $t \in \{0, 1, \ldots, N\}$ and $k \in \{0, 1, \ldots\}$ and to make the parameter estimation algorithm (2.14) have stronger ability in tracking variable parameter, a reset algorithm of $\hat{\theta}_k(t)$ is designed as follows:

$$\hat{\theta}_k(t) = \hat{\theta}_0(t),$$
$$\text{if } \operatorname{sgn}(\hat{\theta}_k(t)) \neq \operatorname{sgn}(\hat{\theta}_0(t)), \text{ or } |\hat{\theta}_k(t)| \leq \varepsilon, \text{ or } |\Delta u_k(t)| \leq \varepsilon \tag{2.15}$$

where $\hat{\theta}_0(t)$ is the initial value of $\hat{\theta}_k(t)$, and $\varepsilon$ is a small positive constant.

Then, the CFIDL-based DDILC (CFIDL-DDILC) is constructed by (2.16), (2.17), and (2.18), shown as follows:

$$u_k(t) = u_{k-1}(t) + \frac{\rho\hat{\theta}_k(t)e_{k-1}(t+1)}{\lambda + |\hat{\theta}_k(t)|^2} \tag{2.16}$$

$$\hat{\theta}_k(t) = \hat{\theta}_{k-1}(t) + \frac{\eta\Delta u_{k-1}(t)(\Delta y_{k-1}(t+1) - \hat{\theta}_{k-1}(t)\Delta u_{k-1}(t))}{\mu + |\Delta u_{k-1}(t)|^2} \tag{2.17}$$

$$\hat{\theta}_k(t) = \hat{\theta}_0(t),$$
$$\text{if } \operatorname{sgn}(\hat{\theta}_k(t)) \neq \operatorname{sgn}(\hat{\theta}_0(t)), \text{ or } |\hat{\theta}_k(t)| \leq \varepsilon, \text{ or } |\Delta u_{k-1}(t)| \leq \varepsilon. \tag{2.18}$$

**Remark 2.7** One can see that the learning control law (2.16) is most similar to the traditional P-type ILC method. However, the learning gain of the CFIDL-DDILC (2.16)–(2.18) is time-iteration-varying and can be tuned iteratively by using the IPPD parameter updating law (2.17) as well as the reset algorithm (2.18). Therefore, the proposed CFIDL-DDILC performs better than the traditional P-type ILC whose learning gain remains unchanged.

**Remark 2.8** There are four controller parameters: $\rho$, $\lambda$, $\eta$, and $\mu$, in the proposed method (2.16)–(2.18). In theory, large values of $\rho$ and $\eta$ mean a faster convergence because they are present in the numerator, but an overshoot may occur. While the smaller values of $\lambda$ and $\mu$ mean a faster convergence because they are present in the denominator. Generally, one can fix the values of $\lambda$ and $\mu$ within a proper scope, and then tune the values of $\rho$ and $\eta$ by trial and error to achieve a good control performance.

**Remark 2.9** Similar to the norm-optimal ILC methods, the proposed CFIDL-DDILC is also designed according to an optimal index function and achieves a monotonic convergence along the iteration direction. However, CFIDL-DDILC is developed for nonlinear systems rather than the exactly known linear systems. So, it is a data-driven control approach, while norm-optimal ILC is model-based.

## 2.4 Convergence Analysis

For the rigorous analysis of the following discussion, we give another assumption on the IPPD parameter as follows:

**Assumption 2.3** The IPPD parameter $\theta_k(t)$ satisfies the assumption that $\theta_k(t) > \ell$ (or $\theta_k(t) < -\ell$), $\forall t \in \{0, \ldots, N\}$ and $\forall k \in \{0, 1, \ldots\}$, where $\ell$ is an arbitrary positive constant. Without loss of generality, we consider $\theta_k(t) > \ell$ in this chapter.

**Remark 2.10** This assumption is similar to the limitation of control input direction. In fact, many practical applications satisfy this assumption such as temperature control systems, pressure control systems, and so on.

**Theorem 2.2** *For a class of nonlinear nonaffine discrete-time systems (2.1), if assumptions 2.1, 2.2, and 2.3 hold, then the proposed CFIDL-DDILC method consisting of the learning control law (2.16), the IPPD parameter updating law (2.17), and the reset algorithm (2.18), guarantees that*

*(i) the IPPD parameter estimation $\hat{\theta}_k(t)$ is bounded over the finite time interval $t \in \{0, 1, \ldots, N\}$, $\forall k$;*
*(ii) the tracking error converges to zero pointwisely over the finite time interval $t \in \{0, 1, \ldots, N\}$ as k approaches to infinity, i.e.,* $\lim_{k\to\infty} e_k(t+1) = 0$;
*(iii) the system output $y_k(t)$ and the control input signal $u_k(t)$ are bounded, $\forall t \in \{0, 1, \ldots, N\}$, and $\forall k$.*

***Proof*** The proof consists of three parts. Part 1 derives the boundedness of $\hat{\theta}_k(t)$. Part 2 proves the almost perfect tracking performance. The boundedness of system output and control input is shown in Part 3.

*Part 1. The boundedness of* $\hat{\theta}_k(t)$

When $|\Delta u_k(t)| \le \varepsilon$, by virtue of (2.18), $\hat{\theta}_k(t)$ is clearly bounded.

When $|\Delta u_k(t)| > \varepsilon$, subtracting $\theta_k(t)$ from both sides of (2.17), we have

$$\tilde{\theta}_k(t) = \tilde{\theta}_{k-1}(t) + \theta_k(t) - \theta_{k-1}(t) - \frac{\eta \Delta u_{k-1}(t)(\Delta y_{k-1}(t+1) - \hat{\theta}_{k-1}(t)\Delta u_{k-1}(t))}{\mu + |\Delta u_{k-1}(t)|^2} \tag{2.19}$$

where $\tilde{\theta}_k(t) = \theta_k(t) - \hat{\theta}_k(t)$.

Let $\Delta\theta_k(t) = \theta_k(t) - \theta_{k-1}(t)$. Using the relationship of (2.3), we can rewrite (2.19) as

$$\begin{aligned}\tilde{\theta}_k(t) &= \tilde{\theta}_{k-1}(t) + \Delta\theta_k(t) - \frac{\eta \Delta u_{k-1}(t)(\theta_{k-1}(t) - \hat{\theta}_{k-1}(t))\Delta u_{k-1}(t)}{\mu + |\Delta u_{k-1}(t)|^2} \\ &= \tilde{\theta}_{k-1}(t)\left(1 - \frac{\eta \Delta u_{k-1}(t)^2}{\mu + |\Delta u_{k-1}(t)|^2}\right) + \Delta\theta_k(t).\end{aligned} \tag{2.20}$$

Consider the following inequality:

$$0 < \left| 1 - \frac{\eta \Delta u_{k-1}(t)^2}{\mu + |\Delta u_{k-1}(t)|^2} \right| < 1. \tag{2.21}$$

Solving inequality (2.21), we have

$$\eta \in \left( 0, 2 + \frac{2\mu}{|\Delta u_{k-1}(t)|^2} \right). \tag{2.22}$$

Since $\mu > 0$, $0 < \eta < 2$, it is apparent that inequality (2.21) can be guaranteed $\forall t \in \{0, 1, \ldots, N\}$ and $\forall k \in \{0, 1, \ldots\}$. Hence, we can find a constant $d_1 = \sup_{k\in\{0,1,\ldots\}} \max_{t\in\{0,1,\ldots,N\}} |1 - \eta \Delta u_{k-1}(t)^2/(\mu + |\Delta u_{k-1}(t)|^2)|$ satisfying $0 < d_1 < 1$ for all $t \in \{0, 1, \ldots, N\}$ and $k \in \{0, 1, \ldots\}$.

Taking norms on both sides of (2.20), we have

$$\begin{aligned} |\tilde{\theta}_k(t)| &\le \left| 1 - \frac{\eta \Delta u_{k-1}(t)^2}{\mu + |\Delta u_{k-1}(t)|^2} \right| |\tilde{\theta}_{k-1}(t)| + |\Delta\theta_k(t)| \\ &\le d_1 |\tilde{\theta}_{k-1}(t)| + 2b \\ &\le d_1^k |\tilde{\theta}_0(t)| + \frac{2b}{1 - d_1}. \end{aligned} \tag{2.23}$$

Hence, $\tilde{\theta}_k(t)$ is bounded because $\hat{\theta}_0(t)$ is chosen bounded and $\theta_k(t)$ is also bounded according to Theorem 2.1. Therefore, the boundedness of $\hat{\theta}_k(t)$ can be guaranteed $\forall t \in \{0, \ldots, N\}$, and $\forall k$.

*Part 2. The convergence of output tracking error $e_k(t)$*

By virtue of (2.7), the tracking error dynamics is

$$\begin{aligned} e_k(t+1) &= y_d(t+1) - y_k(t+1) \\ &= y_d(t+1) - y_{k-1}(t+1) - \theta_k(t)\Delta u_k(t) \\ &= e_{k-1}(t+1) - \theta_k(t)\Delta u_k(t). \end{aligned} \tag{2.24}$$

Substituting control law (2.16) into (2.24), we obtain

$$e_k(t+1) = \left( 1 - \frac{\rho\theta_k(t)\hat{\theta}_k(t)}{\lambda + |\hat{\theta}_k(t)|^2} \right) e_{k-1}(t+1). \tag{2.25}$$

According to Assumption 2.3, the value of $\hat{\theta}_0(t)$ at the initial iteration is selected such that its sign is the same as that of $\theta_k(t)$. Then, the reset algorithm (2.18) implies that $\theta_k(t)\hat{\theta}_k(t) > 0$.

Let $\lambda_{\min} = b^2/16$. According to inequality $\alpha^2 + \beta^2 \geq 2\alpha\beta$, and selecting $\lambda > \lambda_{\min}$, one has

$$0 < \frac{\theta_k(t)\hat{\theta}_k(t)}{\lambda + \left|\hat{\theta}_k(t)\right|^2} \leq \frac{|\theta_k(t)||\hat{\theta}_k(t)|}{2\sqrt{\lambda}|\hat{\theta}_k(t)|} < \frac{b}{2\sqrt{\lambda_{\min}}} = 2. \tag{2.26}$$

In terms of (2.26), by selecting $\rho > 0$ and $\lambda > \lambda_{\min}$, one has

$$\left|1 - \frac{\rho\theta_k(t)\hat{\theta}_k(t)}{\lambda + \left|\hat{\theta}_k(t)\right|^2}\right| \leq d_2 < 1 \tag{2.27}$$

where $0 < d_2 < 1$ is a constant.

Then, taking norms for both sides of (2.25), yields

$$\begin{aligned} |e_k(t+1)| &\leq \left|1 - \frac{\rho\theta_k(t)\hat{\theta}_k(t)}{\lambda + |\hat{\theta}_k(t)|^2}\right| |e_{k-1}(t+1)| \\ &\leq d_2|e_{k-1}(t+1)| \\ &\leq \cdots \leq d_2^k|e_0(t+1)| \end{aligned} \tag{2.28}$$

which implies that the tracking error $e_k(t)$, $t \in \{0, 1, \ldots, N\}$, asymptotically converges to zero along the iteration direction.

*Part 3. Boundedness of system output and control signal*

Since $y_d(t)$ is bounded, the convergence of $e_k(t)$ implies that $y_k(t)$ is also bounded. From learning law (2.16), we get

$$\Delta u_k(t) = \frac{\rho\hat{\theta}_k(t)}{\lambda + |\hat{\theta}_k(t)|^2} e_{k-1}(t+1). \tag{2.29}$$

With the use of $(\sqrt{\lambda})^2 + |\hat{\theta}_k(t)|^2 \geq 2\sqrt{\lambda}\hat{\theta}_k(t)$ and $\lambda > \lambda_{\min}$, one can obtain from (2.29) that

$$\begin{aligned} |\Delta u_k(t)| &= \left|\frac{\rho\hat{\theta}_k(t)e_{k-1}(t+1)}{\lambda + \left|\hat{\theta}_k(t)\right|^2}\right| \leq \left|\frac{\rho\hat{\theta}_k(t)}{2\sqrt{\lambda}\hat{\theta}_k(t)}\right| |e_{k-1}(t+1)| \\ &\leq \left|\frac{\rho}{2\sqrt{\lambda_{\min}}}\right| |e_{k-1}(t+1)| = M|e_{k-1}(t+1)| \end{aligned} \tag{2.30}$$

where $M$ is a constant.

We know that

$$
\begin{aligned}
|u_k(t)| &= |u_k(t) - u_0(t) + u_0(t)| \\
&\le |u_k(t) - u_0(t)| + |u_0(t)| \\
&= |u_k(t) - u_{k-1}(t) + u_{k-1}(t) - \cdots + u_1(t) - u_0(t)| + |u_0(t)| \\
&\le |\Delta u_k(t)| + |\Delta u_{k-1}(t)| + \cdots + |\Delta u_1(t)| + |u_0(t)|. \qquad (2.31)
\end{aligned}
$$

According to (2.28), (2.30), and (2.31), we have

$$
\begin{aligned}
|u_k(t)| &\le Me_{k-1}(t+1) + Me_{k-2}(t+1) + \cdots + Me_0(t+1) + |u_0(t)| \\
&\le \frac{Me_0(t+1)}{1-d_2} + |u_0(t)|. \qquad (2.32)
\end{aligned}
$$

Clearly, the values of initial input $u_0(t)$ and tracking error $e_0(t+1)$ can be chosen to be bounded, so the control input $u_k(t)$ is bounded $\forall t \in \{0, 1, \cdots, N\}$ and $\forall k$ according to (2.32). □

**Remark 2.11** One sees that the initial condition, which is a critical assumption in traditional ILC analysis, is not used in the convergence analysis of the proposed CFIDL-DDILC method. It is a significant breakthrough in ILC theory since the identical initial condition is hard to satisfy in practical applications.

## 2.5 Simulation Study

Consider a nonlinear nonaffine SISO system

$$
y_k(t+1) = \begin{cases} \frac{y_k(t)}{1+y_k(t)^2} + u_k(t)^3, \ 0 \le t < 50 \\ \frac{y_k(t)y_k(t-1)y_k(t-2)u_k(t-1)(y_k(t-2)-1)+a(t)u_k(t)}{1+y_k(t-1)^2+y_k(t-2)^2}, \ 50 \le t \le 100 \end{cases} \qquad (2.33)
$$

where $a(t) = 1 + \text{round}(t/50)$ is a varying parameter. It is notable that the structure, orders, and parameters of the controlled system are time-iteration-varying.

The desired trajectory is given as follows:

$$
y_d(t+1) = \begin{cases} 0.5 \times (-1)^{\text{round}(t/10)}, \ 0 \le t \le 30 \\ 0.5\sin(t\pi/10) + 0.3\cos(t\pi/10), \ 30 < t \le 70 \\ 0.5 \times (-1)^{\text{round}(t/10)}, \ 70 < t \le 100 \end{cases} \qquad (2.34)
$$

By choosing $\eta = 0.5$, $\rho = 0.8$, $\lambda = 1$, $\mu = 0.5$, $\hat{\theta}_0(t) = 1$, $y_k(0) = 0$, $u_0(t) = 0$, the simulation results are shown in Figs. 2.1 and 2.2. Figure 2.1 shows the output tracking performance at the 100th iteration, where the vertical axis denotes the system

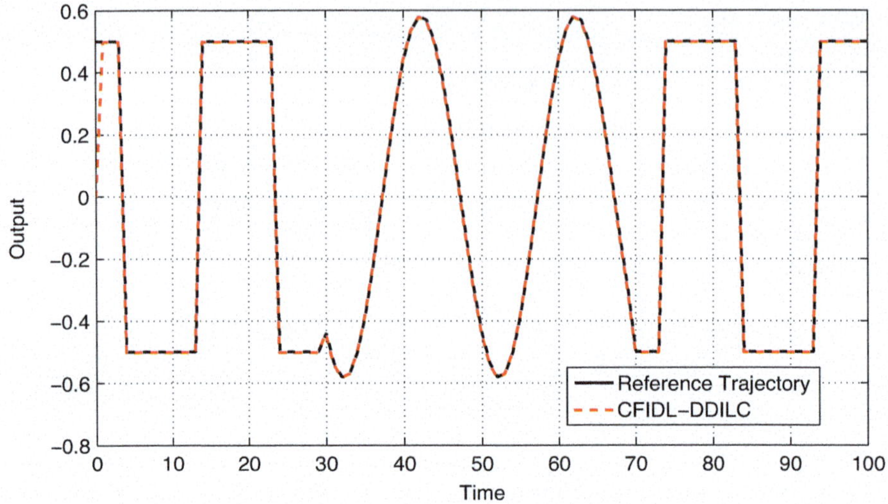


**Fig. 2.1** Output tracking performance with identical initial conditions

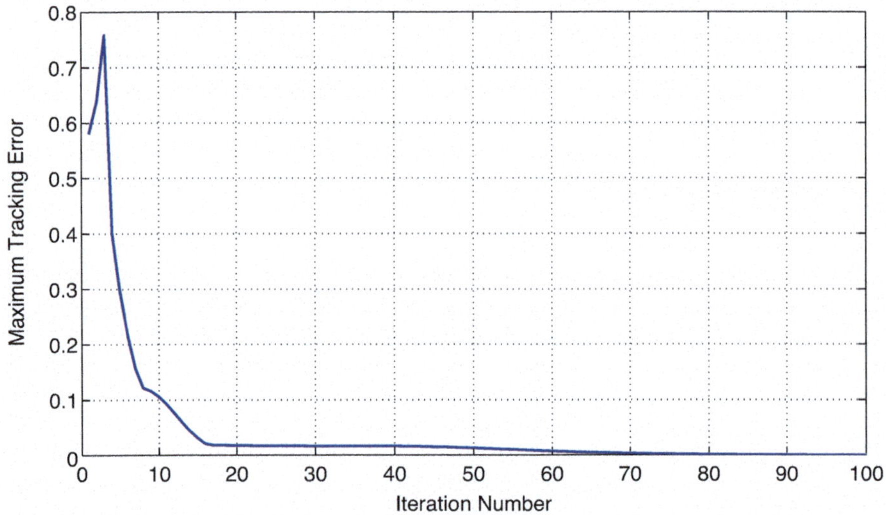


**Fig. 2.2** Convergence of maximum tracking error with identical initial conditions

output and the horizontal axis denotes the time instant. Figure 2.2 shows the convergence performance of tracking error, where the Y-axis denotes the maximum absolute value of the output tracking error, defined as $e_{\max}(k) = \max_{t\in\{1,\ldots,100\}}\{|e_k(t)|\}$, and the X-axis denotes the iteration numbers. One can see that the effectiveness of the presented CFIDL-DDILC is confirmed by the simulation results.

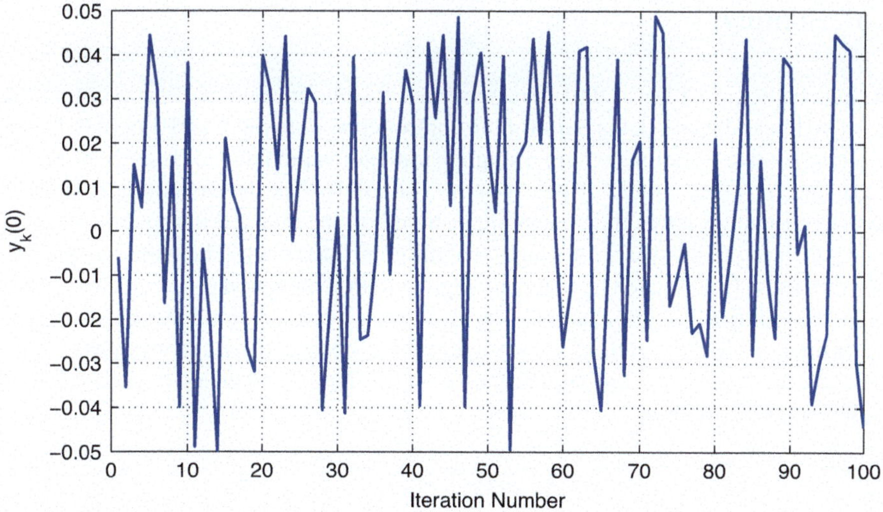


**Fig. 2.3** Profile of initial states

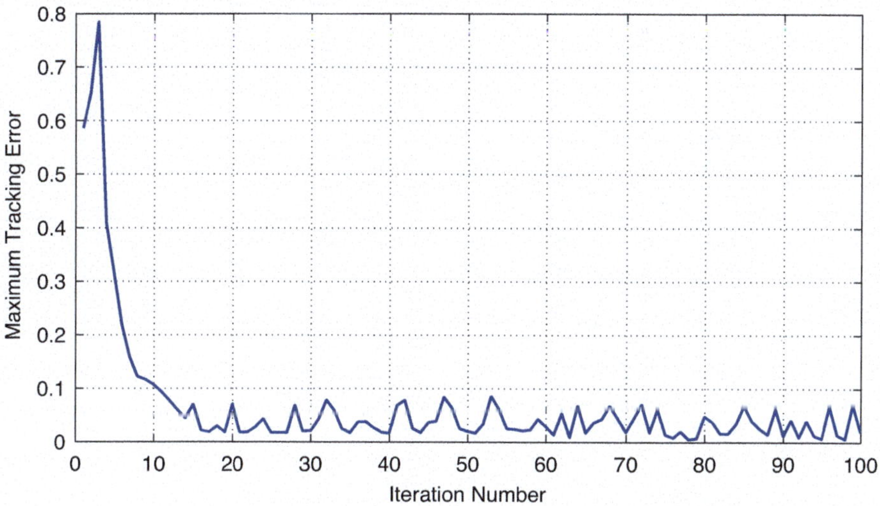


**Fig. 2.4** Convergence of maximum tracking error with nonrepetitive initial states

To show the robustness of CFIDL-DDILC against nonrepetitive iteration-varying initial states, we set $y_k(0)$ be varying randomly over the interval of $[-0.05, 0.05]$, as shown in Fig. 2.3. Under the same simulation condition, the convergence performance of maximum output tracking error is shown in Fig. 2.4. It is seen that the proposed method can also guarantee a bounded convergence although the initial states are varying randomly.

## 2.6 Summary

In this chapter, a CFIDL-DDILC is presented for nonlinear nonaffine systems. An iterative linear data model is built for the controlled system input and output between two consecutive iterations where a single parameter is included to describe the nonlinear uncertain dynamics of the system. The DDILC is developed by design an objective function about the control input based on the attained iterative linear data model instead of any model information of the controlled plant. The unknown parameter is iteratively estimated by using an improved projection updating law conducting along the iteration direction, and thus such an iterative adaptation mechanism makes the proposed CFIDL-DDILC robust against the nonrepetitive uncertainties. Both the theoretical analysis and simulations confirm the effectiveness of the proposed CFIDL-DDILC method.

## References

Ahn HS, Chen YQ, Moore KL (2007) Iterative learning control: brief survey and categorization. IEEE Trans Syst Man Cybern C Appl Rev 37(6):1099–1121

Amann N, Owens H, Rogers E (1996) Iterative learning control for discrete time systems with exponential rate of convergence. Proc Inst Elect Eng Control Theory Appl 143(2):217–224

Arimoto S, Kawamura S, Miyazaki F (1984) Bettering operation of robots by learning. J Robot Syst 1(2):123–140

Billings SA (2013) Nonlinear system identification: NARMAX methods in the time, frequency, and spatio-temporal domains. Wiley, Hoboken

Bristow DA, Tharayil M, Alleyne AG (2006) A survey of iterative learning control: a learning-based method for high-performance tracking control. IEEE Control Syst Mag 26(3):96–114

Chi R, Hou Z (2007) Dual-stage optimal iterative learning control for nonlinear non-affine discrete-time systems. Acta Automatica Sinica 33(10):1061–1065

Chi R, Hou Z, Xu J (2008) Adaptive ILC for a class of discrete-time systems with iteration-varying trajectory and random initial condition. Automatica 44:2207–2213

Chi R, Hou Z, Huang B, Jin S (2015) A unified data-driven design framework of optimality-based generalized iterative learning control. Comp Chem Eng 77:10–23

Gunnarsson S, Norrlof M (2001) On the design of ILC algorithms using optimization. Automatica 37(12):2011–2016

Hou Z, Chi R, Gao H (2016) An overview of dynamic-linearization-based data-driven control and applications. IEEE Trans Ind Electron 64(5):4076–4090

Hou Z, Jin S (2013) Model free adaptive control: theory and applications. CRC Press

Hwang DH, Bien Z, Oh SR (1991) Iterative learning control method for discrete-time dynamic systems. IEE Proceedings-D 138(2):139–144

Janssens P, Pipeleers G, Swevers J (2013) A data-driven constrained norm-optimal iterative learning control framework for LTI systems. IEEE Trans Control Syst Technol 21(2):546–551

Lee JH, Lee KS, Kim WC (2000) Model-based iterative learning control with a quadratic criterion for time-varying linear systems. Automatica 36(5):641–657

Narendra KS, Gallman PG (1966) An iterative method for the identification of nonlinear systems using the hammerstein model. IEEE Trans Automat Contr 11(3):546–550

Park KH, Bien Z, Hwang DH (1999) A study on robustness of PID-type iterative learning controller against initial state error. Int J Syst Sci 30(1):49–59

Radac MB, Precup RE, Petriu EM, Preitl S, Dragos CA (2013) Data-driven reference trajectory tracking algorithm and experimental validation. IEEE Trans Ind Inform 9(4):2327–2336

Rotariu I, Steinbuch M, Ellenbroek R (2008) Adaptive iterative learning control for high precision motion systems. IEEE Trans Control Syst Technol 16(5):1075–1082

Saab SS (1994) On the P-type learning control. IEEE Trans Automat Control 39:2298–2302

Sun HQ, Alleyne AG (2014) A computationally efficient norm optimal iterative learning control ap-proach for LTV systems. Automatica 50:141–148

Sun M, Wang D (2002) Iterative learning control with initial rectifying action. Automatica 38:1177–1182

Tayebi A (2004) Adaptive iterative learning control for robot manipulators. Automatica 40(7):1195–1203

Volckaert M, Diehl M, Swevers J (2013) Generalization of norm optimal ILC for nonlinear systems with constraints. Mech Syst Signal Process 39:280–296

Xu J, Xu J (2004) On iterative learning from different tracking tasks in the presence of time-varying uncertainties. IEEE Trans Syst Man Cybern B Cybern 34(1):589–597

Yin S, Li X, Gao H, Kaynak O (2014) Data-based techniques focused on modern industry: an overview. IEEE Trans Ind Electron 62(1):657–667

# Chapter 3
# DDILC with Partial Form and Full Form-Based Iterative Dynamic Linearizations

## 3.1 Introduction

Linearization of a nonlinear system is an intuitive way to simplify the control system design and analysis. However, most existing linearizations (Slotine and Li 1991; Cheng and Li 2009; Chen and Narendra 2004; Wen and Hill 1990; Xi and Wang 1996; Guardabassi and Savaresi 1997; Deng et al. 2008; Dumont and Fu 1993) are conducted over a small operating range of system and the original nonlinear system is only approximated by the linearized system (Wen and Hill 1990). Furthermore, some of them need an exactly known mathematical model of the controlled plant or have too many parameters. As a result, the unmodeled/mismatched dynamics in the linearization process unavoidably exist and affect the control performance (Hou et al. 2016). To solve these problems, a dynamic linearization (DL) method has been developed for an unknown discrete-time nonlinear system (Hou and Jin 2013, 2011; Chi et al. 2022; Lin et al. 2020) requiring no model information. The resulted linear data model by the DL method is simple with quite few parameters which can be easily updated by using the I/O data merely. Therefore, the DL method (Hou and Jin 2013, 2011; Chi et al. 2022; Lin et al. 2020) is control-design-oriented and can be directly used for control system design by incorporating the well-known model-based control design and analysis methods.

In terms of the similar idea, a compact form iterative dynamic linearization (CFIDL) method (Chi and Hou 2007) is developed in Chap. 2 for repetitive nonlinear systems and it builds an iterative linear data model to describe the dynamic relationship of the system input and output increments between two consecutive iterations. Different from the traditional linearization methods (Slotine and Li 1991; Cheng and Li 2009; Chen and Narendra 2004; Wen and Hill 1990; Xi and Wang 1996; Guardabassi and Savaresi 1997; Deng et al. 2008; Dumont and Fu 1993), the CFIDL is presented without requiring any prior knowledge of the mathematical model, system orders, or time-delay of the controlled nonlinear plant. Further, the CFIDL describes an iterative increment relationship between two iterative opera-



R. Chi et al., *Data-Driven Iterative Learning Control for Discrete-Time Systems*, Intelligent Control and Learning Systems 2,
https://doi.org/10.1007/978-981-19-5950-9_3

tions of the system input and output, and thus it does not have any special physical meaning, but only virtually exists in the computer.

It is worth pointing out that the CFIDL method proposed in Chap. 2 transforms the nonlinear repetitive system into a linear time-iteration-varying dynamic data model, which has only one scalar parameter $\theta_k(t)$. All possible complicated behavior characteristics, such as nonlinearities, time-iteration-varying parameters, time-iteration-varying structures, etc., of the original dynamic nonlinear system are compressed into a single time-iteration-varying scalar parameter $\theta_k(t)$. Therefore, the behavior of IPPD $\theta_k(t)$ may be too complicated to be estimated. On the other hand, the CFIDL method in essence merely considers the iteration-varying dynamic relationship between the iterative increment of the system output at the next time instant and the iterative increment of the control input at the current time instant. In fact, the iterative change of the system output may be also related to iterative changes of the control inputs at the previous time instants besides the iterative change of the control input at the current time instant.

Based on this observation, this chapter proposes a new iterative dynamic linearization method called partial-form iterative dynamic linearization (PFIDL) which considers the influences on the system output iterative increment at the next time instant imposed by the control input iterative increments within a fixed-length moving time window of the current time. Furthermore, the iterative change of the system output at the next time instant is also related to the iterative change of system outputs at the present and previous time instants. Therefore, we further propose a new full form iterative dynamic linearization (FFIDL) method to consider all the influences on the system output iterative increment at the next time instant imposed by both the control input iterative increments and the system output iterative increments within input-related/output-related fixed-length moving time windows at the current time instant. By following the similar procedure in Chap. 2, this chapter presents two DDILC methods based on the PFIDL-based and FFIDL-based linear data models, respectively. The main contributions of this chapter are summarized as follows.

(a) Theoretically speaking, the possible complicated behavior of the original system may be well captured, dispersed and lowered by introducing multiple parameters in the PFIDL and the FFIDL-based data models, rather than a single parameter in the CFIDL data model. Further, the FFIDL method performs better than the PFIDL since more parameters are included to cope with the possible complicated behavior of the original system.
(b) Compared with the CFIDL-DDILC, the two proposed PFIDL-DDILC and FFIDL-DDILC methods utilize more control knowledge in the learning control law, thus enhancing the control performance as a direct result.
(c) The convergence analysis of the proposed PFIDL-DDILC becomes more complex and difficult due to the additional input variables introduced wherein, which are coupled with the error variables strongly. Therefore, more mathematical tools such as the special radius property are introduced for the analysis of PFIDL-DDILC analysis.

(d) The proposed PFIDL-DDILC and FFIDL-DDILC methods can enjoy more adjustable degrees of freedom and more design flexibility by using different parameters. The two proposed DDILC methods provide an opportunity to select a proper control strategy for a complex nonlinear system by considering the trade-off between the control performance and the computation burden.

This chapter is organized as follows. Section 3.2 presents the PFIDL-based DDILC with detail analysis and simulation study. Section 3.3 proposes the FFIDL-based DDILC and shows the simulation study. Section 3.4 gives some conclusions.

## 3.2 PFIDL-Based DDILC

### 3.2.1 *Problem Formulation*

Consider a class of nonlinear nonaffine discrete-time systems,

$$\begin{aligned} y_k(t+1) = f(&y_k(t), y_k(t-1), \ldots, y_k(t-n_y), \\ &u_k(t), u_k(t-1), \ldots, u_k(t-n_u)) \end{aligned} \tag{3.1}$$

where $y_k(t)$ and $u_k(t)$ are system output and input, respectively, where $y_k(t) = 0$ and $u_k(t) = 0$, $\forall t < 0$; $n_y$ and $n_u$ are two unknown orders; $t \in \{0, 1, \ldots, N\}$ is time instant where $N \in Z^+$ represents the endpoint; $k \in \{0, 1, \ldots\}$ indicates the iteration number; $f(\cdots) : R^{n_u+n_y+2} \mapsto R$ is an unknown nonlinear scalar function, and $f(0, 0) = 0$.

Denote $\boldsymbol{U}_k(t) \in R^L$ as a vector consisting of all control input signals within a moving time window $[t-L+1, \ldots, t]$,

$$\boldsymbol{U}_k(t) = [u_k(t), \ldots, u_k(t-L+1)]^{\mathrm{T}}$$

with $\boldsymbol{U}_k(t) = \boldsymbol{0}_L$ for $t < 0$ and $k < 0$. The integer $L$ is called control input linearization length constant (LLC), and $\boldsymbol{0}_L$ is a zero vector of dimension $L$.

For discrete-time nonlinear system (3.1), two assumptions are made as follows.

**Assumption 3.1** The partial derivatives of $f(\cdot)$ with respect to the variables from the $(n_y+2)$th variable to the $(n_y+L+1)$th variable, namely, $u_k(t), \ldots, u_k(t-L+1)$, are continuous.

**Assumption 3.2** System (3.1) satisfies the generalized *Lipschitz* condition, that is,

$$|y_{k_1}(t+1) - y_{k_2}(t+1)| \leq b\|\boldsymbol{U}_{k_1}(t) - \boldsymbol{U}_{k_2}(t)\| \tag{3.2}$$

for $\boldsymbol{U}_{k_1}(t) \neq \boldsymbol{U}_{k_2}(t)$ with $k_1 \neq k_2$, $k_1$, $k_2 \geq 0$, where $y_{k_i}(t+1) = f(y_{k_i}(t), \ldots, y_{k_i}(t-n_y), u_{k_i}(t), \ldots, u_{k_i}(t-n_u))$, $i = 1, 2$, and $b$ is a positive constant.

Denote $\Delta \boldsymbol{U}_k(t) = \boldsymbol{U}_k(t) - \boldsymbol{U}_{k-1}(t)$. The PFIDL of system (3.1) is presented in the following theorem.

**Theorem 3.1** *Consider nonlinear system (3.1) satisfying assumptions 3.1 and 3.2. For any fixed L, if $\|\Delta \boldsymbol{U}_k(t)\| \neq 0$, then there exists an iteration-dependent time varying parameter vector $\boldsymbol{\theta}_k(t)$, called iteration-related pseudo gradient (IPG), such that system (3.1) can be transformed into the following PFIDL data model in iteration domain,*

$$\Delta y_k(t+1) = \boldsymbol{\theta}_k^{\mathrm{T}}(t) \Delta \boldsymbol{U}_k(t), \tag{3.3}$$

*where $\boldsymbol{\theta}_k(t) = [\theta_{1,k}(t) \ \cdots \ \theta_{L,k}(t)]^{\mathrm{T}}$, and $\|\boldsymbol{\theta}_k(t)\| \leq b$.*

***Proof*** It follows from (3.1) that

$$\begin{aligned}
\Delta y_k(t+1) = & \ f(y_k(t), \ldots, y_k(t-n_y), u_k(t), \ldots, u_k(t-n_u)) \\
& - f(y_k(t), \ldots, y_k(t-n_y), u_{k-1}(t), u_k(t-1), \ldots, u_k(t-n_u)) \\
& + f(y_k(t), \ldots, y_k(t-n_y), u_{k-1}(t), u_k(t-1), \ldots, u_k(t-n_u)) \\
& - f(y_{k-1}(t), \ldots, y_{k-1}(t-n_y), u_{k-1}(t), \ldots, u_{k-1}(t-n_u))
\end{aligned} \tag{3.4}$$

By virtue of Assumption 3.1 and differential mean value theorem, (3.4) can be rewritten as

$$\begin{aligned}
\Delta y_k(t+1) = & \ \frac{\partial f}{\partial u_k(t)} \Delta u_k(t) \\
& + f(y_k(t), \ldots, y_k(t-n_y), u_{k-1}(t), u_k(t-1), \ldots, u_k(t-n_u)) \\
& - f(y_{k-1}(t), \ldots, y_{k-1}(t-n_y), u_{k-1}(t), \ldots, u_{k-1}(t-n_u))
\end{aligned} \tag{3.5}$$

where $\partial f / \partial u_k(t)$ represents the partial derivative value of function $f$ with respect to $u_k(t)$ at some point in the interval $[u_k(t), u_{k-1}(t)]$.

From system (3.1), we have

$$\begin{aligned}
y_k(t) = & \ f(y_k(t-1), y_k(t-2), \ldots, y_k(t-n_y-1), \\
& u_k(t-1), u_k(t-2), \ldots, u_k(t-n_u-1))
\end{aligned} \tag{3.6}$$

Substituting (3.6) into (3.5), and defining

$$\begin{aligned}
& \psi_1(y_k(t-1), \ldots, y_k(t-n_y-1), y_{k-1}(t), \ldots, y_{k-1}(t-n_y), \\
& \quad u_k(t-1), \ldots, u_k(t-n_u-1), u_{k-1}(t), \ldots, u_{k-1}(t-n_u)) \\
& = f(f(y_k(t-1), y_k(t-2), \ldots, y_k(t-n_y-1), \\
& \quad u_k(t-1), u_k(t-2), \ldots, u_k(t-n_u-1)), \\
& \quad y_k(t-1), \ldots, y_k(t-n_y), u_{k-1}(t), u_k(t-1), \ldots, u_k(t-n_u)) \\
& \quad - f(y_{k-1}(t), \ldots, y_{k-1}(t-n_y), u_{k-1}(t), \ldots, u_{k-1}(t-n_u))
\end{aligned} \tag{3.7}$$

Equation (3.5) can be rewritten as

$$\begin{aligned}
\Delta y_k(t+1) &= \frac{\partial f}{\partial u_k(t)}\Delta u_k(t) \\
&\quad +\psi_1(y_k(t-1),\ldots,y_k(t-n_y-1),y_{k-1}(t),\ldots,y_{k-1}(t-n_y), \\
&\quad u_k(t-1),\ldots,u_k(t-n_u-1),u_{k-1}(t),\ldots,u_{k-1}(t-n_u)) \\
&\quad -\psi_1(y_k(t-1),\ldots,y_k(t-n_y-1),y_{k-1}(t),\ldots,y_{k-1}(t-n_y), \\
&\quad u_{k-1}(t-1),u_k(t-2),\ldots,u_k(t-n_u-1),u_{k-1}(t),\ldots,u_{k-1}(t-n_u)) \\
&\quad +\psi_1(y_k(t-1),\ldots,y_k(t-n_y-1),y_{k-1}(t),\ldots,y_{k-1}(t-n_y), \\
&\quad u_{k-1}(t-1),u_k(t-2),\ldots,u_k(t-n_u-1),u_{k-1}(t),\ldots,u_{k-1}(t-n_u)) \\
&= \frac{\partial f}{\partial u_k(t)}\Delta u_k(t)+\frac{\partial \psi_1}{\partial u_k(t-1)}\Delta u_k(t-1) \\
&\quad +\psi_2(y_k(t-2),\ldots,y_k(t-n_y-2),y_{k-1}(t),\ldots,y_{k-1}(t-n_y), \\
&\quad u_k(t-2),\ldots,u_k(t-n_u-2),u_{k-1}(t),\ldots,u_{k-1}(t-n_u))
\end{aligned} \tag{3.8}$$

where $\partial\psi_1/\partial u_k(t-1)$ represents the partial derivative value of function $\psi_1$ with respect to $u_k(t-1)$ at some point in the interval $[u_k(t-1), u_{k-1}(t-1)]$, and

$$\begin{aligned}
&\psi_2(y_k(t-2),\ldots,y_k(t-n_y-2),y_{k-1}(t),\ldots,y_{k-1}(t-n_y), \\
&\quad u_k(t-2),\ldots,u_k(t-n_u-2),u_{k-1}(t),\ldots,u_{k-1}(t-n_u)) \\
&= \psi_1(f(y_k(t-2),\ldots,y_k(t-n_y-2),u_k(t-2),\ldots,u_k(t-n_u-2)), \\
&\quad y_k(t-2),\ldots,y_k(t-n_y-1),y_{k-1}(t),\ldots,y_{k-1}(t-n_y), \\
&\quad u_{k-1}(t-1),u_k(t-2),\ldots,u_k(t-n_u-1),u_{k-1}(t),\ldots,u_{k-1}(t-n_u))
\end{aligned}$$

Analogously, from (3.8) with a fixed $L$, we have

$$\begin{aligned}
\Delta y_k(t+1) = {} & \frac{\partial f}{\partial u_k(t)}\Delta u_k(t)+\frac{\partial \psi_1}{\partial u_k(t-1)}\Delta u_k(t-1) \\
& +\cdots+\frac{\partial \psi_{L-1}}{\partial u_k(t-L+1)}\Delta u_k(t-L+1) \\
& +\psi_L(y_k(t-L),\ldots,y_k(t-L-n_y),y_{k-1}(t),\ldots,y_{k-1}(t-n_y), \\
& \quad u_k(t-L),\ldots,u_k(t-L-n_u),u_{k-1}(t),\ldots,u_{k-1}(t-n_u))
\end{aligned} \tag{3.9}$$

where

$$\begin{aligned}
&\psi_i(y_k(t-i),\ldots,y_k(t-i-n_y),y_{k-1}(t),\ldots,y_{k-1}(t-n_y), \\
&\quad u_k(t-i),\ldots,u_k(t-i-n_u),u_{k-1}(t),\ldots,u_{k-1}(t-n_u)) \\
&= \psi_{i-1}(f(y_k(t-i),\ldots,y_k(t-i-n_y),u_k(t-i),\ldots,u_k(t-n_u-i)), \\
&\quad y_k(t-i),\ldots,y_k(t-i-n_y+1),y_{k-1}(t),\ldots,y_{k-1}(t-n_y), \\
&\quad u_{k-1}(t-i+1),u_k(t-i),\ldots,u_k(t-n_u-i+1),u_{k-1}(t),\ldots,u_{k-1}(t-n_u))
\end{aligned}$$

for $i = 2, \ldots, L$.

Consider following equation with a column vector $\varsigma_k(t)$,

$$\begin{aligned}&\psi_L(y_k(t-L), \ldots, y_k(t-L-n_y), y_{k-1}(t), \ldots, y_{k-1}(t-n_y),\\&\quad u_k(t-L), \ldots, u_k(t-L-n_u), u_{k-1}(t), \ldots, u_{k-1}(t-n_u))\\&= \varsigma_k^{\mathrm{T}}(t)\Delta \boldsymbol{U}_k(t)\end{aligned} \tag{3.10}$$

Since $\|\Delta \boldsymbol{U}_k(t)\| \neq 0$, there at least exists a solution $\varsigma_k(t)$ to (3.10).

Let

$$\begin{aligned}\boldsymbol{\theta}_k(t) &= [\theta_{1,k}(t), \ldots, \theta_{L,k}(t)]^{\mathrm{T}}\\&= \left[\frac{\partial f}{\partial u_k(t)} \cdots \frac{\partial f}{\partial u_k(t-L+1)}\right]^{\mathrm{T}} + \varsigma_k(t)\end{aligned} \tag{3.11}$$

Equation (3.9) can be written as

$$\Delta y_k(t+1) = \boldsymbol{\theta}_k^{\mathrm{T}}(t)\Delta \boldsymbol{U}_k(t) \tag{3.12}$$

This gives the main conclusion of Theorem 3.1. And the secondary concern of the boundedness of IPG is guaranteed directly by using Assumption 3.2 and (3.3).□

**Remark 3.1** Similar to IPPD, the IPG, $\boldsymbol{\theta}_k(t)$, is also merely related to I/O signals of the controlled system; it can be regarded as a set of differential signals in a certain sense. In particular, the PFIDL data model (3.3) becomes the CFIDL data model (2.3) when $L = 1$.

**Remark 3.2** Compared with CFIDL, the PFIDL method synthetically takes all the relationships between the system output change at the next time instant $t+1$ and the control input changes within a fixed length moving time window $[t-L+1, \ldots, t]$ into consideration. Therefore, the PFIDL method can use additional parameters to capture and disperse the possible complicated behavior of the original system, rather than imprudently compressing all these uncertain factors into the single time-iteration-varying scalar parameter.

**Remark 3.3** Although the dimension of IPG in the PFIDL-based linear data model increases compared with the IPPD in the CFIDL-based linear data model, the dynamic complexity of its every element is reduced consequently. So, $\boldsymbol{\theta}_k(t)$ can be estimated more easily than the IPPD.

**Remark 3.4** Note that the IPG in PFIDL data model (3.3) is not unique and there at least exists one bounded IPG, $\boldsymbol{\theta}_k(t)$, such that this linear data model holds. Further, we can obtain different PFIDL-based linear data models by selecting different control input LLC, $L$. Thus, the flexibility of the dynamic linearization approach of equivalently describing original nonlinear system can be enhanced via a proper selection of IPG and $L$.

### 3.2.2 Controller Design

Given a desired trajectory $y_d(t)$, $t \in \{0, 1, \ldots, N\}$, the control target is to find a sequence of appropriate control inputs $u_k(t)$ such that the tracking error $e_k(t+1) = y_d(t+1) - y_k(t+1)$ converges to zero as the iteration number $k$ approaches infinity.

Rewrite (3.3) as

$$y_k(t+1) = y_{k-1}(t+1) + \boldsymbol{\theta}_k^{\mathrm{T}}(t)\Delta \boldsymbol{U}_k(t), \tag{3.13}$$

Define an index function of the control input as

$$J(u_k(t)) = |e_k(t+1)|^2 + \lambda |u_k(t) - u_{k-1}(t)|^2 \tag{3.14}$$

where $\lambda > 0$ is a weighting factor.

According to (3.13) and the definition of $e_k(t+1)$, we can rewrite $J(u_k(t))$ as follows,

$$\begin{aligned} J(u_k(t)) = {} & |y_d(t+1) - y_{k-1}(t+1) - \boldsymbol{\theta}_k^{\mathrm{T}}(t)(\boldsymbol{U}_k(t) - \boldsymbol{U}_{k-1}(t))|^2 \\ & + \lambda |u_k(t) - u_{k-1}(t)|^2 \end{aligned} \tag{3.15}$$

Using the optimal condition $\partial J / \partial u_k(t) = 0$, we have

$$\begin{aligned} u_k(t) = {} & u_{k-1}(t) + \frac{\rho_1 \theta_{1,k}(t) e_{k-1}(t+1)}{\lambda + |\theta_{1,k}(t)|^2} \\ & - \frac{\sum_{i=2}^{L} \rho_i \theta_{1,k}(t) \theta_{i,k}(t) \Delta u_k(t-i+1)}{\lambda + |\theta_{1,k}(t)|^2} \end{aligned} \tag{3.16}$$

where $0 < \rho_i \leq 1$, $i \in \{1, \ldots, L\}$ is a step-size factor.

Since $\boldsymbol{\theta}_k(t)$ is unknown, the following cost function of parameter estimation is used to derive the estimator,

$$J(\hat{\boldsymbol{\theta}}_k(t)) = |\Delta y_{k-1}(t+1) - \hat{\boldsymbol{\theta}}_k^{\mathrm{T}}(t)\Delta \boldsymbol{U}_{k-1}(t)|^2 + \mu \|\hat{\boldsymbol{\theta}}_k(t) - \hat{\boldsymbol{\theta}}_{k-1}(t)\|^2 \tag{3.17}$$

where $\mu > 0$ is a weighting factor; $\hat{\boldsymbol{\theta}}_k(t)$ is the estimation value of $\boldsymbol{\theta}_k(t)$.

The minimization of above criterion function (3.17) gives the estimation algorithm

$$\hat{\boldsymbol{\theta}}_k(t) = \hat{\boldsymbol{\theta}}_{k-1}(t) + \frac{\eta \Delta \boldsymbol{U}_{k-1}(t)}{\mu + \|\Delta \boldsymbol{U}_{k-1}(t)\|^2}(\Delta y_{k-1}(t+1) - \hat{\boldsymbol{\theta}}_{k-1}^{\mathrm{T}}(t)\Delta \boldsymbol{U}_{k-1}(t)) \tag{3.18}$$

where $0 < \eta < 2$ is a step-size factor.

Correspondingly, the learning control law (3.16) becomes

$$
\begin{aligned}
u_k(t) = u_{k-1}(t) &+ \frac{\rho_1 \hat{\theta}_{1,k}(t) e_{k-1}(t+1)}{\lambda + |\hat{\theta}_{1,k}(t)|^2} \\
&- \frac{\sum_{i=2}^{L} \rho_i \hat{\theta}_{1,k}(t) \hat{\theta}_{i,k}(t) \Delta u_k(t-i+1)}{\lambda + |\hat{\theta}_{1,k}(t)|^2}
\end{aligned} \tag{3.19}
$$

To make the parameter estimation algorithm (3.18) have a strong ability to track time-iteration-varying parameter, a reset algorithm is presented as follows,

$$
\begin{aligned}
&\hat{\boldsymbol{\theta}}_k(t) = \hat{\boldsymbol{\theta}}_0(t), \\
&\text{if } \operatorname{sgn}(\hat{\theta}_{1,k}(t)) \neq \operatorname{sgn}(\hat{\theta}_{1,0}(t)), \text{ or } \|\hat{\boldsymbol{\theta}}_k(t)\| \le \varepsilon, \text{ or } \|\Delta \boldsymbol{U}_{k-1}(t)\| \le \varepsilon
\end{aligned} \tag{3.20}
$$

where $\varepsilon$ is a small positive constant, and $\hat{\boldsymbol{\theta}}_0(t)$ is the initial value of $\hat{\boldsymbol{\theta}}_k(t)$.

Then, the PFIDL-based DDILC (PFIDL-DDILC) is constructed by (3.18), (3.19), and (3.20), shown as follows for convenience,

$$
\begin{aligned}
u_k(t) = u_{k-1}(t) &+ \frac{\rho_1 \hat{\theta}_{1,k}(t) e_{k-1}(t+1)}{\lambda + |\hat{\theta}_{1,k}(t)|^2} \\
&- \frac{\sum_{i=2}^{L} \rho_i \hat{\theta}_{1,k}(t) \hat{\theta}_{i,k}(t) \Delta u_k(t-i+1)}{\lambda + |\hat{\theta}_{1,k}(t)|^2}
\end{aligned} \tag{3.21}
$$

$$
\begin{aligned}
\hat{\boldsymbol{\theta}}_k(t) = \hat{\boldsymbol{\theta}}_{k-1}(t) &+ \frac{\eta \Delta \boldsymbol{U}_{k-1}(t)}{\mu + \|\Delta \boldsymbol{U}_{k-1}(t)\|^2} \\
&\times (\Delta y_{k-1}(t+1) - \hat{\boldsymbol{\theta}}_{k-1}^{\mathrm{T}}(t) \Delta \boldsymbol{U}_{k-1}(t))
\end{aligned} \tag{3.22}
$$

$$
\begin{aligned}
&\hat{\boldsymbol{\theta}}_k(t) = \hat{\boldsymbol{\theta}}_0(t), \\
&\text{if } \operatorname{sgn}(\hat{\theta}_{1,k}(t)) \neq \operatorname{sgn}(\hat{\theta}_{1,0}(t)), \text{ or } \|\hat{\boldsymbol{\theta}}_k(t)\| \le \varepsilon, \text{ or } \|\Delta \boldsymbol{U}_{k-1}(t)\| \le \varepsilon
\end{aligned} \tag{3.23}
$$

**Remark 3.5** The PFIDL-DDILC scheme for nonlinear system (3.1) is constructed by integrating learning control law (3.21), parameter updating law (3.22), and reset algorithm (3.23). The proposed scheme has $L$-dimensional vector to be updated iteratively, and it is quite different from the traditional adaptive control system, in which there are usually $2n$ model parameters to be estimated online, where $n$ denotes the order of the controlled plant.

**Remark 3.6** The proposed PFIDL-DDILC only utilizes I/O data for the controller design and analysis. It is noteworthy that the learning gain in control law (3.21) is nonlinear and consists of $\hat{\boldsymbol{\theta}}_k(t)$, therefore it can be updated in real-time by the use of (3.22) and (3.23). This is quite different from standard ILC, where the learning gain is fixed, and cannot be tuned automatically and iteratively.

### 3.2.3 Convergence Analysis

Before proceeding to the analysis of the proposed PFIDL-DDILC scheme (3.21)–(3.23), one assumption and two lemmas are given as follows.

**Assumption 3.3** The first element of IPG $\theta_{1,k}(t) > \ell$ (or $\theta_{1,k}(t) < -\ell$), $\forall t \in \{0, 1, \ldots, N\}$, $\forall k$, where $\ell$ is a positive constant. Without loss of generality, we consider $\theta_{1,k}(t) > \ell$.

**Lemma 3.1** (Jury (1964)) *Let* $A = \begin{bmatrix} a_1 & a_2 & \cdots & a_{L-1} & a_L \\ 1 & 0 & \cdots & 0 & 0 \\ 0 & 1 & \cdots & 0 & 0 \\ \vdots & \vdots & \vdots & \vdots & \vdots \\ 0 & 0 & \cdots & 1 & 0 \end{bmatrix}$. *If* $\sum_{i=1}^{L} |a_i| < 1$, *then* $s(A) < 1$, *where* $s(A)$ *is the spectral radius of* $A$.

**Lemma 3.2** (Huang (1984)) $A \in R^{n \times n}$, *and* $s(A)$ *represents the spectral radius of* $A$. *Then,* $\forall \delta > 0$, *there always exists a proper matrix norm* $\|A\|_v$ *on the normed vector space* $v$, *such that* $\|A\|_v \leq s(A) + \delta$.

The convergence property of the proposed PFIDL-DDILC scheme is shown as the following theorem.

**Theorem 3.2** *For a class of nonlinear nonaffine discrete-time systems (3.1), if assumptions 3.1, 3.2, and 3.3 hold, then the proposed PFIDL-DDILC method (3.21)–(3.23) guarantees that:*

*(i) the parameter estimation* $\hat{\theta}_k(t)$ *is bounded,* $\forall t \in \{0, 1, \ldots, N\}$, $\forall k$;
*(ii) the tracking error converges to zero pointwisely over the finite time interval* $t \in \{0, 1, \ldots, N\}$ *as* $k$ *approaches to infinity, i.e.,* $\lim_{k\to\infty} e_k(t+1) = 0$;
*(iii) the system output* $y_k(t)$ *and the control input signal* $u_k(t)$ *are bounded,* $\forall t \in \{0, 1, \ldots, N\}$, $\forall k$.

***Proof*** *Part 1. The boundedness of* $\hat{\theta}_k(t)$

If $\text{sgn}(\hat{\theta}_{1,k}(t)) \neq \text{sgn}(\hat{\theta}_{1,0}(t))$, or $\|\hat{\theta}_k(t)\| \leq \varepsilon$ or $\|\Delta U_{k-1}(t)\| \leq \varepsilon$, the boundedness of $\hat{\theta}_k(t)$ is derived clearly according to the reset algorithm (3.23).

In other case, define IPG estimation error as

$$\tilde{\theta}_k(t) = \theta_k(t) - \hat{\theta}_k(t) \tag{3.24}$$

Subtracting $\theta_k(t)$ from both sides of (3.22) and using (3.3), we have

$$\tilde{\theta}_k(t) = \left(I - \frac{\eta \Delta U_{k-1}(t) \Delta U_{k-1}^{\mathrm{T}}(t)}{\mu + \|\Delta U_{k-1}(t)\|^2}\right) \tilde{\theta}_{k-1}(t) + \theta_k(t) - \theta_{k-1}(t) \tag{3.25}$$

Taking norms on both sides of (3.25), yields

$$\|\tilde{\boldsymbol{\theta}}_k(t)\| \leq \left\| \left( I - \frac{\eta \Delta \boldsymbol{U}_{k-1}(t) \Delta \boldsymbol{U}_{k-1}^{\mathrm{T}}(t)}{\mu + \|\Delta \boldsymbol{U}_{k-1}(t)\|^2} \right) \tilde{\boldsymbol{\theta}}_{k-1}(t) \right\| + 2b \tag{3.26}$$

For the first term of right side of (3.26), one has

$$\begin{aligned} \left\| \left( I - \frac{\eta \Delta \boldsymbol{U}_{k-1}(t) \Delta \boldsymbol{U}_{k-1}^{\mathrm{T}}(t)}{\mu + \|\Delta \boldsymbol{U}_{k-1}(t)\|^2} \right) \tilde{\boldsymbol{\theta}}_{k-1}(t) \right\|^2 &= \|\tilde{\boldsymbol{\theta}}_{k-1}(t)\|^2 + \frac{\|\eta \Delta \boldsymbol{U}_{k-1}^{\mathrm{T}}(t) \tilde{\boldsymbol{\theta}}_{k-1}(t)\|^2}{\mu + \|\Delta \boldsymbol{U}_{k-1}(t)\|^2} \\ &\quad \times \left( -2 + \frac{\eta \|\Delta \boldsymbol{U}_{k-1}(t)\|^2}{\mu + \|\Delta \boldsymbol{U}_{k-1}(t)\|^2} \right) \end{aligned} \tag{3.27}$$

Since $0 < \eta < 2$, $\mu > 0$, one has,

$$\left( -2 + \frac{\eta \|\Delta \boldsymbol{U}_{k-1}(t)\|^2}{\mu + \|\Delta \boldsymbol{U}_{k-1}(t)\|^2} \right) < 0 \tag{3.28}$$

Then, one can obtain from (3.27) that,

$$\left\| \left( I - \frac{\eta \Delta \boldsymbol{U}_{k-1}(t) \Delta \boldsymbol{U}_{k-1}^{\mathrm{T}}(t)}{\mu + \|\Delta \boldsymbol{U}_{k-1}(t)\|^2} \right) \tilde{\boldsymbol{\theta}}_{k-1}(t) \right\|^2 \leq d_1 \|\tilde{\boldsymbol{\theta}}_{k-1}(t)\|^2 \tag{3.29}$$

where $0 < d_1 < 1$ is a constant.

In terms of (3.26) and (3.29)

$$\begin{aligned} \|\tilde{\boldsymbol{\theta}}_k(t)\| &\leq d_1 \|\tilde{\boldsymbol{\theta}}_{k-1}(t)\| + 2b \\ &\leq d_1^k \|\tilde{\boldsymbol{\theta}}_0(t)\| + \frac{2b}{1 - d_1} \end{aligned} \tag{3.30}$$

Note that, we only need the existence of constant $d_1$ instead of its exact value. Hence, $\tilde{\boldsymbol{\theta}}_k(t)$ is bounded because $\hat{\boldsymbol{\theta}}_0(t)$ is chosen bounded and $\boldsymbol{\theta}_k(t)$ is also bounded according to Theorem 3.1. Therefore, the boundedness of $\hat{\boldsymbol{\theta}}_k(t)$ can be guaranteed $\forall t \in \{0, 1, \ldots, N\}$, and $\forall k$.

*Part 2. The convergence of output tracking error* $e_k(t)$

Since $\boldsymbol{\theta}_k(t)$ and $\hat{\boldsymbol{\theta}}_k(t)$ are bounded, there exists $\lambda_{\min} > 0$, such that following inequalities (3.31)–(3.34) hold when $\lambda > \lambda_{\min}$,

$$\left| \frac{\hat{\theta}_{1,k}(t)}{\lambda + |\hat{\theta}_{1,k}(t)|^2} \right| \leq \left| \frac{\hat{\theta}_{1,k}(t)}{2\sqrt{\lambda} |\hat{\theta}_{1,k}(t)|} \right| < \frac{1}{2\sqrt{\lambda_{\min}}} = M_1 < \frac{0.5}{b} \tag{3.31}$$

$$0 < M_2 \leq \left| \frac{\theta_{1,k}(t) \hat{\theta}_{1,k}(t)}{\lambda + |\hat{\theta}_{1,k}(t)|^2} \right| \leq b \left| \frac{\hat{\theta}_{1,k}(t)}{2\sqrt{\lambda} |\hat{\theta}_{1,k}(t)|} \right| < \frac{b}{2\sqrt{\lambda_{\min}}} < 0.5 \tag{3.32}$$

$$\|\boldsymbol{\theta}_k(t)\|_v \le M_3 \tag{3.33}$$

$$\left(\sum_{i=2}^{L}\left|\frac{\hat{\theta}_{1,k}(t)\hat{\theta}_{i,k}(t)}{\lambda+|\hat{\theta}_{1,k}(t)|^2}\right|\right)^{\frac{1}{L-1}} \le M_4 \tag{3.34}$$

where $M_1$, $M_2$, $M_3$, and $M_4$ are positive constants.

Selecting $\rho_{\max} = \max_{i\in\{2,\ldots,L\}}\{\rho_i\}$, such that

$$\begin{aligned}\sum_{i=2}^{L}\rho_i\left|\frac{\hat{\theta}_{1,k}(t)\hat{\theta}_{i,k}(t)}{\lambda+|\hat{\theta}_{1,k}(t)|^2}\right| &\le \rho_{\max}\sum_{i=2}^{L}\left|\frac{\hat{\theta}_{1,k}(t)\hat{\theta}_{i,k}(t)}{\lambda+|\hat{\theta}_{1,k}(t)|^2}\right| \\ &\le \rho_{\max}M_4^{L-1} < 1\end{aligned} \tag{3.35}$$

Let $\boldsymbol{A}_k(t) = \begin{bmatrix} -\frac{\rho_2\hat{\theta}_{1,k}(t)\hat{\theta}_{2,k}(t)}{\lambda+|\hat{\theta}_{1,k}(t)|^2} \cdots -\frac{\rho_L\hat{\theta}_{1,k}(t)\hat{\theta}_{L,k}(t)}{\lambda+|\hat{\theta}_{1,k}(t)|^2} & 0 \\ I & \boldsymbol{0}\end{bmatrix}$ and $\boldsymbol{C} = [1\,0\,\cdots\,0]^{\mathrm{T}}$, then Eq. (3.21) can be rewritten as

$$\Delta\boldsymbol{U}_k(t) = \boldsymbol{A}_k(t)\Delta\boldsymbol{U}_k(t-1) + \frac{\rho_1\hat{\theta}_{1,k}(t)}{\lambda+|\hat{\theta}_{1,k}(t)|^2}\boldsymbol{C}e_{k-1}(t+1) \tag{3.36}$$

The characteristic equation of $\boldsymbol{A}_k(t)$ is

$$z^L + \frac{\rho_2\hat{\theta}_{1,k}(t)\hat{\theta}_{2,k}(t)}{\lambda+|\hat{\theta}_{1,k}(t)|^2}z^{L-1} + \cdots + \frac{\rho_L\hat{\theta}_{1,k}(t)\hat{\theta}_{L,k}(t)}{\lambda+|\hat{\theta}_{1,k}(t)|^2}z = 0 \tag{3.37}$$

From (3.35) and Lemma 3.1, we have $|z| < 1$. Thus, the following equality holds

$$\begin{aligned}|z|^{L-1} &\le \sum_{i=2}^{L}\rho_i\left|\frac{\hat{\theta}_{1,k}(t)\hat{\theta}_{i,k}(t)}{\lambda+|\hat{\theta}_{1,k}(t)|^2}\right||z|^{L-i} \le \sum_{i=2}^{L}\rho_i\left|\frac{\hat{\theta}_{1,k}(t)\hat{\theta}_{i,k}(t)}{\lambda+|\hat{\theta}_{1,k}(t)|^2}\right| \\ &\le \rho_{\max}M_4^{L-1}\end{aligned} \tag{3.38}$$

Inequality (3.38) implies that

$$|z| \le \rho_{\max}^{\frac{1}{L-1}}M_4 < 1 \tag{3.39}$$

According to Lemma 3.2, there exists an arbitrary small positive constant $\delta$ such that the following inequality holds,

$$\|\boldsymbol{A}_k(t)\|_v \le s(\boldsymbol{A}_k(t)) + \delta \le \rho_{\max}^{\frac{1}{L-1}}M_4 + \delta < 1 \tag{3.40}$$

where $s(\boldsymbol{A}_k(t))$ denotes the spectral radius of $\boldsymbol{A}_k(t)$, and $\|\boldsymbol{A}_k(t)\|_v$ denotes the consistent matrix norm of $\boldsymbol{A}_k(t)$.

Let $d_2 = \rho_{\max}^{\frac{1}{L-1}} M_4 + \delta$. Taking norms on both sides of Eq. (3.36), we have

$$
\begin{aligned}
\|\Delta \boldsymbol{U}_k(t)\|_v &\le \|\boldsymbol{A}_k(t)\|_v \|\Delta \boldsymbol{U}_k(t-1)\|_v + \rho_1 \left| \frac{\hat{\theta}_{1,k}(t)}{\lambda + |\hat{\theta}_{1,k}(t)|^2} \right| |e_{k-1}(t+1)| \\
&< d_2 \|\Delta \boldsymbol{U}_k(t-1)\|_v + \rho_1 M_1 |e_{k-1}(t+1)| \\
&\ \ \vdots \\
&< d_2^{t+1} \|\Delta \boldsymbol{U}_k(-1)\|_v + d_2^t \rho_1 M_1 |e_{k-1}(1)| + \cdots + \rho_1 M_1 |e_{k-1}(t+1)| \\
&= \rho_1 M_1 \sum_{i=0}^{t} d_2^{t-i} |e_{k-1}(i+1)|
\end{aligned}
\tag{3.41}
$$

In terms of (3.3) and (3.36), we have

$$
\begin{aligned}
e_k(t+1) &= y_d(t+1) - y_k(t+1) \\
&= y_d(t+1) - y_{k-1}(t+1) - \boldsymbol{\theta}_k^{\mathrm{T}}(t) \Delta \boldsymbol{U}_k(t) \\
&= \left( 1 - \frac{\rho_1 \theta_{1,k}(t) \hat{\theta}_{1,k}(t)}{\lambda + |\hat{\theta}_{1,k}(t)|^2} \right) e_{k-1}(t+1) \\
&\quad - \boldsymbol{\theta}_k^{\mathrm{T}}(t) \boldsymbol{A}_k(t) \Delta \boldsymbol{U}_k(t-1)
\end{aligned}
\tag{3.42}
$$

Taking norms on both sides of Eq. (3.42), we have

$$
\begin{aligned}
|e_k(t+1)| &\le \left| 1 - \frac{\rho_1 \theta_{1,k}(t) \hat{\theta}_{1,k}(t)}{\lambda + |\hat{\theta}_{1,k}(t)|^2} \right| |e_{k-1}(t+1)| \\
&\quad + d_2 \|\boldsymbol{\theta}_k^{\mathrm{T}}(t)\|_v \|\Delta \boldsymbol{U}_k(t-1)\|_v
\end{aligned}
\tag{3.43}
$$

According to Assumption 3.3, the value of $\hat{\theta}_{1,0}(t)$ at the initial iteration is selected such that its sign is the same as that of $\theta_{1,k}(t)$. Then, the reset algorithm (3.23) implies that $\theta_{1,k}(t)\hat{\theta}_{1,k}(t) > 0$. Selecting $0 < \rho_1 \le 1$ such that

$$
\begin{aligned}
\left| 1 - \rho_1 \frac{\theta_{1,k}(t) \hat{\theta}_{1,k}(t)}{\lambda + |\hat{\theta}_{1,k}(t)|^2} \right| &= \left| 1 - \rho_1 \left| \frac{\theta_{1,k}(t) \hat{\theta}_{1,k}(t)}{\lambda + |\hat{\theta}_{1,k}(t)|^2} \right| \right| \\
&\le 1 - \rho_1 M_2 < 1
\end{aligned}
$$

Let $d_3 = 1 - \rho_1 M_2$. Using (3.33), (3.43) can be rewritten as

$$
|e_k(t+1)| \le d_3 |e_{k-1}(t+1)| + d_2 M_3 \|\Delta \boldsymbol{U}_k(t-1)\|_v \tag{3.44}
$$

Substituting (3.41) into (3.44), yields

$$|e_k(t+1)| < d_3|e_{k-1}(t+1)| + \rho_1 M_1 M_3 \sum_{i=0}^{t-1} d_2^{t-i}|e_{k-1}(i+1)| \tag{3.45}$$

Assume $e_{\max}(k) = e_k(\tau_1)$, then the following inequality holds,

$$\begin{aligned}|e_{\max}(k)| = |e_k(\tau_1)| &\le d_3|e_{k-1}(\tau_1)| + \rho_1 M_1 M_3 \sum_{i=0}^{\tau_1-2} d_2^{\tau_1-1-i}|e_{k-1}(i+1)| \\ &\le d_3|e_{\max}(k-1)| + \rho_1 M_1 M_3 \sum_{i=0}^{\tau_1-2} d_2^{\tau_1-1-i}|e_{\max}(k-1)| \\ &< (d_3 + \rho_1 M_1 M_3 \frac{d_2}{1-d_2})|e_{\max}(k-1)| \end{aligned} \tag{3.46}$$

Selecting $0 < \rho_1 \le 1, \ldots, 0 < \rho_L \le 1$ such that

$$\begin{aligned} d_3 + \rho_1 M_1 M_3 \frac{d_2}{1-d_2} &= 1 - \rho_1 M_2 + \rho_1 M_1 M_3 \frac{d_2}{1-d_2} \\ &= 1 - \rho_1(M_2 + M_1 M_3) + \frac{\rho_1 M_1 M_3}{1-d_2} \\ &= 1 - \rho_1(M_2 + M_1 M_3) \\ &\quad + \frac{\rho_1 M_1 M_3}{1 - \rho_{\max}^{\frac{1}{L-1}} M_4 - \delta} < 1. \end{aligned} \tag{3.47}$$

Let $d_4 = d_3 + \rho_1 M_1 M_3 \frac{d_2}{1-d_2}$. It is obviously that

$$\begin{aligned} \lim_{k\to\infty} |e_{\max}(k)| &< \lim_{k\to\infty} d_4|e_{\max}(k-1)| \\ &< \cdots < \lim_{k\to\infty} d_4^k|e_{\max}(0)| \end{aligned} \tag{3.48}$$

Therefore, one can obtain from (3.48) that $\lim_{k\to\infty} |e_{\max}(k)| = 0$. Then, $\lim_{k\to\infty} e_k(t) = 0$ can be obtained directly based on the definition of $e_{\max}(k)$.

*Part 3. Boundedness of control signal and system output*

Since $y_d(t)$ is bounded, the convergence of $e_k(t)$ implies that $y_k(t)$ is also bounded. We know that

$$\begin{aligned} \|\boldsymbol{U}_k(t)\|_v &= \|\boldsymbol{U}_k(t) - \boldsymbol{U}_0(t) + \boldsymbol{U}_0(t)\|_v \\ &\le \|\boldsymbol{U}_k(t) - \boldsymbol{U}_0(t)\|_v + \|\boldsymbol{U}_0(t)\|_v \\ &= \|\boldsymbol{U}_k(t) - \boldsymbol{U}_{k-1}(t) + \boldsymbol{U}_{k-1}(t) - \cdots + \boldsymbol{U}_1(t) - \boldsymbol{U}_0(t)\|_v + \|\boldsymbol{U}_0(t)\|_v \\ &\le \|\Delta\boldsymbol{U}_k(t)\|_v + \|\Delta\boldsymbol{U}_{k-1}(t)\|_v + \cdots + \|\Delta\boldsymbol{U}_1(t)\|_v + \|\boldsymbol{U}_0(t)\|_v \end{aligned} \tag{3.49}$$

From (3.41) and (3.49), we have

$$\begin{aligned}
\|\boldsymbol{U}_k(t)\|_v &\le \sum_{h=1}^{k} \|\Delta \boldsymbol{U}_h(t)\|_v + \|\boldsymbol{U}_0(t)\|_v \\
&< \rho_1 M_1 \sum_{h=1}^{k} \sum_{i=0}^{t} d_2^{t-i} |e_{h-1}(i+1)| + \|\boldsymbol{U}_0(t)\|_v \\
&< \frac{\rho_1 M_1}{1-d_2} \sum_{h=1}^{k} e_{\max}(h-1) + \|\boldsymbol{U}_0(t)\|_v \\
&< \frac{\rho_1 M_1 e_{\max}(0)}{(1-d_2)(1-d_4)} + \|\boldsymbol{U}_0(t)\|_v
\end{aligned} \tag{3.50}$$

Since $e_{\max}(0)$ and $\boldsymbol{U}_0(t)$ in the first iteration are given bounded, (3.50) implies that $u_k(t)$ is bounded. □

### 3.2.4 Simulation Study

Consider a nonlinear nonaffine SISO system

$$y_k(t+1) = \begin{cases} \frac{y_k(t)}{1+y_k(t)^2} + u_k(t)^3, \ 0 \le t < 50 \\ \frac{y_k(t)y_k(t-1)y_k(t-2)u_k(t-1)(y_k(t-2)-1)+a(t)u_k(t)}{1+y_k(t-1)^2+y_k(t-2)^2}, \ 50 \le t \le 100 \end{cases} \tag{3.51}$$

where $a(t) = 1 + \text{round}(t/50)$ is a varying parameter. It is notable that the structure, orders and parameters of the controlled system are time-varying.

The desired trajectory is given as follows,

$$y_d(t+1) = \begin{cases} 0.5 \times (-1)^{\text{round}(t/10)}, \ 0 \le t \le 30 \\ 0.5\sin(t\pi/10) + 0.3\cos(t\pi/10), \ 30 < t \le 70 \\ 0.5 \times (-1)^{\text{round}(t/10)}, \ 70 < t \le 100 \end{cases} \tag{3.52}$$

By choosing $\eta = 0.5$, $\rho_1 = 1$, $\rho_2 = 0.6$, $\lambda = 1$, $\mu = 0.5$, $L = 2$, $\hat{\boldsymbol{\theta}}_0(t) = 1$, $u_0(t) = 0$, and $y_k(0) = 0$, the simulation results are shown in Figs. 3.1 and 3.2. Figure 3.1 shows the output tracking performance at the 100th iteration, where the vertical axis denotes the system output and the horizontal axis denotes the time instants. Figure 3.2 shows the convergence performance of tracking error, where the Y-axis denotes the maximum absolute value of the output tracking error, defined as $e_{\max}(k) = \max_{t\in\{1,\ldots,100\}}\{|e_k(t)|\}$, and the X-axis denotes the iteration numbers. One can see that the effectiveness of the presented PFIDL-DDILC is confirmed by the simulation results.

For the comparison, the CFIDL-DDILC proposed in Chap. 2 is also applied. Selecting $\eta = 0.5$, $\rho = 0.8$, $\lambda = 1$, $\mu = 0.5$, using CFIDL-DDILC (2.16)–(2.18), the convergence performance of maximum tracking error is shown in Fig. 3.2. It is

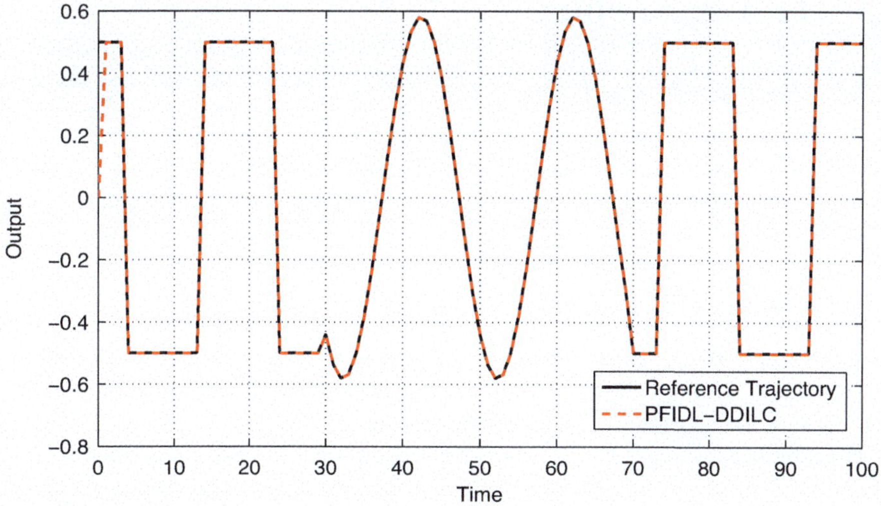


**Fig. 3.1** Profile of system output

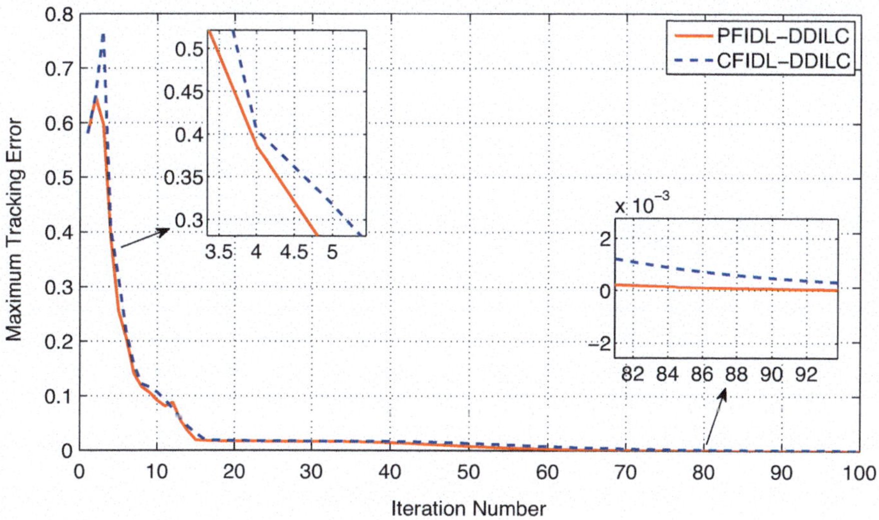


**Fig. 3.2** Convergence of maximum tracking error

seen that the proposed PFIDL-DDILC outperforms CFIDL-DDILC because more control input information is utilized in the control system design.

## 3.3 FFIDL-Based DDILC

### 3.3.1 Problem Formulation

Denote $\boldsymbol{H}_k(t)$ as a vector consisting of all control inputs signals within an input-related moving time window $[t - L_u + 1, \ldots, t]$ and all system output signals within a output-related moving time window $[t - L_y + 1, \ldots, t]$, i.e.,

$$\boldsymbol{H}_k(t) = [y_k(t), \ldots, y_k(t - L_y + 1), u_k(t), \ldots, u_k(t - L_u + 1)]^{\mathrm{T}} \tag{3.53}$$

where $\boldsymbol{H}_k(t) = \mathbf{0}_{L_y+L_u}$ for $k < 0$ and $t < 0$. The two integers $L_y$ and $L_u$ $(0 \leq L_y \leq n_y,\ 1 \leq L_u \leq n_u)$ are called *pseudo-orders* of the system, which are similar to the definition of the constant $L$ in the PFIDL method.

For nonlinear system (3.1), another two assumptions are made as follows.

**Assumption 3.4** The partial derivatives of $f(\cdots)$ with respect to all variables are continuous.

**Assumption 3.5** System (3.1) satisfies the generalized *Lipschitz* condition, that is,

$$|y_{k_1}(t+1) - y_{k_2}(t+1)| \leq b\|\boldsymbol{H}_{k_1}(t) - \boldsymbol{H}_{k_2}(t)\|$$

for $\boldsymbol{H}_{k_1}(t) \neq \boldsymbol{H}_{k_2}(t)$ with any $k_1 \neq k_2, k_1, k_2 \geq 0$, where $y_{k_i}(t+1) = f(y_{k_i}(t), \ldots, y_{k_i}(t-n_y), u_{k_i}(t), \ldots, u_{k_i}(t-n_u))$, $i = 1, 2$, and $b$ is a positive constant.

Denote $\Delta\boldsymbol{H}_k(t) = \boldsymbol{H}_k(t) - \boldsymbol{H}_{k-1}(t)$. The FFIDL of system (3.1) is shown in the following theorem in detail.

**Theorem 3.3** *Consider nonlinear system (3.1) satisfying assumptions 3.4 and 3.5. For any* $0 \leq L_y \leq n_y$ *and* $1 \leq L_u \leq n_u$, *if* $\|\Delta\boldsymbol{H}_k(t)\| \neq 0$, *there exists a time-iteration-varying vector* $\boldsymbol{\vartheta}_k(t) \in R^{L_y+L_u}$, *called the iteration-related pseudo gradient (IPG), such that system (3.1) can be transformed into the following FFIDL data model,*

$$\Delta y_k(t+1) = \boldsymbol{\vartheta}_k^{\mathrm{T}}(t)\Delta\boldsymbol{H}_k(t) \tag{3.54}$$

*where* $\boldsymbol{\vartheta}_k(t) = [\vartheta_{1,k}(t), \ldots, \vartheta_{L_y,k}(t), \vartheta_{L_y+1,k}(t), \ldots, \vartheta_{L_y+L_u,k}(t)]^{\mathrm{T}}$, *and* $\|\boldsymbol{\vartheta}_k(t)\| \leq b$.

***Proof*** It follows from (3.1) that

$$\begin{aligned}\Delta y_k(t+1) &= f(y_k(t),\ldots,y_k(t-n_y),u_k(t),\ldots,u_k(t-n_u))\\&\quad - f(y_{k-1}(t),\ldots,y_{k-1}(t-n_y),u_{k-1}(t),\ldots,u_{k-1}(t-n_u))\\&= f(y_k(t),\ldots,y_k(t-L_y+1),y_k(t-L_y),\ldots,y_k(t-n_y),\\&\quad u_k(t),\ldots,u_k(t-L_u+1),u_k(t-L_u),\ldots,u_k(t-n_u))\\&\quad - f(y_{k-1}(t),\ldots,y_{k-1}(t-L_y+1),y_k(t-L_y),\ldots,y_k(t-n_y),\\&\quad u_{k-1}(t),\ldots,u_{k-1}(t-L_u+1),u_k(t-L_u),\ldots,u_k(t-n_u))\\&\quad + f(y_{k-1}(t),\ldots,y_{k-1}(t-L_y+1),y_k(t-L_y),\ldots,y_k(t-n_y),\\&\quad u_{k-1}(t),\ldots,u_{k-1}(t-L_u+1),u_k(t-L_u),\ldots,u_k(t-n_u))\\&\quad - f(y_{k-1}(t),\ldots,y_{k-1}(t-n_y),u_{k-1}(t),\ldots,u_{k-1}(t-n_u))\end{aligned}\tag{3.55}$$

Denote

$$\begin{aligned}\psi_k(t) = {}& f(y_{k-1}(t),\ldots,y_{k-1}(t-L_y+1),y_k(t-L_y),\ldots,y_k(t-n_y),\\& u_{k-1}(t),\ldots,u_{k-1}(t-L_u+1),u_k(t-L_u),\ldots,u_k(t-n_u))\\& - f(y_{k-1}(t),\ldots,y_{k-1}(t-n_y),u_{k-1}(t),\ldots,u_{k-1}(t-n_u))\end{aligned}$$

By virtue of Assumption 3.4 and *Differential Mean Value Theorem*, (3.55) can be rewritten as

$$\begin{aligned}\Delta y_k(t+1) = {}& \frac{\partial f}{\partial y_k(t)}\Delta y_k(t)+\cdots+\frac{\partial f}{\partial y_k(t-L_y+1)}\Delta y_k(t-L_y+1)\\&+\frac{\partial f}{\partial u_k(t)}\Delta u_k(t)+\cdots+\frac{\partial f}{\partial u_k(t-L_u+1)}\Delta u_k(t-L_u+1)\\&+\psi_k(t)\end{aligned}\tag{3.56}$$

where $\partial f/\partial y_k(t-i)$, $0\le i\le L_y-1$ and $\partial f/\partial u_k(t-j)$, $0\le j\le L_u-1$ denote the partial derivatives of $f(\cdots)$ with respect to $y_k(i)$ and $u_k(j)$ at some point in the interval $[y_k(i), y_{k-1}(i)]$ and $[u_k(j), u_{k-1}(j)]$, respectively.

Consider the following equation with a vector-valued variable $\boldsymbol{\varsigma}_k(t)$

$$\psi_k(t) = \boldsymbol{\varsigma}_k^{\mathrm{T}}(t)\Delta\boldsymbol{H}_k(t)\tag{3.57}$$

Since $\|\Delta\boldsymbol{H}_k(t)\|\neq 0$, there at least exists a solution $\boldsymbol{\varsigma}_k(t)$ to (3.57).

Let

$$\begin{aligned}\boldsymbol{\vartheta}_k(t) &= [\vartheta_{1,k}(t)\ \cdots\ \vartheta_{L_y,k}(t)\ \vartheta_{L_y+1,k}(t)\ \cdots\ \vartheta_{L_y+L_u,k}(t)]^{\mathrm{T}}\\&= \left[\frac{\partial f}{\partial y_k(t)}\ \cdots\ \frac{\partial f}{\partial y_k(t-L_y+1)}\ \frac{\partial f}{\partial u_k(t)}\ \cdots\ \frac{\partial f}{\partial u_k(t-L_u+1)}\right]^{\mathrm{T}}\\&\quad+\boldsymbol{\varsigma}_k(t)\end{aligned}\tag{3.58}$$

Equation (3.56) can be written as the following linear data model (LDM)

$$\Delta y_k(t+1) = \boldsymbol{\vartheta}_k^{\mathrm{T}}(t)\Delta \boldsymbol{H}_k(t) \tag{3.59}$$

This gives the main conclusion of Theorem 3.3. And the boundedness of $\boldsymbol{\vartheta}_k(t)$ is guaranteed directly by using Assumption 3.5 and (3.54). □

**Remark 3.7** In particular, the FFIDL-based LDM (3.54) becomes the PFIDL-based LDM (3.3) when $L_y = 0$ and $L_u = L$, and the FFIDL-based LDM (3.54) is transformed into the CFIDL-based LDM (2.3) when $L_y = 0$ and $L_u = 1$. Therefore, the FFIDL proposed in this section can be regarded as the most general dynamic linearization method. In practice, we can obtain different LDMs by selecting different pseudo-orders.

**Remark 3.8** Generally, the FFIDL-based LDM with a higher pseudo-order is needed for a complicated system, and the one with a lower pseudo-order is suitable to a simple system. Obviously, compared with the PFIDL, the FFIDL method considers all the relationships between the system output iterative change at time instant $t+1$, and both the control input iterative change within an input-related fixed-length moving time window $[t - L_u + 1, \ldots, t]$ and the system output iterative change within an output-related fixed length moving time window $[t - L_y + 1, \ldots, t]$.

**Remark 3.9** The FFIDL method not only enhances the applicability of the dynamic linearization method but also reduces the complexity of dynamics of IPG element. However, it may lead to a heavy computational burden for the controller design.

**Remark 3.10** The pseudo-orders $L_y$ and $L_u$ are adjustable. If the orders of the system $n_y$ and $n_u$ are known, it is reasonable to choose pseudo-orders satisfying $L_y = n_y$ and $L_u = n_u$. In practical applications, the values of $n_y$ and $n_u$ are generally unknown, and sometimes $n_y$ and $n_u$ are time-iteration-varying, so they are difficult to be determined. In such circumstances, one may select pseudo-orders of the FFIDL data model to be approximate values of the true orders of the system. On the other hand, when the orders of the plant are large, it is necessary to select pseudo orders in a proper size so as to derive a dynamic linearization data model with lower orders, which can help to reduce the computational burden in practice. As we know, a high-order nonlinear model must lead to a high order nonlinear controller, which definitely renders unavoidable difficulties in the design and application of the controller. Thus, with a proper selection of the pseudo-orders, the simplification procedure for a complex system model or for a complex controller can be avoided.

### 3.3.2 Controller Design

For a desired trajectory $y_d(t), t \in \{0, 1, \ldots, N\}$, the objective is to design a sequence of appropriate control inputs $u_k(t)$ such that the tracking error $e_k(t+1) = y_d(t+1) - y_k(t+1)$ converges to zero as the iteration number $k$ approaches infinity.

Rewrite (3.54) as

$$y_k(t+1) = y_{k-1}(t+1) + \boldsymbol{\vartheta}_k^{\mathrm{T}}(t)\Delta \boldsymbol{H}_k(t) \tag{3.60}$$

Define an index function of control input as

$$J(u_k(t)) = |e_k(t+1)|^2 + \lambda |u_k(t) - u_{k-1}(t)|^2 \tag{3.61}$$

where $\lambda > 0$ is a weighting factor.

According to (3.60) and the definition of $e_k(t+1)$, $J(u_k(t))$ can be rewritten as follows,

$$\begin{aligned} J(u_k(t)) = {} & |y_d(t+1) - y_{k-1}(t+1) - \boldsymbol{\vartheta}_k^{\mathrm{T}}(t)(\boldsymbol{H}_k(t) - \boldsymbol{H}_{k-1}(t))|^2 \\ & + \lambda |u_k(t) - u_{k-1}(t)|^2 \end{aligned} \tag{3.62}$$

Using the optimal condition $\partial J / \partial u_k(t) = 0$, we have

$$\begin{aligned} u_k(t) = {} & u_{k-1}(t) + \frac{\rho_{L_y+1}\vartheta_{L_y+1,k}(t) e_{k-1}(t+1)}{\lambda + \vartheta_{L_y+1,k}(t)^2} - \frac{\vartheta_{L_y+1,k}(t) \sum_{l=1}^{L_y} \rho_l \vartheta_{l,k}(t) \Delta y_k(t-l+1)}{\lambda + \vartheta_{L_y+1,k}(t)^2} \\ & - \frac{\vartheta_{L_y+1,k}(t) \sum_{l=L_y+2}^{L_y+L_u} \rho_l \vartheta_{l,k}(t) \Delta u_k(t-l+L_y+1)}{\lambda + \vartheta_{L_y+1,k}(t)^2} \end{aligned} \tag{3.63}$$

where $0 < \rho_l \leq 1, l \in \{1, \ldots, L_y + L_u\}$ is the steps-size factor.

Since $\boldsymbol{\vartheta}_k(t)$ is unknown, the following cost function of parameter estimation is used to derive estimator,

$$J(\hat{\boldsymbol{\vartheta}}_k(t)) = |\Delta y_{k-1}(t+1) - \hat{\boldsymbol{\vartheta}}_k^{\mathrm{T}}(t)\Delta \boldsymbol{H}_{k-1}(t)|^2 + \mu \|\hat{\boldsymbol{\vartheta}}_k(t) - \hat{\boldsymbol{\vartheta}}_{k-1}(t)\|^2 \tag{3.64}$$

where $\mu > 0$ is a weighting factor, $\hat{\boldsymbol{\vartheta}}_k(t)$ is the estimation value of $\boldsymbol{\vartheta}_k(t)$.

The minimization of above criterion function gives the estimation algorithm,

$$\hat{\boldsymbol{\vartheta}}_k(t) = \hat{\boldsymbol{\vartheta}}_{k-1}(t) + \frac{\eta \Delta \boldsymbol{H}_{k-1}(t)(\Delta y_{k-1}(t+1) - \hat{\boldsymbol{\vartheta}}_{k-1}^{\mathrm{T}}(t)\Delta \boldsymbol{H}_{k-1}(t))}{\mu + \|\Delta \boldsymbol{H}_{k-1}(t)\|^2} \tag{3.65}$$

where $0 < \eta < 2$ is a step-size factor.

Correspondingly, the learning control law (3.63) is rewritten as,

$$\begin{aligned} u_k(t) = {} & u_{k-1}(t) + \frac{\rho_{L_y+1}\hat{\vartheta}_{L_y+1,k}(t) e_{k-1}(t+1)}{\lambda + \hat{\vartheta}_{L_y+1,k}(t)^2} - \frac{\hat{\vartheta}_{L_y+1,k}(t) \sum_{l=1}^{L_y} \rho_l \hat{\vartheta}_{l,k}(t) \Delta y_k(t-l+1)}{\lambda + \hat{\vartheta}_{L_y+1,k}(t)^2} \\ & - \frac{\hat{\vartheta}_{L_y+1,k}(t) \sum_{l=L_y+2}^{L_y+L_u} \rho_l \hat{\vartheta}_{l,k}(t) \Delta u_k(t-l+L_y+1)}{\lambda + \hat{\vartheta}_{L_y+1,k}(t)^2} \end{aligned} \tag{3.66}$$

where $\hat{\vartheta}_{l,k}(t)$ is the $l$th element of $\hat{\boldsymbol{\vartheta}}_k(t)$, $l \in \{1, \ldots, L_y + L_u\}$.

In order to endow the parameter estimation algorithm (3.65) with a strong ability to track time-iteration-varying parameter, we present a reset algorithm as follows,

$$\hat{\boldsymbol{\vartheta}}_k(t) = \hat{\boldsymbol{\vartheta}}_0(t), \text{ if } \|\hat{\boldsymbol{\vartheta}}_k(t)\| \le \varepsilon, \text{ or } \|\Delta \boldsymbol{H}_{k-1}(t)\| \le \varepsilon \\ \text{or } \operatorname{sgn}(\hat{\vartheta}_{L_y+1,k}(t)) \ne \operatorname{sgn}(\hat{\vartheta}_{L_y+1,0}(t)) \quad (3.67)$$

where $\varepsilon$ is a small positive constant, $\hat{\boldsymbol{\vartheta}}_0(t)$ is the initial value of $\hat{\boldsymbol{\vartheta}}_k(t)$.

Then, the FFIDL-based DDILC (FFIDL-DDILC) is constructed by (3.65), (3.66), and (3.67), shown as follows for facilitating the readers,

$$u_k(t) = u_{k-1}(t) + \frac{\rho_{L_y+1}\hat{\vartheta}_{L_y+1,k}(t)e_{k-1}(t+1)}{\lambda + \hat{\vartheta}_{L_y+1,k}(t)^2} - \frac{\hat{\vartheta}_{L_y+1,k}(t)\sum_{l=1}^{L_y}\rho_l\hat{\vartheta}_{l,k}(t)\Delta y_k(t-l+1)}{\lambda + \hat{\vartheta}_{L_y+1,k}(t)^2} \\ - \frac{\hat{\vartheta}_{L_y+1,k}(t)\sum_{l=L_y+2}^{L_y+L_u}\rho_l\hat{\vartheta}_{l,k}(t)\Delta u_k(t-l+L_y+1)}{\lambda + \hat{\vartheta}_{L_y+1,k}(t)^2} \quad (3.68)$$

$$\hat{\boldsymbol{\vartheta}}_k(t) = \hat{\boldsymbol{\vartheta}}_{k-1}(t) + \frac{\eta\Delta\boldsymbol{H}_{k-1}(t)(\Delta y_{k-1}(t+1) - \hat{\boldsymbol{\vartheta}}_{k-1}^{\mathrm{T}}(t)\Delta\boldsymbol{H}_{k-1}(t))}{\mu + \|\Delta\boldsymbol{H}_{k-1}(t)\|^2} \quad (3.69)$$

$$\hat{\boldsymbol{\vartheta}}_k(t) = \hat{\boldsymbol{\vartheta}}_0(t), \text{ if } \|\hat{\boldsymbol{\vartheta}}_k(t)\| \le \varepsilon, \\ \text{or } \|\Delta\boldsymbol{H}_{k-1}(t)\| \le \varepsilon, \text{ or } \operatorname{sgn}(\hat{\vartheta}_{L_y+1,k}(t)) \ne \operatorname{sgn}(\hat{\vartheta}_{L_y+1,0}(t)) \quad (3.70)$$

### 3.3.3 Simulation Study

Consider a nonlinear nonaffine SISO system

$$y_k(t+1) = \begin{cases} \frac{y_k(t)}{1+y_k(t)^2} + u_k(t)^3, & 0 \le t < 50 \\ \frac{y_k(t)y_k(t-1)y_k(t-2)u_k(t-1)(y_k(t-2)-1)+a(t)u_k(t)}{1+y_k(t-1)^2+y_k(t-2)^2}, & 50 \le t \le 100 \end{cases} \quad (3.71)$$

where $a(t) = 1 + \text{round}(t/50)$ is a varying parameter. It is notable that the structure, orders and parameters of the controlled system are time-varying.

The desired trajectory is given as follows,

$$y_d(t+1) = \begin{cases} 0.5 \times (-1)^{\text{round}(t/10)}, & 0 \le t \le 30 \\ 0.5\sin(t\pi/10) + 0.3\cos(t\pi/10), & 30 < t \le 70 \\ 0.5 \times (-1)^{\text{round}(t/10)}, & 70 < t \le 100 \end{cases} \quad (3.72)$$

By choosing $\eta = 0.5$, $\rho_1 = 1$, $\rho_2 = 0.4$, $\rho_3 = 0.1$, $\lambda = 0.1$, $\mu = 0.1$, $L_y = 1$, $L_u = 2$, $\hat{\boldsymbol{\vartheta}}_0(t) = 1$, $u_0(t) = 0$, and $y_k(0) = 0$, and applying the FFIDL-DDILC method (3.68)–(3.70), the results are shown in Figs. 3.3 and 3.4. Figure 3.3 shows the output tracking performance at the 100th iteration, where the vertical axis

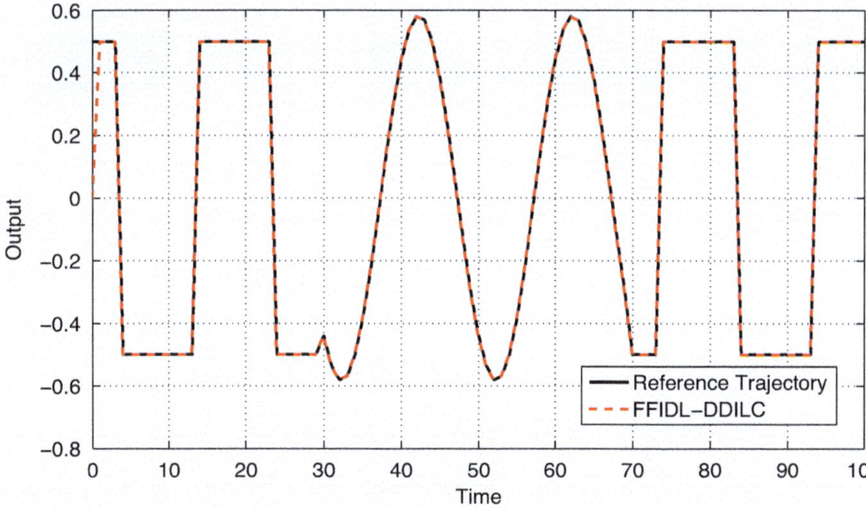


**Fig. 3.3** Profile of system output

denotes the system output and the horizontal axis denotes the time instant. Figure 3.4 shows the convergence performance of tracking error, where the Y-axis denotes the maximum absolute value of the output tracking error, defined as $e_{\max}(k) = \max_{t\in\{1,\ldots,100\}}\{|e_k(t)|\}$, and the X-axis denotes the iteration numbers. One can see that the effectiveness of the presented FFIDL-DDILC is confirmed by the simulation results.

For the comparison, the CFIDL-DDILC in Chap. 2 and PFIDL-DDILC in Sect. 3.2 are also applied. Selecting $\eta = 0.5$, $\rho = 0.8$, $\lambda = 1$, and $\mu = 0.5$ for CFIDL-DDILC (2.16)–(2.18), the convergence performance of maximum tracking error is shown as the dotted line in Fig. 3.4. Selecting $\eta = 0.5$, $\rho_1 = 1$, $\rho_2 = 0.6$, $\lambda = 1$, $\mu = 0.5$, and $L = 2$ for PFIDL-DDILC (3.21)–(3.23), the convergence performance of maximum tracking error is shown as the dashed line in Fig. 3.4. It is seen that the proposed FFIDL-DDILC outperforms CFIDL-DDILC and PFIDL-DDILC because more control input and output information is utilized in the control system design.

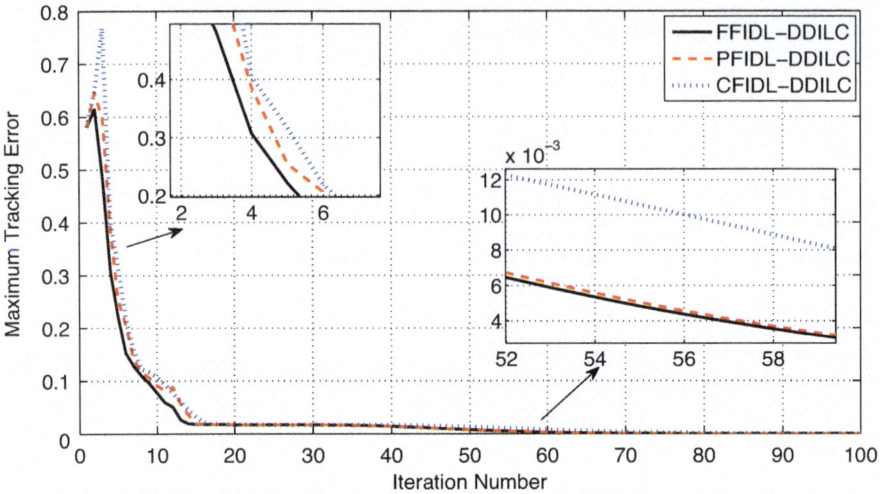


**Fig. 3.4** Convergence of maximum tracking error

## 3.4 Summary

In order to deal with the heavy complex nonlinearity and uncertainty in the controlled plant, this chapter develops a PFIDL method and a FFIDL method by considering much more influences on the system output iterative increment at the next time instant imposed by the iterative increments of the system input and output within a fixed-length moving time window of the current time. On this basis, two DDILC methods are studied in this chapter, including PFIDL-DDILC and FFIDL-DDILC. The proposed methods merely utilize the I/O data of the controlled plant, and no model information of the plant is involved. The theoretical results are verified by both mathematical analysis and simulation results. It is worth pointing out that the CFIDL-DDILC in Chap. 2 is designed for simple SISO discrete-time nonlinear systems and has a simple form of controller with only one scalar IPPD parameter updated iteratively. While, the two DDILC methods proposed in this chapter have an $L$-dimensional and an $(L_y + L_u)$-dimensional IPG vectors that are updated from iteration to iteration, respectively. One can select a proper DDILC method for a nonlinear plant by carefully considering the trade-off between the computation burden and the simplicity of the control system design. The convergence analysis of FFIDL-DDILC is still open because much more parameters and input and output variables are included, which needs further investigation in the future.

## References


Chen L, Narendra KS (2004) Identification and control of a nonlinear discrete-time system based on its linearization: a unified framework. IEEE Trans Neural Netw 15(3):663–673

Cheng D, Li Z (2009) A survey on linearization of nonlinear systems. J Shandong Univ (Eng Sci) 39(2):26–36

Chi R, Hou Z (2007) Dual-stage optimal iterative learning control for nonlinear non-affine discrete-time systems. Acta Automatica Sinica 33(10):1061–1065

Chi R, Li H, Shen D, Hou Z, Huang B (2022) Enhanced P-type control: Indirect adaptive learning from set-point updates. IEEE Trans Automat Control. https://doi.org/10.1109/TAC.2022.3154347

Deng H, Li HX, Wu YH (2008) Feedback-linearization-based neural adaptive control for unknown nonaffine nonlinear discrete-time systems. IEEE Trans Neural Netw 19(9):1615–1625

Dumont GA, Fu Y (1993) Nonlinear adaptive control via laguerre expansion of volterra kernels. Int J Adapt Control Signal Process 7(5):367–382

Guardabassi GO, Savaresi SM (1997) Approximate feedback linearization of discrete-time nonlinear systems using virtual input direct design. Syst Control Lett 32(2):63–74

Hou Z, Jin S (2011) A novel data-driven control approach for a class of discrete-time nonlinear systems. IEEE Trans Control Syst Technol 19(6):1549–1558

Hou Z, Chi R, Gao H (2016) An overview of dynamic-linearization-based data-driven control and applications. IEEE Trans Ind Electron 64(5):4076–4090

Hou Z, Jin S (2013) Model free adaptive control: theory and applications. Springer, Berlin

Huang L (1984) Linear algebra system and control theory. Science Press, Beijing

Jury EI (1964) Theory and application of the z-transform method. Wiley, New York

Lin N, Chi R, Huang B, Hou Z (2020) Event-triggered nonlinear iterative learning control. IEEE Trans Neural Netw Learn Syst 32(11):5118–5128

Slotine JJE, Li W (1991) Applied nonlinear control. Prentice Hall, Englewood Cliffs

Wen C, Hill DJ (1990) Adaptive linear control of nonlinear systems. IEEE Trans Automat Control 35(11):1253–1257

Xi Y, Wang F (1996) Nonlinear multi-model predictive control. IFAC Proc Vol 29(1):2359–2364

# Chapter 4
# DDILC with State-Transition-Based Iterative Dynamic Linearization

## 4.1 Introduction

In general, there are three alternate formulations of the ILC systems for the controller design and analysis. The first one is based on the frequency response of the plant and the learning filter (Amann et al. 1994). The second formulation is based on two-dimensional (2D) system theory (Dijkstra and Bosgra 2002). The third method is to formulate the ILC problem using a lifted domain representation (Hätönen 2004) where the 2D ILC system is reformulated as 1D system that evolves only in the trial domain. The lifted representation with supervector approach is often used in the design of norm optimization-based ILC (Amann et al. 1996; Gunnarsson and Norrlof 2001; Mishra et al. 2011; Rogers et al. 2010; Owens et al. 2013). It is common to consider all the signals of input, output, and tracking error over the entire finite time interval as supervectors and to rewrite linear discrete-time systems as matrices. Therefore, a state transition technique is usually employed to map the ILC input signals to the system outputs with a lower triangular Toeplitz matrix constructed using the finite time impulse responses of the system where the initial states are set as zeros straightforwardly (Hakvoort et al. 2008).

Using the lifted system representation, one can easily formulate the ILC problem as an optimization problem with convex quadratic objective and affine constraints (Boyd and Vandenberghe 2004). This formulation captures many ILC design problems including those with truncated objective functions and input saturation, rate, and state constraints. Recently, therefore, many excellent results of lifted ILC have been studied including the robustness of lifted ILC laws in van de Wijdeven et al. (2009), the robust lifted ILC laws obtained by inverting lifted system's dynamics (Harte et al. 2005), the linear matrix inequalities (LMIs) framework for design of lifted ILC laws (Yang et al. 2008), the design methodology for lifted ILC using the convex optimization techniques (Lee et al. 2000), the lifted ILC with quadratic cost design using weighting matrices (van de Wijdeven et al. 2009; Bristow 2008). However, most existing lifted ILC schemes are limited to linear discrete-time systems

 
R. Chi et al., *Data-Driven Iterative Learning Control for Discrete-Time Systems*, Intelligent Control and Learning Systems 2,
https://doi.org/10.1007/978-981-19-5950-9_4

that hinders their effective applications to practical complex processes with stronger nonlinearities and uncertainties. Meanwhile, it is unreasonable to set the initial states to be zeros due to the initial shifts occurring inevitably in practice.

On the other hand, the CFIDL, PFIDL, and FFIDL proposed in Chaps. 2 and 3 require that the solutions of the algebra equations must exist under the condition that the iterative input increments between two consecutive iterations are not zeros. Although it is reasonable in mathematics, the values of these solutions may become extremely large when the input is too small at some time instant. As a result, the values of IPPD and IPG will become too large to estimate from a practical viewpoint.

Based on this observation, this chapter proposes a state transition-based iterative dynamic linearization (STIDL) method in which the above condition is removed. First, the system output is reformulated only using the initial state and the input vector from zero time instant to the current time instant by referring to the state transition technique. Then, a lifted STIDL method is developed by making a difference in the system output between two consecutive iterations and incorporating the differential mean value theorem. Consequently, a lifted DDILC is proposed on the basis of the lifted linear data model obtained via the lifted STIDL method. The convergence of the proposed lifted STIDL-based DDILC (STIDL-DDILC) is shown rigorously.

It is worth pointing out that in the trial domain representation, dimensions of the lifted system matrices (system matrices in the trial domain) are proportional to the number of sampling steps in the trial. This implies that for large trial lengths the lifted system matrices have extremely large dimensions, and thus the execution of lifted ILC over a long time interval is not only time-consuming but also costly in memory requirement (Hakvoort et al. 2008; Rice and Verhaegen 2010). Several researches have recognized the problem of high computational complexity of the lifted ILC (Rice and Verhaegen 2010; Hakvoort et al. 2009; Barton et al. 2010; Haber et al. 2012; Sun and Alleyne 2014; Freeman 2012). However, they are mainly developed under a linear system framework depending on the exact model information of the controlled plant though some attempts exist for nonlinear systems.

Therefore, this chapter further presents a nonlifted STIDL-DDILC method by using a nonlifted STIDL where the supervector formulation of system input and output is no longer required, and the computational complexity is reduced consequently.

The main contributions of this chapter are summarized as (a) Compared with the CFIDL, PFIDL, and FFIDL methods, the proposed STIDL relaxes the conditions on the controlled system, therefore extending its wide applications. (b) Although it is similar to the PFIDL in the form, the STIDL considers all the influences imposed by the iterative input increments from zero time instant to the current time instant and thus the dimension of the input vector in the STIDL is varying with time instants, instead of a fixed one as that in the PFIDL. (c) The proposed lifted STIDL-DDILC and nonlifted STIDL-DDILC are both data-driven methods without requiring any model information of the controlled nonlinear plant, except for the I/O data only. (d) The lifted STIDL-DDILC is structured with a compact matrix form which enables all the exogenous constraints can be easily included in the cost function by rewriting them in a lifted form. The nonlifted STIDL-DDILC can well address the problem of computational complexity caused by large operation length and small sampling period. So, the two proposed STIDL-DDILC methods provide an alternative for the practical applications on the trade-off between control performance and computation complexity.

This chapter is organized as follows. Section 4.2 presents the state transition-based IDL method. Section 4.3 proposes a lifted STIDL-DDILC method with analysis and simulation study. Section 4.4 proposes a nonlifted STIDL-DDILC scheme with rigorous analysis and simulation verification. Section 4.5 gives some conclusions.

## 4.2 State-Transition-Based Iterative Dynamic Linearization

Consider a class of nonlinear nonaffine discrete-time systems,

$$\begin{aligned} y_k(t+1) = f(&y_k(t), y_k(t-1), \cdots, y_k(t-n_y), \\ &u_k(t), u_k(t-1), \ldots, u_k(t-n_u)) \end{aligned} \tag{4.1}$$

where $y_k(t)$ and $u_k(t)$ are system output and input, respectively, where $y_k(t) = 0$ and $u_k(t) = 0$, $\forall t < 0$; $n_y$ and $n_u$ are two unknown orders; $t \in \{0, 1, \ldots, N\}$ is time instant where $N \in Z^+$ represents the endpoint; $k \in \{0, 1, \ldots\}$ indicates the iteration number; $f(\cdots) : R^{n_u+n_y+2} \mapsto R$ is an unknown nonlinear scalar function, $f(0, 0) = 0$.

Denote $\boldsymbol{u}_k(t) \in R^{t+1}$ as a vector,

$$\boldsymbol{u}_k(t) = [u_k(0), u_k(1), \ldots, u_k(t)]^{\mathrm{T}}$$

with $\boldsymbol{u}_k(t) = \mathbf{0}_{t+1}$ for $t \leq 0$ and $k \leq 0$.

For discrete-time nonlinear system (4.1), two assumptions are made as follows.

**Assumption 4.1** The initial state $y_k(0)$ at every iteration $k$ is identical, that is $y_k(0) = y_0$, $\forall k$.

**Assumption 4.2** System (4.1) satisfies globally *Lipschitz* condition, that is

$$\begin{aligned} &|f(y_{k_1}(t), \ldots, y_{k_1}(t-n_y), u_{k_1}(t), \ldots, u_{k_1}(t-n_u)) \\ &\quad - f(y_{k_2}(t), \ldots, y_{k_2}(t-n_y), u_{k_2}(t), \ldots, u_{k_2}(t-n_u))| \\ &\leq b_1|y_{k_1}(t) - y_{k_2}(t)| + \cdots + b_{n_y+1}|y_{k_1}(t-n_y) - y_{k_2}(t-n_y)| \\ &\quad + b_{n_y+2}|u_{k_1}(t) - u_{k_2}(t)| + \cdots + b_{n_u+n_y+2}|u_{k_1}(t-n_u) - u_{k_2}(t-n_u)| \end{aligned} \tag{4.2}$$

where $b_l$, $l = 1, \ldots, n_y + n_u + 2$ are Lipschitz constants.

**Remark 4.1** Assumptions 4.1 and 4.2 are standard conditions in the literature of ILC systems. In Assumption 4.2, we only need the existence of the positive Lipschitz constants without requiring their exact values.

Denote $\Delta \boldsymbol{u}_k(t) = \boldsymbol{u}_k(t) - \boldsymbol{u}_{k-1}(t)$. The nonlifted STIDL of system (4.1) is proposed in the following theorem.

**Theorem 4.1** *Consider nonlinear system (4.1) satisfying Assumptions 4.1 and 4.2. There exists an time-iteration-varying parameter $\boldsymbol{\varphi}_k(t) \in R^{t+1}$, called iterative gradient (IG), such that system (4.1) can be transformed into the following nonlifted state-transition-based iterative dynamic linearization (STIDL) data model,*

$$\Delta y_k(t+1) = \boldsymbol{\varphi}_k(t) \Delta \boldsymbol{u}_k(t) \tag{4.3}$$

*where $\boldsymbol{\varphi}_k(t) = [\frac{\partial g_t}{\partial u_k(0)}, \cdots, \frac{\partial g_t}{\partial u_k(t)}]$, and $\|\boldsymbol{\varphi}_k(t)\| \le b_\varphi$ with $b_\varphi > 0$ being a positive constant.*

***Proof*** On each trial, the relationships between the input and output series can be expressed by the following algebraic functions,

$$\begin{aligned}
y_k(1) &= f(y_k(0), u_k(0)) \\
&= g_0(y_k(0), u_k(0)) \\
y_k(2) &= f(y_k(1), y_k(0), u_k(1), u_k(0)) \\
&= f(g_0(y_k(0), u_k(0)), y_k(0), u_k(1), u_k(0)) \\
&= g_1(y_k(0), u_k(0), u_k(1)) \\
&\ \vdots \\
y_k(t+1) &= f(y_k(t), \ldots, y_k(t-n_y), u_k(t), \ldots, u_k(t-n_u)) \\
&= f(g_{t-1}(y_k(0), u_k(0), \ldots, u_k(t-1)), g_{t-2}(y_k(0), u_k(0), \ldots, u_k(t-2)), \\
&\quad \ldots, g_{t-n_y-1}(y_k(0), u_k(0), \ldots, u_k(t-n_y-1)), u_k(t), \ldots, u_k(t-n_u)) \\
&= g_t(y_k(0), u_k(0), u_k(1), \ldots, u_k(t))
\end{aligned} \tag{4.4}$$

where $g_t(\cdots)$, $t \in \{0, 1, \cdots, N-1\}$ is a compound function of $f(\cdots)$ and has the same properties as that of $f(\cdots)$, that is, $g_t(\cdots)$ also satisfies Assumption 4.2.

From (4.4), we have,

$$\begin{aligned}
\Delta y_k(t+1) &= g_t(y_k(0), u_k(0), u_k(1), \ldots, u_k(t)) \\
&\quad - g_t(y_{k-1}(0), u_{k-1}(0), u_{k-1}(1), \ldots, u_{k-1}(t)) \\
&= \frac{\partial g_t}{\partial y_k(0)} \Delta y_k(0) + \frac{\partial g_t}{\partial u_k(0)} \Delta u_k(0) + \cdots + \frac{\partial g_t}{\partial u_k(t)} \Delta u_k(t)
\end{aligned} \tag{4.5}$$

where $\partial g_t / \partial y_k(0)$ and $\partial g_t / \partial u_k(l)$, $0 \le l \le t$ denote the partial derivatives of $g_t(\cdots)$ with respect to $y_k(0)$ and $u_k(l)$ at some point in the interval $[y_k(0), y_{k-1}(0)]$ and $[u_k(l), u_{k-1}(l)]$, respectively.

Owing to Assumption 4.1, Equation (4.5) becomes,

$$\begin{aligned}
\Delta y_k(t+1) &= \frac{\partial g_t}{\partial u_k(0)} \Delta u_k(0) + \cdots + \frac{\partial g_t}{\partial u_k(t)} \Delta u_k(t) \\
&= \boldsymbol{\varphi}_k(t) \Delta \boldsymbol{u}_k(t)
\end{aligned} \tag{4.6}$$

where $\boldsymbol{\varphi}_k(t) = \left[\frac{\partial g_t}{\partial u_k(0)}, \ldots, \frac{\partial g_t}{\partial u_k(t)}\right]$.

This gives the conclusion of Theorem 4.1, and the boundedness of $\boldsymbol{\varphi}_k(t)$ is guaranteed directly by Assumption 4.2. □

Define $\boldsymbol{U}_k = [u_k(0), u_k(1), \ldots, u_k(N-1)]^{\mathrm{T}} \in R^N$, $\boldsymbol{Y}_k = [y_k(1), y_k(2), \ldots, y_k(N)]^{\mathrm{T}}$ and $\boldsymbol{g}(\cdot) = [g_0(\cdot), g_1(\cdot), \ldots, g_{N-1}(\cdot)]^{\mathrm{T}} : R^N \to R^N$; $\Delta \boldsymbol{Y}_k = \boldsymbol{Y}_k - \boldsymbol{Y}_{k-1}$, $\Delta \boldsymbol{U}_k = \boldsymbol{U}_k - \boldsymbol{U}_{k-1}$. The lifted STIDL is given in the following theorem.

**Theorem 4.2** *Consider nonlinear system (4.1) satisfying assumptions 4.1 and 4.2. There exists an time-iteration-varying parameter $\boldsymbol{\Phi}_k \in R^{N\times N}$, such that system (4.1) can be transformed into the following lifted STIDL data model,*

$$\Delta \boldsymbol{Y}_k = \boldsymbol{\Phi}_k \Delta \boldsymbol{U}_k \tag{4.7}$$

*with bounded* $\boldsymbol{\Phi}_k = \begin{bmatrix} \frac{\partial g_0}{\partial u_k(0)} & 0 & \cdots & 0 \\ \frac{\partial g_1}{\partial u_k(0)} & \frac{\partial g_1}{\partial u_k(1)} & \cdots & 0 \\ \vdots & \vdots & \ddots & \vdots \\ \frac{\partial g_{N-1}}{\partial u_k(0)} & \frac{\partial g_{N-1}}{\partial u_k(1)} & \cdots & \frac{\partial g_{N-1}}{\partial u_k(N-1)} \end{bmatrix} \in R^{N\times N}$*, and* $\|\boldsymbol{\Phi}_k\| \le b_{\Phi}$*, where* $b_{\Phi} > 0$ *is a constant.*

***Proof*** According to (4.4), one can obtain

$$\boldsymbol{Y}_k = \boldsymbol{g}(y_k(0), \boldsymbol{U}_k^{\mathrm{T}}). \tag{4.8}$$

From (4.8), one has

$$\Delta \boldsymbol{Y}_k = \frac{\partial \boldsymbol{g}}{\partial y_k(0)}(y_k(0) - y_{k-1}(0)) + \frac{\partial \boldsymbol{g}}{\partial \boldsymbol{U}_k^{\mathrm{T}}}(\boldsymbol{U}_k - \boldsymbol{U}_{k-1}). \tag{4.9}$$

Owing to Assumption 4.1, (4.9) becomes

$$\Delta \boldsymbol{Y}_k = \boldsymbol{\Phi}_k(\boldsymbol{U}_k - \boldsymbol{U}_{k-1}) \tag{4.10}$$

where $\boldsymbol{\Phi}_k = \partial \boldsymbol{g}/\partial \boldsymbol{U}_k^{\mathrm{T}}$, and $\frac{\partial \boldsymbol{g}}{\partial \boldsymbol{U}_k^{\mathrm{T}}} = \begin{bmatrix} \frac{\partial g_0}{\partial u_k(0)} & 0 & \cdots & 0 \\ \frac{\partial g_1}{\partial u_k(0)} & \frac{\partial g_1}{\partial u_k(1)} & \cdots & 0 \\ \vdots & \vdots & \ddots & \vdots \\ \frac{\partial g_{N-1}}{\partial u_k(0)} & \frac{\partial g_{N-1}}{\partial u_k(1)} & \cdots & \frac{\partial g_{N-1}}{\partial u_k(N-1)} \end{bmatrix} \in R^{N\times N}$. The boundedness of $\boldsymbol{\Phi}_k(t)$ is guaranteed directly by Assumption 4.2. □

**Remark 4.2** Compared with the CFIDL, PFIDL, and FFIDL-based LDMs, the IG in STIDL-based LDM (4.3) denotes the system gradient vector, instead of a pseudo one. It reflects the I/O dynamic relationship of the control system in the iteration domain.

**Remark 4.3** The existence conditions of the solutions of equations (2.6), (3.10), and (3.57) are no longer required for the STIDL by a trade-off that the initial states are identical for all iterations which however is most common in the ILC design.

**Remark 4.4** Similar to PFIDL, the STIDL method built the relationship between the system output iterative change at the next time instant $t+1$ and all the control input iterative changes within a time-varying moving time window $[0, \ldots, t]$. However, the vector dimension of IG is time-varying which is different from the fixed dimension of the IPG when the LLC is selected artificially. Consequently, the STIDL can be more flexible in dealing with dynamic complexity of the systems that evolve with increasing time.

## 4.3 Lifted STIDL-Based DDILC

### *4.3.1 Controller Design*

The control objective is to track a desired trajectory $\boldsymbol{Y}_d = [y_d(1), y_d(2), \ldots, y_d(N)]^{\mathrm{T}} \in R^N$ such that the tracking error $\boldsymbol{E}_k = \boldsymbol{Y}_d - \boldsymbol{Y}_k$ converges to zero as the iteration number $k$ goes to infinity, i.e., $\lim_{k\to\infty} \boldsymbol{E}_k = 0$.

Consider an objective function as follows:

$$J(\boldsymbol{U}_k) = \|\boldsymbol{E}_k\|^2 + \lambda\|\boldsymbol{U}_k - \boldsymbol{U}_{k-1}\|^2 \tag{4.11}$$

where $\lambda > 0$ is a weighting factor.

According to (4.7), one can rewrite (4.11) as

$$\begin{aligned} J(\boldsymbol{U}_k) &= \|\boldsymbol{Y}_d - \boldsymbol{Y}_{k-1} - \boldsymbol{\Phi}_k(\boldsymbol{U}_k - \boldsymbol{U}_{k-1})\|^2 + \lambda\|\boldsymbol{U}_k - \boldsymbol{U}_{k-1}\|^2 \\ &= \|\boldsymbol{E}_{k-1} - \boldsymbol{\Phi}_k(\boldsymbol{U}_k - \boldsymbol{U}_{k-1})\|^2 + \lambda\|\boldsymbol{U}_k - \boldsymbol{U}_{k-1}\|^2. \end{aligned} \tag{4.12}$$

Differentiating (4.12) with respect to $\boldsymbol{U}_k$, and setting it to zero, gives

$$\Delta\boldsymbol{U}_k = (\lambda \boldsymbol{I} + \boldsymbol{\Phi}_k^{\mathrm{T}}\boldsymbol{\Phi}_k)^{-1}\boldsymbol{\Phi}_k^{\mathrm{T}}\boldsymbol{E}_{k-1}. \tag{4.13}$$

Note that control algorithm (4.13) requires calculating the matrix inversion, which is difficult when the dimensions of the system input and output are large. Hence, to give a simplified control algorithm, the following matrix inverse lemma is introduced.

**Lemma 4.1** (Horn RA and Johnson CR (2012)) *Suppose a nonsingular matrix* $\boldsymbol{A}$ *has a known inverse* $\boldsymbol{A}^{-1}$ *and consider*

$$\boldsymbol{B} = \boldsymbol{A} + \boldsymbol{X}\boldsymbol{Y},$$

*where* $\boldsymbol{X}$ *is* $n$*-by-*1, $\boldsymbol{Y}$ *is* 1*-by-*$n$. *If* $\boldsymbol{B}$ *is nonsingular, then*

$$\boldsymbol{B}^{-1} = \boldsymbol{A}^{-1} - \frac{1}{1 + \boldsymbol{Y}\boldsymbol{A}^{-1}\boldsymbol{X}}\boldsymbol{A}^{-1}\boldsymbol{X}\boldsymbol{Y}\boldsymbol{A}^{-1}.$$

According to Lemma 4.1, it is easy to get

$$\begin{aligned}
&\left(\lambda I + \boldsymbol{\Phi}_k^{\mathrm{T}}\boldsymbol{\Phi}_k\right)^{-1}\boldsymbol{\Phi}_k^{\mathrm{T}} \\
&= \left[(\lambda I)^{-1} - \frac{(\lambda I)^{-1}\boldsymbol{\Phi}_k^{\mathrm{T}}\boldsymbol{\Phi}_k(\lambda I)^{-1}}{1 + \boldsymbol{\Phi}_k(\lambda I)^{-1}\boldsymbol{\Phi}_k^{\mathrm{T}}}\right]\boldsymbol{\Phi}_k^{\mathrm{T}} \\
&= (\lambda I)^{-1}\boldsymbol{\Phi}_k^{\mathrm{T}} - \frac{(\lambda I)^{-1}\boldsymbol{\Phi}_k^{\mathrm{T}}\boldsymbol{\Phi}_k\boldsymbol{\Phi}_k^{\mathrm{T}}}{\lambda + \boldsymbol{\Phi}_k\boldsymbol{\Phi}_k^{\mathrm{T}}} \\
&= \frac{\boldsymbol{\Phi}_k^{\mathrm{T}} + (\lambda I)^{-1}\|\boldsymbol{\Phi}_k\|^2\boldsymbol{\Phi}_k^{\mathrm{T}} - (\lambda I)^{-1}\|\boldsymbol{\Phi}_k\|^2\boldsymbol{\Phi}_k^{\mathrm{T}}}{\lambda + \|\boldsymbol{\Phi}_k\|^2} \\
&= \frac{\boldsymbol{\Phi}_k^{\mathrm{T}}}{\lambda + \|\boldsymbol{\Phi}_k\|^2}.
\end{aligned} \tag{4.14}$$

Then, the simplified control law is derived from (4.13) and (4.14) as follows,

$$\boldsymbol{U}_k = \boldsymbol{U}_{k-1} + \frac{\rho\boldsymbol{\Phi}_k^{\mathrm{T}}\boldsymbol{E}_{k-1}}{\lambda + \|\boldsymbol{\Phi}_k\|^2} \tag{4.15}$$

where the step-size factor $\rho > 0$ is added to make the control algorithm more general and it will be used in the convergence analysis.

Since $\boldsymbol{\Phi}_k$ is unknown, an estimation algorithm is developed in this section. Note that $\boldsymbol{\Phi}_k$ is a lower triangular matrix and all of the upper triangular elements are zeros, hence we only need to estimate the nonzero elements of $\boldsymbol{\Phi}_k$.

Define an objective function of $\boldsymbol{\varphi}_k(t)$ as

$$\begin{aligned}
J(\boldsymbol{\varphi}_k(t)) = \|\Delta y_{k-1}(t+1) - \boldsymbol{\varphi}_k(t)(\boldsymbol{u}_{k-1}(t) - \boldsymbol{u}_{k-2}(t))\|^2 \\
+\mu\|\boldsymbol{\varphi}_k(t) - \hat{\boldsymbol{\varphi}}_{k-1}(t)\|^2
\end{aligned} \tag{4.16}$$

where $\mu > 0$ is a weighting factor; $\boldsymbol{\varphi}_k(t)$ is the vector consisting of nonzero elements of the $(t+1)$th row of $\boldsymbol{\Phi}_k$, $t \in \{0, 1, \ldots, N-1\}$.

Minimizing (4.16) with respect to $\boldsymbol{\varphi}_k(t)$ gives the following estimation algorithm,

$$\hat{\boldsymbol{\varphi}}_k(t) = \hat{\boldsymbol{\varphi}}_{k-1}(t) + \frac{\eta(\Delta y_{k-1}(t+1) - \hat{\boldsymbol{\varphi}}_{k-1}(t)\Delta\boldsymbol{u}_{k-1}(t))\Delta\boldsymbol{u}_{k-1}^{\mathrm{T}}(t)}{\mu + \|\Delta\boldsymbol{u}_{k-1}(t)\|^2} \tag{4.17}$$

where $\hat{\boldsymbol{\varphi}}_k(t)$ is the estimation of $\boldsymbol{\varphi}_k(t)$, and $0 < \eta < 2$ is a step-size factor which is added to make the estimation algorithm more general and flexible.

By integrating the control algorithm (4.15) and the parameter estimation algorithm (4.17), the proposed lifted STIDL-DDILC scheme is constructed as follows:

$$\boldsymbol{U}_k = \boldsymbol{U}_{k-1} + \frac{\rho \hat{\boldsymbol{\Phi}}_k^{\mathrm{T}} \boldsymbol{E}_{k-1}}{\lambda + \|\hat{\boldsymbol{\Phi}}_k\|^2} \tag{4.18}$$

$$\hat{\boldsymbol{\varphi}}_k(t) = \hat{\boldsymbol{\varphi}}_{k-1}(t) + \frac{\eta(\Delta y_{k-1}(t+1) - \hat{\boldsymbol{\varphi}}_{k-1}(t)\Delta \boldsymbol{u}_{k-1}(t))\Delta \boldsymbol{u}_{k-1}^{\mathrm{T}}(t)}{\mu + \|\Delta \boldsymbol{u}_{k-1}(t)\|^2} \tag{4.19}$$

$$\hat{\boldsymbol{\varphi}}_k(t) = \hat{\boldsymbol{\varphi}}_0(t), \text{ if } \mathrm{sgn}(\hat{\phi}_k(i)) \neq \mathrm{sgn}(\hat{\phi}_0(i)) \text{ or } \|\hat{\boldsymbol{\varphi}}_k(t)\| \leq \varepsilon,\ i = 0, 1, \ldots, t \tag{4.20}$$

where $\varepsilon$ is a small positive constant, $\hat{\boldsymbol{\varphi}}_0(t)$ is the initial vector value of $\hat{\boldsymbol{\varphi}}_k(t)$, and $\hat{\phi}_k(i)$ is the $i$th element of vector $\hat{\boldsymbol{\varphi}}_k(t)$, $i \in \{0, 1, \ldots, t\}$.

**Remark 4.5** A reset algorithm (4.20) is used to strengthen the ability of parameter estimation algorithm (4.19) in tracking time-iteration-varying parameter. $\hat{\boldsymbol{\varphi}}_0(t)$ is selected such that sign of all the elements are the same as that of $\boldsymbol{\varphi}_k(t)$.

**Remark 4.6** Note that the presented control law (4.18) calculates the learning gain through estimate algorithm (4.19) and reset algorithm (4.20) by using the measured I/O data, which is easily implemented with little computation burden.

**Remark 4.7** Compared with the traditional optimal ILCs (Lee et al. 2000; Bristow 2008), the proposed lifted STIDL-DDILC is a data-driven control strategy. The controller design and analysis only depends on the I/O data without requiring any model information of the plant either explicitly or implicitly.

### 4.3.2 Convergence Analysis

Before proceeding to the convergence analysis, another assumption is given as follows.

**Assumption 4.3** The partial gradient vector $\boldsymbol{\varphi}_k(t), t \in \{0, 1, \ldots, N-1\}$, is nonsingular and all the signs of its elements are known and unchanged, that is, $\phi_k(i) > \ell$(or $\phi_k(i) < -\ell$), $\forall i \in \{0, 1, \ldots, t\}$, where $\ell$ is a positive constant.

**Remark 4.8** Note that Assumption 4.3 is similar to the constraint on control input direction in traditional model-based control methods.

The properties of the lifted STIDL-DDILC (4.18)–(4.20) are provided by the following theorem.

**Theorem 4.3** *For a class of nonlinear nonaffine discrete-time systems (4.1), if Assumptions 4.1, 4.2, and 4.3 hold, then the presented lifted STIDL-DDILC (4.18)–(4.20) guarantees that*

(i) *the parameter estimation* $\hat{\boldsymbol{\Phi}}_k$ *is bounded,* $\forall k$*;*
(ii) *the tracking error* $\boldsymbol{E}_k$ *converges to zero as* $k$ *approaches to infinity, i.e.,* $\lim_{k\to\infty} \boldsymbol{E}_k = 0$*;*
(iii) *the system output* $\boldsymbol{Y}_k$ *and the control input signal* $\boldsymbol{U}_k$ *are bounded,* $\forall k$.

***Proof*** *Part 1. The boundedness of estimation value* $\hat{\boldsymbol{\Phi}}_k$

Subtracting $\boldsymbol{\varphi}_k(t)$ from both sides of (4.19) and using (4.3), we have

$$\begin{aligned}
\tilde{\boldsymbol{\varphi}}_k(t) &= \tilde{\boldsymbol{\varphi}}_{k-1}(t) - \frac{\eta(\Delta y_{k-1}(t+1) - \hat{\boldsymbol{\varphi}}_{k-1}(t)\Delta \boldsymbol{u}_{k-1}(t))\Delta \boldsymbol{u}_{k-1}^{\mathrm{T}}(t)}{\mu + \|\Delta \boldsymbol{u}_{k-1}(t)\|^2} \\
&\quad + \boldsymbol{\varphi}_k(t) - \boldsymbol{\varphi}_{k-1}(t) \\
&= \tilde{\boldsymbol{\varphi}}_{k-1}(t) - \frac{\eta(\boldsymbol{\varphi}_{k-1}^{\mathrm{T}}(t)\Delta \boldsymbol{u}_{k-1}(t) - \hat{\boldsymbol{\varphi}}_{k-1}(t)\Delta \boldsymbol{u}_{k-1}(t))\Delta \boldsymbol{u}_{k-1}^{\mathrm{T}}(t)}{\mu + \|\Delta \boldsymbol{u}_{k-1}(t)\|^2} \\
&\quad + \boldsymbol{\varphi}_k(t) - \boldsymbol{\varphi}_{k-1}(t) \\
&= \tilde{\boldsymbol{\varphi}}_{k-1}(t) - \frac{\eta\tilde{\boldsymbol{\varphi}}_{k-1}(t)\Delta \boldsymbol{u}_{k-1}(t)\Delta \boldsymbol{u}_{k-1}^{\mathrm{T}}(t)}{\mu + \|\Delta \boldsymbol{u}_{k-1}(t)\|^2} + \boldsymbol{\varphi}_k(t) - \boldsymbol{\varphi}_{k-1}(t) \\
&= \tilde{\boldsymbol{\varphi}}_{k-1}(t)\left(I - \frac{\eta\Delta \boldsymbol{u}_{k-1}(t)\Delta \boldsymbol{u}_{k-1}^{\mathrm{T}}(t)}{\mu + \|\Delta \boldsymbol{u}_{k-1}(t)\|^2}\right) + \boldsymbol{\varphi}_k(t) - \boldsymbol{\varphi}_{k-1}(t)
\end{aligned} \tag{4.21}$$

where $\tilde{\boldsymbol{\varphi}}_k(t) = \boldsymbol{\varphi}_k(t) - \hat{\boldsymbol{\varphi}}_k(t)$ denotes the parameter estimate error.

According to Assumption 4.2, $\|\boldsymbol{\varphi}_k(t)\| \le b_\varphi$ is bounded, which leads to $\|\boldsymbol{\varphi}_k(t) - \boldsymbol{\varphi}_{k-1}(t)\| \le 2b_\varphi$. Taking norms on both sides of Eq. (4.21), yields

$$\|\tilde{\boldsymbol{\varphi}}_k(t)\| \le \left\|\tilde{\boldsymbol{\varphi}}_{k-1}(t)\left(I - \frac{\eta\Delta \boldsymbol{u}_{k-1}(t)\Delta \boldsymbol{u}_{k-1}^{\mathrm{T}}(t)}{\mu + \|\Delta \boldsymbol{u}_{k-1}(t)\|^2}\right)\right\| + 2b_\varphi. \tag{4.22}$$

Squaring the first term of the right side of inequality (4.22) leads to

$$\begin{aligned}
&\left\|\tilde{\boldsymbol{\varphi}}_{k-1}(t)\left(I - \frac{\eta\Delta \boldsymbol{u}_{k-1}(t)\Delta \boldsymbol{u}_{k-1}^{\mathrm{T}}(t)}{\mu + \|\Delta \boldsymbol{u}_{k-1}(t)\|^2}\right)\right\|^2 \\
&= \|\tilde{\boldsymbol{\varphi}}_{k-1}(t)\|^2 + \left(-2 + \frac{\eta\|\Delta \boldsymbol{u}_{k-1}(t)\|^2}{\mu + \|\Delta \boldsymbol{u}_{k-1}(t)\|^2}\right)\frac{\eta\|\Delta \boldsymbol{u}_{k-1}^{\mathrm{T}}(t)\tilde{\boldsymbol{\varphi}}_{k-1}(t)\|^2}{\mu + \|\Delta \boldsymbol{u}_{k-1}(t)\|^2}.
\end{aligned} \tag{4.23}$$

Since $0 < \eta < 2$ and $\mu > 0$, then

$$\left(-2 + \frac{\eta\|\Delta \boldsymbol{u}_{k-1}(t)\|^2}{\mu + \|\Delta \boldsymbol{u}_{k-1}(t)\|^2}\right) < 0. \tag{4.24}$$

So Eqs. (4.23) and (4.24) together imply that

$$\left\| \tilde{\boldsymbol{\varphi}}_{k-1}(t)\left(I-\frac{\eta\Delta \boldsymbol{u}_{k-1}(t)\Delta \boldsymbol{u}_{k-1}^{\mathrm{T}}(t)}{\mu+\|\Delta \boldsymbol{u}_{k-1}(t)\|^2}\right)\right\|^2 < \|\tilde{\boldsymbol{\varphi}}_{k-1}(t)\|^2.$$

According to (4.22), it is obvious that there exists a positive constant $0 < d_1 < 1$ such that

$$\|\tilde{\boldsymbol{\varphi}}_k(t)\| \le d_1\|\tilde{\boldsymbol{\varphi}}_{k-1}(t)\| + 2b_\varphi \le \cdots \le d_1^k\|\tilde{\boldsymbol{\varphi}}_0(t)\| + \frac{2b_\varphi}{1-d_1}. \tag{4.25}$$

Hence, $\tilde{\boldsymbol{\varphi}}_k(t)$ is bounded because $\boldsymbol{\varphi}_k(t)$ is bounded according to Theorem 4.1, and $\hat{\boldsymbol{\varphi}}_0(t)$ is a bounded value. Therefore, the boundedness of $\hat{\boldsymbol{\varphi}}_k(t)$ is evident, $\forall t \in \{0, \cdots, N\}$, and $\forall k$. Consequently, conclusion (i) of Theorem 4.3 is obtained directly, i.e., $\hat{\boldsymbol{\Phi}}_k$ is bounded.

*Part 2. The convergence of tracking error $\boldsymbol{E}_k$*

From Eq. (4.7) and control (4.18), we have

$$\begin{aligned}\boldsymbol{E}_k &= \boldsymbol{Y}_d - \boldsymbol{Y}_{k-1} - \boldsymbol{\Phi}_k(\boldsymbol{U}_k - \boldsymbol{U}_{k-1})\\ &= \boldsymbol{E}_{k-1} - \frac{\rho\boldsymbol{\Phi}_k\hat{\boldsymbol{\Phi}}_k^{\mathrm{T}}}{\lambda+\|\hat{\boldsymbol{\Phi}}_k\|^2}\boldsymbol{E}_{k-1}\\ &= \left(I - \frac{\rho\boldsymbol{\Phi}_k\hat{\boldsymbol{\Phi}}_k^{\mathrm{T}}}{\lambda+\|\hat{\boldsymbol{\Phi}}_k\|^2}\right)\boldsymbol{E}_{k-1}.\end{aligned} \tag{4.26}$$

The reset algorithm (4.20) together with Assumption 4.3 implies that $\boldsymbol{\Phi}_k\hat{\boldsymbol{\Phi}}_k^{\mathrm{T}}$ is positive definite. Furthermore, both of $\boldsymbol{\Phi}_k$ and $\hat{\boldsymbol{\Phi}}_k$ have been shown to be bounded, so by properly selecting controller parameters $\rho$ and $\lambda$, one can guarantee that there exists a positive constant $0 < d_2 < 1$ such that

$$0 < \left\| I - \frac{\rho\boldsymbol{\Phi}_k\hat{\boldsymbol{\Phi}}_k^{\mathrm{T}}}{\lambda+\left\|\hat{\boldsymbol{\Phi}}_k\right\|^2}\right\| \le d_2 < 1. \tag{4.27}$$

Hence, from (4.26) and (4.27),

$$\|\boldsymbol{E}_k\| \le \left\| I - \frac{\rho\boldsymbol{\Phi}_k\hat{\boldsymbol{\Phi}}_k^{\mathrm{T}}}{\lambda+\left\|\hat{\boldsymbol{\Phi}}_k\right\|^2}\right\| \|\boldsymbol{E}_{k-1}\| \le d_2\|\boldsymbol{E}_{k-1}\| \le \cdots \le d_2^k\|\boldsymbol{E}_0\|. \tag{4.28}$$

So, conclusion (ii) of Theorem 4.3 is obtained from (4.28), i.e., $\lim_{k\to\infty}\boldsymbol{E}_k = 0$.

*Part 3. The boundedness of system output and control input*

It is obvious that the boundedness of $\boldsymbol{Y}_d$ and $\boldsymbol{E}_k$ leads to the boundedness of $\boldsymbol{Y}_k$. Using the property $a^2 + b^2 \ge 2ab$, one obtains that

$$\left\| \frac{\rho \hat{\boldsymbol{\Phi}}_k^{\mathrm{T}}}{\lambda + \left\| \hat{\boldsymbol{\Phi}}_k \right\|^2} \right\| \leq \left| \frac{\rho \left\| \hat{\boldsymbol{\Phi}}_k \right\|}{2\sqrt{\lambda} \left\| \hat{\boldsymbol{\Phi}}_k \right\|} \right| = \frac{\rho}{2\sqrt{\lambda}}. \tag{4.29}$$

It is clear that

$$\boldsymbol{U}_k = \boldsymbol{U}_k - \boldsymbol{U}_{k-1} + \boldsymbol{U}_{k-1} - \cdots + \boldsymbol{U}_1 - \boldsymbol{U}_0 + \boldsymbol{U}_0 = \boldsymbol{U}_0 + \sum_{j=1}^{k} \Delta \boldsymbol{U}_j. \tag{4.30}$$

In view of Eqs. (4.18), (4.29) and (4.30), and using the Schwarz inequality, we have

$$\begin{aligned} \|\boldsymbol{U}_k\| &\leq \|\boldsymbol{U}_0\| + \sum_{j=1}^{k} \left\| \Delta \boldsymbol{U}_j \right\| \\ &\leq \|\boldsymbol{U}_0\| + \frac{\rho}{2\sqrt{\lambda}} \sum_{j=1}^{k} \left\| \boldsymbol{E}_{j-1} \right\| \\ &\leq \|\boldsymbol{U}_0\| + \frac{\rho}{2\sqrt{\lambda}} \sum_{j=1}^{k} d_2^{j-1} \|\boldsymbol{E}_0\| \\ &\leq \|\boldsymbol{U}_0\| + \frac{\rho}{(1-d_2)\, 2\sqrt{\lambda}} \|\boldsymbol{E}_0\|. \end{aligned} \tag{4.31}$$

Because both the initial tracking error $\boldsymbol{E}_0$ and initial control input $\boldsymbol{U}_0$ are given bounded, the inequality (4.31) leads to a bounded control input sequence directly. Thus the conclusion (iii) of Theorem 4.3 is obtained. □

### 4.3.3 Simulation Study

***Example 1***. Consider a nonlinear nonaffine SISO system

$$y_k(t+1) = \frac{y_k(t)}{1 + y_k(t)^3} + u_k(t)^3 \tag{4.32}$$

where $t \in \{0, 1, \ldots, 20\}$ is the time instant; $k \in \{0, 1, \ldots\}$ is the iteration number.

The desired trajectory is given as follows:

$$y_d(t) = \sin(t\pi/10) + 0.5\sin(t\pi/15),\ t \in \{0, 1, \ldots, 20\}. \tag{4.33}$$

Set $\rho = 3.5,\ \eta = 1.5,\ \lambda = 2,\ \mu = 1,\ y_k(0) = 0,\ u_0(t) = 0,\ \hat{\varphi}_0(t) = 0.5,\ t \in \{0, 1, \ldots, 20\}$. Applying the proposed lifted STIDL-DDILC scheme (4.18)–(4.20),

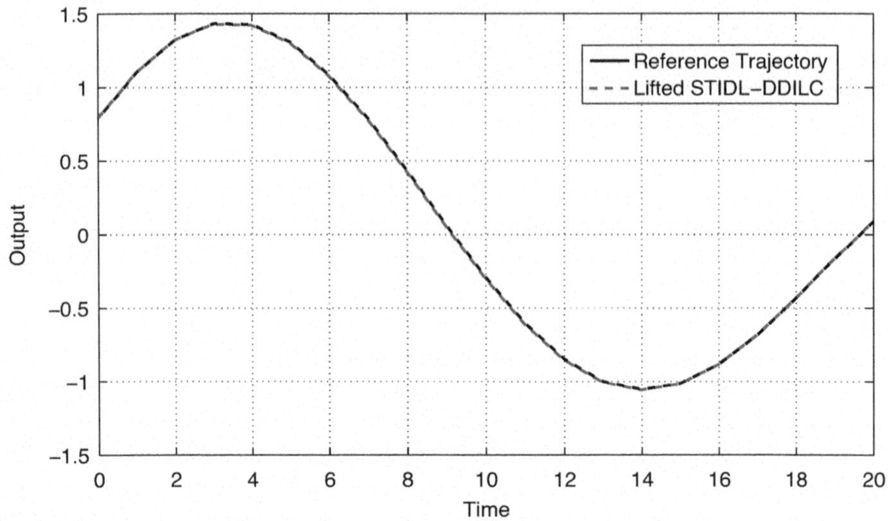


**Fig. 4.1** Profile of system output

the results are shown in Figs. 4.1 and 4.2. Figure 4.1 shows the output tracking performance at the 200th iteration, where the vertical axis denotes the system output and the horizontal axis denotes the time instant. Figure 4.2 shows the convergence performance of tracking error, where the Y-axis denotes the maximum absolute value of the output tracking error, defined as $e_{\max}(k) = \max_{t\in\{1,\ldots,20\}}\{|e_k(t)|\}$, and the X-axis denotes the iteration numbers. One can see that the effectiveness of the presented lifted STIDL-DDILC is confirmed by the simulation results.

***Example 2***. A typical nonlinear batch reactor (Logsdon and Biegler 1989) is considered with the well-tested dynamical equations as below

$$
\begin{aligned}
\frac{dx_1}{dt} &= -k_1 \exp\left(\frac{-E_1}{uT_{ref}}\right) x_1^2 \\
\frac{dx_2}{dt} &= -k_1 \exp\left(\frac{-E_1}{uT_{ref}}\right) x_1^2 - k_2 \exp\left(\frac{-E_2}{uT_{ref}}\right) x_2 \\
y &= x_2
\end{aligned}
$$

where $x_1$ and $x_2$ are the dimensionless concentrations of the raw material and the product, respectively; $u = T/T_{\text{ref}}$ is the dimensionless temperature of the reactor with $T_{\text{ref}}$ being the reference temperature; the reactor temperature $T$ is limited to $298\ \text{K} \le T \le 398\ \text{K}$; the final time $t_f$ is 1.0 h and the sampling time $h = t_f/N = 0.1$ with $N = 10$.

The system parameters are given as $k_1 = 4.0 \times 10^3$, $k_2 = 6.2 \times 10^5$, $E_1 = 2.5 \times 10^3$ K, $E_2 = 5.0 \times 10^3$ K, and $T_{ref} = 348$ K. The desired product reference trajectory is $y_d(t) = 0.61 \sin(\pi t/2N)$, $t \in \{0, 1, \ldots, 10\}$.

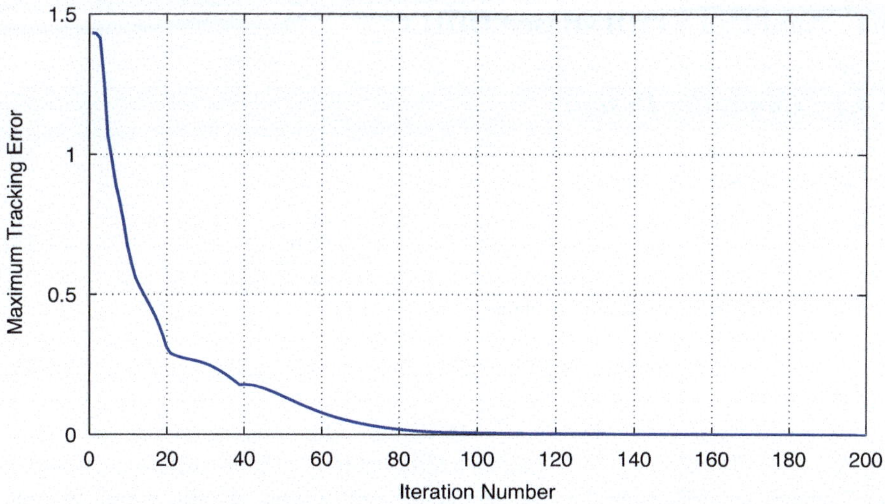


**Fig. 4.2** Convergence of tracking error

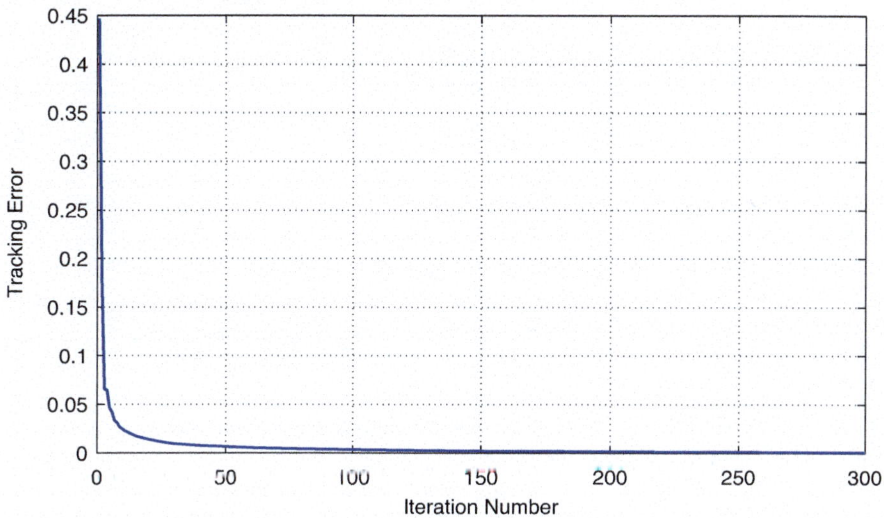


**Fig. 4.3** Convergence of tracking error

The controller parameters are set as $x_1(0) = 1$, $x_2(0) = 0$; $u_0(t) = 0.5$, $\hat{\varphi}_0(t) = 0.01$, for $t \in \{0, 1, \ldots, 10\}$; $\rho = 0.2$, $\eta = 0.1$, $\lambda = 0.01$, $\mu = 0.5$. Using the proposed lifted STIDL-DDILC (4.18)–(4.20), the convergence performance of tracking error is shown in Fig. 4.3. It is seen that a satisfied control performance can be guaranteed.

## 4.4 Nonlifted STIDL-Based DDILC

### 4.4.1 Controller Design

According to (4.3), we have

$$y_k(t+1) = y_{k-1}(t+1) + \boldsymbol{\varphi}_k(t)\Delta \boldsymbol{u}_k(t). \tag{4.34}$$

Consider a cost function as follows:

$$J\left(u_k(t)\right) = |e_k(t+1)|^2 + \lambda \left|u_k(t) - u_{k-1}(t)\right|^2 \tag{4.35}$$

where $e_k(t+1) = y_d(t+1) - y_k(t+1)$ denotes the tracking error, and $\lambda > 0$ denotes a weighting factor. Note that an additional term of input difference is added for the purpose of limiting the change rate of control input signals, which is often required in practical applications.

Substituting (4.34) into (4.35), yields

$$\begin{aligned} J\left(u_k(t)\right) &= \left| e_{k-1}(t+1) - \sum_{i=0}^{t-1} \phi_k(i)\Delta u_k(i) - \phi_k(t)\Delta u_k(t) \right|^2 \\ &\quad + \lambda \left|u_k(t) - u_{k-1}(t)\right|^2 \end{aligned} \tag{4.36}$$

where $u_k(i)$ is the $(i+1)$th element of $\boldsymbol{u}_k(t)$; $\Delta u_k(i) = u_k(i) - u_{k-1}(i), i \in \{0, 1, \ldots, t\}$.

Differentiating (4.36) with respect to $u_k(t)$ and setting it to zero, we have

$$u_k(t) = u_{k-1}(t) + \frac{\rho_t \phi_k(t) e_{k-1}(t+1)}{\lambda + |\phi_k(t)|^2} - \frac{\phi_k(t)\sum_{i=0}^{t-1} \rho_i \phi_k(i)\Delta u_k(i)}{\lambda + |\phi_k(t)|^2} \tag{4.37}$$

where $\rho_i > 0$, $i \in \{0, 1, \cdots, t\}$, denotes the step-size constant added to make the control algorithm (4.37) more flexible.

Since $\phi_k(i)$, $i \in \{0, 1, \ldots, t\}$, is unknown, it needs estimating by designing an objective function of $\boldsymbol{\varphi}_k(t)$, shown as follows

$$\begin{aligned} J\left(\hat{\boldsymbol{\varphi}}_k(t)\right) &= \left\| \Delta y_{k-1}(t+1) - \hat{\boldsymbol{\varphi}}_k(t)\left(\boldsymbol{u}_{k-1}(t) - \boldsymbol{u}_{k-2}(t)\right) \right\|^2 \\ &\quad + \mu \left\| \hat{\boldsymbol{\varphi}}_k(t) - \hat{\boldsymbol{\varphi}}_{k-1}(t) \right\|^2 \end{aligned} \tag{4.38}$$

where $\mu > 0$ is a weighting factor and $\hat{\boldsymbol{\varphi}}_k(t)$ is the estimated value of $\boldsymbol{\varphi}_k(t)$.

Minimizing Eq. (4.38) with respect to $\boldsymbol{\varphi}_k(t)$ gives the following estimation algorithm,

$$\hat{\boldsymbol{\varphi}}_k(t) = \hat{\boldsymbol{\varphi}}_{k-1}(t) + \frac{\eta\left(\Delta y_{k-1}(t+1) - \hat{\boldsymbol{\varphi}}_{k-1}(t)\Delta \boldsymbol{u}_{k-1}(t)\right)\Delta \boldsymbol{u}_{k-1}^{\mathrm{T}}(t)}{\mu + \|\Delta \boldsymbol{u}_{k-1}(t)\|^2} \tag{4.39}$$

where $0 < \eta < 2$ is a step-size factor being included to make the estimation algorithm more general and flexible. The initial value of $\hat{\boldsymbol{\varphi}}_k(t)$ is given as $\hat{\boldsymbol{\varphi}}_0(t)$, which is bounded for all time.

So, the controller becomes

$$u_k(t) = u_{k-1}(t) + \frac{\rho_t \hat{\phi}_k(t) e_{k-1}(t+1)}{\lambda + \left|\hat{\phi}_k(t)\right|^2} - \frac{\hat{\phi}_k(t)\sum_{i=0}^{t-1}\rho_i \hat{\phi}_k(i)\Delta u_k(i)}{\lambda + \left|\hat{\phi}_k(t)\right|^2}. \tag{4.40}$$

To enhance the tracking performance of $\hat{\boldsymbol{\varphi}}_k(t)$ following the time-iteration-varying parameter $\boldsymbol{\varphi}_k(t)$, a reset algorithm is added as

$$\begin{aligned}&\hat{\boldsymbol{\varphi}}_k(t) = \hat{\boldsymbol{\varphi}}_0(t),\\ &\text{if } \operatorname{sgn}\left(\hat{\phi}_k(i)\right) \neq \operatorname{sgn}\left(\hat{\phi}_0(i)\right) \text{ or } \left\|\hat{\boldsymbol{\varphi}}_k(t)\right\| \leq \varepsilon, i \in \{0, 1, \ldots, t\}\end{aligned} \tag{4.41}$$

where $\varepsilon$ is a small positive scalar.

Thus, the proposed nonlifted STIDL-DDILC method is constructed by (4.42)–(4.44),

$$u_k(t) = u_{k-1}(t) + \frac{\rho_t \hat{\phi}_k(t) e_{k-1}(t+1)}{\lambda + \left|\hat{\phi}_k(t)\right|^2} - \frac{\hat{\phi}_k(t)\sum_{i=0}^{t-1}\rho_i \hat{\phi}_k(i)\Delta u_k(i)}{\lambda + \left|\hat{\phi}_k(t)\right|^2} \tag{4.42}$$

$$\hat{\boldsymbol{\varphi}}_k(t) = \hat{\boldsymbol{\varphi}}_{k-1}(t) + \frac{\eta\left(\Delta y_{k-1}(t+1) - \hat{\boldsymbol{\varphi}}_{k-1}(t)\Delta \boldsymbol{u}_{k-1}(t)\right)\Delta \boldsymbol{u}_{k-1}^{\mathrm{T}}(t)}{\mu + \|\Delta \boldsymbol{u}_{k-1}(t)\|^2} \tag{4.43}$$

$$\begin{aligned}&\hat{\boldsymbol{\varphi}}_k(t) = \hat{\boldsymbol{\varphi}}_0(t),\\ &\text{if } \operatorname{sgn}\left(\hat{\phi}_k(i)\right) \neq \operatorname{sgn}\left(\hat{\phi}_0(i)\right) \text{ or } \left\|\hat{\boldsymbol{\varphi}}_k(t)\right\| \leq \varepsilon,\ i \in \{0, 1, \ldots, t\}.\end{aligned} \tag{4.44}$$

**Remark 4.9** Different from the lifted STIDL-DDILC, there is no matrix computation included in learning control law (4.42). Hence the computational complexity can be reduced. Further, the objective function (4.35) is optimized pointwisely and the control law (4.42) is implemented point by point on a real-time basis, which will produce the optimal solution for every point. The effect of the control input at a single time point on other time points is considered by adding the additional control inputs in previous time instants into the control law.

**Remark 4.10** It is seen that the proposed nonlifted STIDL-DDILC (4.42)–(4.44) does not use any model information, such as model structure, model parameters, and so on, except for the I/O data only.

**Remark 4.11** In the practical applications, one can select the same values of $\rho_i$, $i = 1, \ldots, N$, for the simplicity. In general, we can fix the values of $\lambda$ in the given scope, and then tuning the values of $\rho_i$ in the given scope by trials. If the parameters are selected differently under the given scopes, the convergence can be still guaranteed in theory but the convergence rates may be different. The larger values of $\rho_i$ and smaller values of $\lambda$ may generally produce a faster convergence.

### 4.4.2 Convergence Analysis

**Lemma 4.2** (Jury EI (1964)) *Let* $A = \begin{bmatrix} 0 & 1 & 0 & \cdots & 0 \\ 0 & 0 & 1 & \cdots & 0 \\ \vdots & \vdots & \vdots & \ddots & \vdots \\ 0 & 0 & 0 & \cdots & 1 \\ a_1 & a_2 & a_3 & \cdots & a_t \end{bmatrix} \in R^{t\times t}$. *If* $\sum_{i=1}^{t} |a_i| < 1$, *then the spectral radius* $s(A) < 1$.

**Theorem 4.4** *For a class of nonlinear nonaffine discrete-time systems (4.1), if Assumptions 4.1, 4.2, and 4.3 hold, and* $\rho_i^2 \le \min\left\{\lambda/b_{\hat{\phi}}^2 N^2, 4\lambda, 4\lambda/b_{\varphi}^2\right\}$, $i \in \{0, 1, \ldots, t\}$, $\mu > 0$, *and* $\eta \in (0, 2)$, *then the proposed nonlifted STIDL-DDILC scheme (4.42)–(4.44) guarantees that:*

*(i) the parameter estimation* $\hat{\boldsymbol{\varphi}}_k(t)$ *is bounded over the finite time interval* $t \in \{0, 1, \ldots, N\}$, $\forall k$;
*(ii) the output error converges to zero pointwisely over the finite time interval* $t \in \{0, 1, \ldots, N\}$ *as* $k$ *approaches to infinity, i.e.,* $\lim_{k\to\infty} e_k(t) = 0$;
*(iii) the system output* $y_k(t)$ *and the control input signal* $u_k(t)$ *are bounded,* $\forall t \in \{0, 1, \ldots, N\}$, $\forall k$.

***Proof*** *Part 1. The boundedness of estimation value* $\hat{\boldsymbol{\varphi}}_k(t)$

The boundedness of $\hat{\boldsymbol{\varphi}}_k(t)$ can be easily obtained following the similar steps in Theorem 4.3. Therefore, we assume that $\|\hat{\boldsymbol{\varphi}}_k(t)\| \le b_{\hat{\varphi}}$, where $b_{\hat{\varphi}}$ is a positive constant.

*Part 2. The convergence of output tracking error*

Let $\boldsymbol{A}_k(t) = \begin{bmatrix} 0 & 1 & 0 & \cdots & 0 \\ 0 & 0 & 1 & \cdots & 0 \\ \vdots & \vdots & \vdots & \ddots & \vdots \\ 0 & 0 & 0 & \cdots & 1 \\ 0 & -\frac{\rho_0\hat{\phi}_k(t)\hat{\phi}_k(0)}{\lambda+\left|\hat{\phi}_k(t)\right|^2} & -\frac{\rho_1\hat{\phi}_k(t)\hat{\phi}_k(1)}{\lambda+\left|\hat{\phi}_k(t)\right|^2} & \cdots & -\frac{\rho_{t-1}\hat{\phi}_k(t)\hat{\phi}_k(t-1)}{\lambda+\left|\hat{\phi}_k(t)\right|^2} \end{bmatrix} \in R^{(t+1)\times(t+1)}$

and $\boldsymbol{C} = [0 \cdots 0\ 1]^{\mathrm{T}} \in R^{t+1}$, then according to (4.42), one has

$$\Delta \boldsymbol{u}_k(t) = \boldsymbol{A}_k(t)\Delta \boldsymbol{u}_k(t-1) + \frac{\rho_t\hat{\phi}_k(t)}{\lambda + \left|\hat{\phi}_k(t)\right|^2}\boldsymbol{C}e_{k-1}(t+1). \tag{4.45}$$

The characteristic equation of $\boldsymbol{A}_k(t)$ is

$$z^{t+1} + \frac{\rho_{t-1}\hat{\phi}_k(t)\hat{\phi}_k(t-1)}{\lambda + \left|\hat{\phi}_k(t)\right|^2} z^t + \cdots + \frac{\rho_0\hat{\phi}_k(t)\hat{\phi}_k(0)}{\lambda + \left|\hat{\phi}_k(t)\right|^2} z = 0. \tag{4.46}$$

Let $\rho_{\max} = \max_{i=0,\cdots,N-1}\{\rho_i\}$. Since $a^2 + b^2 \geq 2ab$, then

$$\begin{aligned}\sum_{i=0}^{t-1}\left|\frac{\rho_i\hat{\phi}_k(t)\hat{\phi}_k(i)}{\lambda + \left|\hat{\phi}_k(t)\right|^2}\right| &\leq \rho_{\max}\sum_{i=0}^{t-1}\left|\frac{\hat{\phi}_k(t)\hat{\phi}_k(i)}{\lambda + \left|\hat{\phi}_k(t)\right|^2}\right| \\ &\leq \rho_{\max}\sum_{i=0}^{t-1}\frac{\left|\hat{\phi}_k(t)\right|\left|\hat{\phi}_k(i)\right|}{2\sqrt{\lambda}\left|\hat{\phi}_k(t)\right|} = \rho_{\max}\sum_{i=0}^{t-1}\frac{\left|\hat{\phi}_k(i)\right|}{2\sqrt{\lambda}}.\end{aligned} \tag{4.47}$$

According to conclusion (i) of Theorem 4.4, $\left|\hat{\phi}_k(i)\right| < b_{\hat{\phi}}$ is bounded, where $b_{\hat{\phi}}$ is a positive constant. Further, the time instant $t$ is bounded by $N$. Therefore, one can select $\lambda$ and $\rho_{\max}$ properly such that

$$\sum_{i=0}^{t-1}\left|\frac{\rho_i\hat{\phi}_k(t)\hat{\phi}_k(i)}{\lambda + \left|\hat{\phi}_k(t)\right|^2}\right| \leq \rho_{\max}\frac{b_{\hat{\phi}}N}{2\sqrt{\lambda}} \leq d_1 < 0.5 \tag{4.48}$$

where $d_1$ is a positive constant. According to Lemma 4.2, Eq. (4.48) implies that $|z| < 0.5$ and $s\left(\boldsymbol{A}_k(t)\right) < 0.5$ where $s\left(\boldsymbol{A}_k(t)\right)$ is the spectral radius of $\boldsymbol{A}_k(t)$. Hence, according to Lemma 3.2, there exists an arbitrary small positive constant $\delta$ such that the following inequality holds

$$\|\boldsymbol{A}_k(t)\|_v \leq s\left(\boldsymbol{A}_k(t)\right) + \delta \leq d_1 + \delta \leq d_2 < 0.5 \tag{4.49}$$

where $\|\boldsymbol{A}_k(t)\|_v$ is a proper matrix norm of $\boldsymbol{A}_k(t)$, and $d_2$ is a positive constant.

Taking norm on both sides of equation (4.45),

$$\|\Delta\boldsymbol{u}_k(t)\|_v \leq \|\boldsymbol{A}_k(t)\|_v\,\|\Delta\boldsymbol{u}_k(t-1)\|_v + \left|\frac{\rho_t\hat{\phi}_k(t)}{\lambda + \left|\hat{\phi}_k(t)\right|^2}\right| |e_{k-1}(t+1)|. \tag{4.50}$$

By properly selecting $\rho_{\max}$ and $\lambda$, one can guarantee that

$$\left|\frac{\rho_t\hat{\phi}_k(t)}{\lambda + \left|\hat{\phi}_k(t)\right|^2}\right| \leq \frac{\rho_{\max}\left|\hat{\phi}_k(t)\right|}{2\sqrt{\lambda}\left|\hat{\phi}_k(t)\right|} = \frac{\rho_{\max}}{2\sqrt{\lambda}} = d_3 < 1 \tag{4.51}$$

where $d_3$ is a positive constant.

Thereafter, one can derive according to (4.49)–(4.51) that

$$\begin{aligned}\|\Delta \boldsymbol{u}_k(t)\|_v &\le d_2 \|\Delta \boldsymbol{u}_k(t-1)\|_v + d_3 |e_{k-1}(t+1)| \\ &\le d_2^2 \|\Delta \boldsymbol{u}_k(t-2)\|_v + d_2 d_3 |e_{k-1}(t)| + d_3 |e_{k-1}(t+1)| \\ &\le \cdots \le d_2^{t+1} \|\Delta \boldsymbol{u}_k(-1)\|_v + d_3 d_2^t |e_{k-1}(1)| \\ &\quad + d_3 d_2^{t-1} |e_{k-1}(2)| + \cdots + d_3 |e_{k-1}(t+1)| \\ &= d_3 \sum_{i=0}^{t} d_2^{t-i} |e_{k-1}(i+1)| \end{aligned} \tag{4.52}$$

where the input signal $\boldsymbol{u}_k(-1)$ is set as zero for $t < 0$.

By virtue of the definition of tracking error, Eqs. (4.34) and (4.45), we have

$$\begin{aligned} e_k(t+1) &= e_{k-1}(t+1) - \boldsymbol{\varphi}_k(t)\Delta \boldsymbol{u}_k(t) \\ &= \left(1 - \frac{\rho_t \phi_k(t)\hat{\phi}_k(t)}{\lambda + \left|\hat{\phi}_k(t)\right|^2}\right) e_{k-1}(t+1) - \boldsymbol{\varphi}_k(t)\boldsymbol{A}_k(t)\Delta \boldsymbol{u}_k(t-1). \end{aligned} \tag{4.53}$$

Taking norm for the two sides of Eq. (4.53), we have

$$\begin{aligned} |e_k(t+1)| \le & \left|\left(1 - \frac{\rho_t \phi_k(t)\hat{\phi}_k(t)}{\lambda + \left|\hat{\phi}_k(t)\right|^2}\right)\right| |e_{k-1}(t+1)| \\ & + \left\|\boldsymbol{\varphi}_k(t)\right\|_v \|\boldsymbol{A}_k(t)\|_v \|\Delta \boldsymbol{u}_k(t-1)\|_v . \end{aligned} \tag{4.54}$$

The values of the elements of $\hat{\boldsymbol{\varphi}}_0(t)$ at the initial iteration are selected such that their signs are the same as that of the elements of $\boldsymbol{\varphi}_k(t)$. Therefore, according to Assumption 4.3 and the reset algorithm (4.44), it is obvious that there exist suitable $\rho_{\max}$ and $\lambda$ such that

$$0 < \frac{\rho_t \phi_k(t)\hat{\phi}_k(t)}{\lambda + \left|\hat{\phi}_k(t)\right|^2} \le \frac{\rho_t \left|\phi_k(t)\right| \left|\hat{\phi}_k(t)\right|}{2\sqrt{\lambda}\left|\hat{\phi}_k(t)\right|} = \frac{\rho_{\max} b_\varphi}{2\sqrt{\lambda}} = b_\varphi d_3 = d_4 < 1 \tag{4.55}$$

where $d_4$ is a positive scalar.

Since $\left\|\boldsymbol{\varphi}_k(t)\right\| \le b_\varphi$, using (4.49), (4.52), and (4.55), we can rewrite (4.54) as

$$\begin{aligned} |e_k(t+1)| &\le d_5 |e_{k-1}(t+1)| + d_2 b_\varphi \|\Delta \boldsymbol{u}_k(t-1)\|_v \\ &\le d_5 |e_{k-1}(t+1)| + b_\varphi d_3 \sum_{i=0}^{t-1} d_2^{t-i} |e_{k-1}(i+1)| \end{aligned} \tag{4.56}$$

where $0 < d_5 = 1 - d_4 < 1$.

Define $e_{\max,k} = \max_{t\in\{1,\ldots,N\}}\{|e_k(t)|\}$. Assume $e_{\max,k} = |e_k(\tau)|$, where $\tau$ is some time instant over the finite time interval $\{1, \ldots, N\}$. Then according to (4.56), the following inequality holds,

$$\begin{aligned} e_{\max,k} &= |e_k(\tau)| \\ &\leq d_5 |e_{k-1}(\tau)| + b_\varphi d_3 \sum_{i=1}^{\tau} d_2^{\tau-i} |e_{k-1}(i)| \\ &\leq d_5 e_{\max,k-1} + b_\varphi d_3 \sum_{i=1}^{\tau} d_2^{\tau-i} e_{\max,k-1} \\ &\leq \left(d_5 + b_\varphi d_3 \frac{d_2}{1-d_2}\right) e_{\max,k-1}. \end{aligned} \tag{4.57}$$

It is clear that

$$0 < d_5 + b_\varphi d_3 \frac{d_2}{1-d_2} = 1 - d_4 + \frac{d_2 d_4}{1-d_2} \leq d_6 < 1 \tag{4.58}$$

where $0 < d_6 < 1$ is a positive constant.

Inequality (4.56) and the condition (4.57) imply that

$$e_{\max}(k) \leq d_6 e_{\max}(k-1) \leq \cdots \leq d_6^k e_{\max}(0). \tag{4.59}$$

Apparently, $\lim_{k\to\infty} e_{\max}(k) = 0$. Then, conclusion (ii) of Theorem 4.4 is derived directly from (4.59), i.e., $\lim_{k\to\infty} |e_k(t+1)| = 0$, $t \in \{0, \ldots, N-1\}$.

*Part 3. Then boundedness of system output and control input*

The boundedness of both of $y_d(t+1)$ and $e_k(t+1)$ means that $y_k(t+1)$ is bounded.

It is clear that

$$\begin{aligned} \boldsymbol{u}_k(t) &= \boldsymbol{u}_k(t) - \boldsymbol{u}_{k-1}(t) + \boldsymbol{u}_{k-1}(t) - \cdots - \boldsymbol{u}_0(t) + \boldsymbol{u}_0(t) \\ &= \boldsymbol{u}_0(t) + \sum_{j=1}^{k} \Delta \boldsymbol{u}_j(t). \end{aligned} \tag{4.60}$$

Then using the Schwarz inequality, we obtain that

$$\|\boldsymbol{u}_k(t)\|_v \leq \sum_{j=1}^{k} \left\|\Delta \boldsymbol{u}_j(t)\right\|_v + \|\boldsymbol{u}_0(t)\|_v. \tag{4.61}$$

According to (4.52) and (4.59), it is easy to get

$$
\begin{aligned}
\|\boldsymbol{u}_k(t)\|_v &\le d_3 \sum_{j=1}^{k}\sum_{i=0}^{t} d_2^{t-i}\left|e_{j-1}(i+1)\right| + \|\boldsymbol{u}_0(t)\|_v \\
&\le \frac{d_3\left(1-d_2^t\right)}{1-d_2}\sum_{j=1}^{k} e_{\max}(j-1) + \|\boldsymbol{u}_0(t)\|_v \\
&\le \frac{d_3}{1-d_2}\left(1+d_6+\cdots+d_6^k\right)|e_{\max}(0)| + \|\boldsymbol{u}_0(t)\|_v \\
&\le \frac{d_3}{1-d_2}\frac{1}{1-d_6}|e_{\max}(0)| + \|\boldsymbol{u}_0(t)\|_v\,.
\end{aligned}
\tag{4.62}
$$

Because both the initial maximum tracking error $e_{\max}(0)$ and initial control input $u_0(t)$ are bounded, the inequality (4.62) leads to a bounded control input sequence directly. Thus the conclusion (iii) of Theorem 4.4 is obtained. □

### 4.4.3 Simulation Study

In the simulation, the given mathematical models are assumed unknown. They are only used to generate the I/O values of the controlled processes. There is no any model information used in both the proposed nonlifted STIDL-DDILC and the previously proposed lifted STIDL-DDILC. However, the exact model is required for the traditional lifted optimal ILC (Gunnarsson and Norrlof 2001; Mishra et al. 2011; Rogers et al. 2010; Owens et al. 2013).

***Example 1***. A linear discrete-time system is considered as below

$$
\begin{cases}
\boldsymbol{x}_k(t+1) = \boldsymbol{A}\boldsymbol{x}_k(t) + \boldsymbol{B}u_k(t) \\
y_k(t) = \boldsymbol{C}\boldsymbol{x}_k(t)
\end{cases}
\tag{4.63}
$$

where $\boldsymbol{A} = \begin{bmatrix} 0 & 1 \\ -0.5 & -0.5 \end{bmatrix}$, $\boldsymbol{B} = \begin{bmatrix} 0 \\ 1 \end{bmatrix}$, and $\boldsymbol{C} = \begin{bmatrix} 1 & 0 \end{bmatrix}$; $t \in \{0, \ldots, 100\}$; $k$ is the iteration index. The control task is to track the following desired trajectory,

$$
y_d(t) = 10^{-6}(t-1)^3(4-0.03(t-1)), t \in \{0, \ldots, 100\}.
$$

In the simulation, we set $\boldsymbol{x}_k(0) = \begin{bmatrix} 0 & 0 \end{bmatrix}^T$ and $u_0(t) = 0$ for all the following cases.

*Case 1.1*: *Nonlifted STIDL-DDILC Scheme*. In the simulation, the initial value $\hat{\varphi}_0(t) = 1$ for all time $t \in \{0, \ldots, 100\}$. The controller parameters are set as $\rho_i = 1.3$, $i = 1, \ldots, N$, $\eta = 0.1$, $\lambda = 0.1$, and $\mu = 0.01$. Applying the proposed nonlifted STIDL-DDILC (4.42)–(4.44), the tracking error is shown as the solid line in Fig. 4.4, where the X-axis is the iteration number and the Y-axis is the maximum absolute value of tracking error $e_{\max}(k) = \max_{t\in\{1,\ldots,100\}}\{|e_k(t)|\}$. It is clear that the proposed

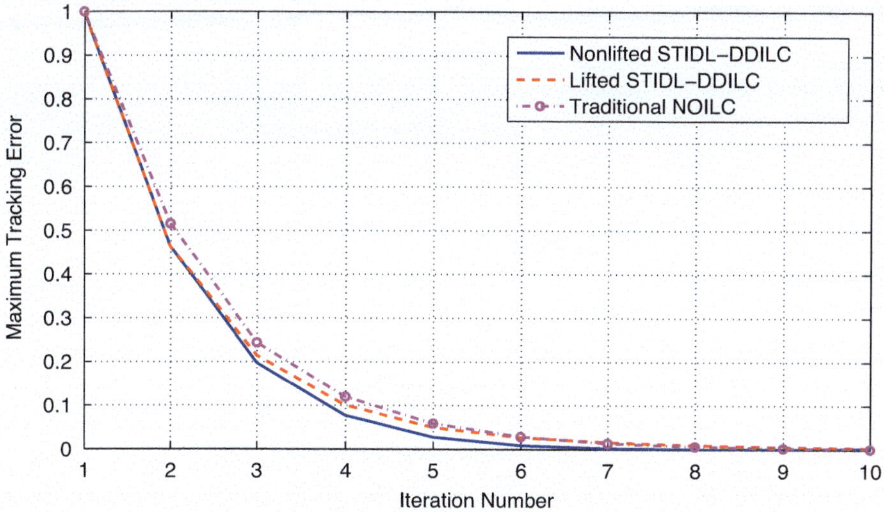


**Fig. 4.4** The convergence of tracking error in Example 1

nonlifted STIDL-DDILC guarantees the convergent property of the tracking error in the iteration direction.

*Case 1.2*: *Lifted STIDL-DDILC Scheme*. For the purpose of comparison, the lifted STIDL-DDILC proposed in Sect. 4.3 is also applied to the system (4.63).

The controller parameters are selected to be the same as that in Case 1.1, i.e., $\hat{\varphi}_0(t) = 1$, $\rho = 1.3$, $\eta = 0.1$, $\lambda = 0.1$, and $\mu = 0.01$. Then the simulation result of the lifted STIDL-DDILC is shown in Fig. 4.4. Obviously, the lifted STIDL-DDILC also warrants that system tracking error approaches to zero iteratively. However, compared with the nonlifted STIDL-DDILC, the nonlifted STIDL-DDILC utilizes more additional control inputs in the previous time instants of the current iteration, helping to improve control performance as shown in Fig. 4.4.

*Case 1.3*: *Traditional Lifted NOILC Scheme*. For the purpose of comparison of the proposed nonlifted STIDL-DDILC approach with the traditional optimal ILC, a standard NOILC is selected as follows:

$$\boldsymbol{U}_{k+1} = L_u \boldsymbol{U}_k + L_d \boldsymbol{E}_k \tag{4.64}$$

where $\boldsymbol{U}_k$ and $\boldsymbol{E}_k$ are two supervectors of the control input and tracking error, respectively; $\boldsymbol{L}_u = \left(\boldsymbol{G}^{\mathrm{T}}\boldsymbol{Q}\boldsymbol{G} + \boldsymbol{R} + \boldsymbol{S}\right)^{-1}\boldsymbol{R}$, and $\boldsymbol{L}_d = \left(\boldsymbol{G}^{\mathrm{T}}\boldsymbol{Q}\boldsymbol{G} + \boldsymbol{R} + \boldsymbol{S}\right)^{-1}\boldsymbol{G}^{\mathrm{T}}\boldsymbol{Q}$ with $\boldsymbol{G} = \begin{bmatrix} CB & CAB & CB \\ \vdots & \vdots & \ddots \\ CA^{N-1}B & CA^{N-2}B & \cdots & CB \end{bmatrix}$; $\boldsymbol{Q}$, $\boldsymbol{R}$, and $\boldsymbol{S}$ are real symmetric positive definite matrices with appropriate dimensions.

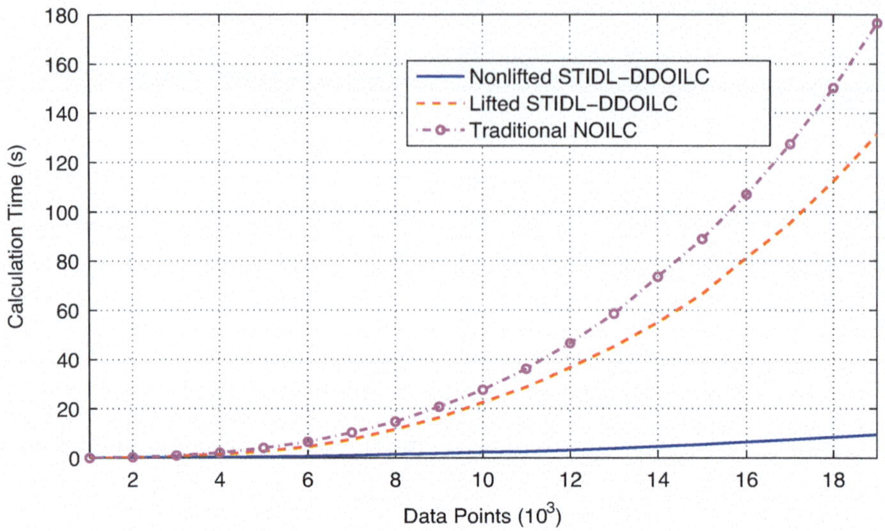


**Fig. 4.5** Computation time

By selecting $Q = 1$, $R = 0.2$, $S = 0$, the simulation result is shown as the dash-dotted line with circles in Fig. 4.4, which shows that the monotonic convergence of the tracking error is guaranteed from iteration to iteration. Although a slightly better performance of the proposed nonlifted STIDL-DDILC is achieved compared with the traditional NOILC (Fig. 4.4), it is difficult to prove in theory that the proposed nonlifted STIDL-DDILC outperforms the traditional NOILC. In general, a better convergence can be achieved by utilizing more control knowledge. Although more additional control inputs in previous time instant of the current iteration have been used in the proposed approach, the traditional NOILC also utilizes more model information, such as the system matrices, which also helps the convergence. Compared with the traditional NOILC, the main contributions of the proposed nonlifted STIDL-DDILC are that (1) no explicit model information of the controlled plant is needed; (2) the computation burden is reduced by avoiding matrix calculation in the control law. The following case is the comparative result of computation complexity.

*Case 1.4: Comparison of the computational complexity.* Besides the error convergence, we now compare the computational effort for calculating the proposed nonlifted STIDL-DDILC, the lifted STIDL-DDILC, and the standard lifted NOILC. The calculation is performed in MATLAB on a Lenovo Thinkpad laptop computer with a 2.40GHz Intel Core i3 process and 2 GB of RAM. Multiple trial lengths are selected and the computation time of the three approaches is shown in Fig. 4.5.

In Fig. 4.5, the computational time for the traditional lifted NOILC approach and the lifted STIDL-DDILC increases dramatically as the trial length increases because the matrix operation is included in the control law, where the matrix dimension increases greatly with the data points. Comparatively, the computation time of the

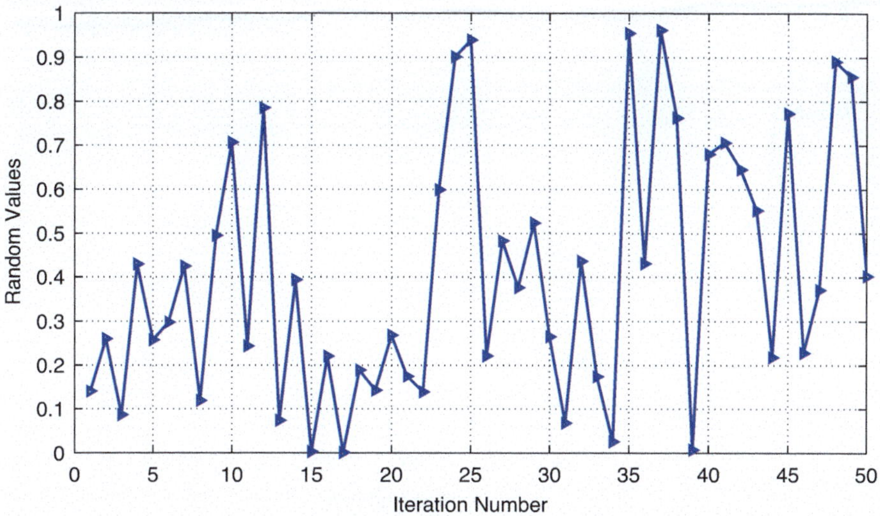


**Fig. 4.6** Random values of $\delta_k$ along iteration axis

lifted STIDL-DDILC is less than that of the traditional lifted NOILC, because there is no matrix inversion computation involved in the control law (4.18). Hence, as illustrated in Fig. 4.5, the computation time of the proposed nonlifted STIDL-DDILC is not a problem because it does not need to construct or calculate large matrices in its implementation.

***Example 2***. A typical nonlinear batch reactor (Logsdon and Biegler 1989) is considered with the well-tested dynamical equations as below

$$\frac{dx_{1,k}}{dt} = -k_1 \exp\left(\frac{-E_1}{u_k T_{ref}}\right) x_{1,k}^2 + 0.01\delta_k$$
$$\frac{dx_{2,k}}{dt} = -k_1 \exp\left(\frac{-E_1}{u_k T_{ref}}\right) x_{1,k}^2 - k_2 \exp\left(\frac{-E_2}{u_k T_{ref}}\right) x_{2,k} + 0.01\delta_k$$
$$y_k = x_{2,k} + 0.001\delta_k$$

where $x_{1,k}$ and $x_{2,k}$ are the dimensionless concentrations of the raw material and the product, respectively; $u_k = T_k/T_{ref}$ is the dimensionless temperature of the reactor with $T_{\rm ref}$ being the reference temperature; the reactor temperature $T_k$ is limited to 298 K $\leq T_k \leq$ 398 K; subscript $k$ is added to denote the batch number of the reaction. The disturbance $\delta_k$ varies randomly over [0, 1], as shown in Fig. 4.6. The final time $t_f$ is 1.0 h and the sampling time $h = t_f/N = 0.1$ with $N = 10$.

The system parameters are given as $k_1 = 4.0 \times 10^3$, $k_2 = 6.2 \times 10^5$, $E_1 = 2.5 \times 10^3$ K, $E_2 = 5.0 \times 10^3$ K, and $T_{ref} = 348$ K. The desired product reference trajectory is $y_d(t) = 0.61$. The initial states vary randomly as $x_{1,k}(0) = 1 + 0.001\delta_k$ and $x_{2,k}(0) = 0.001\delta_k$.

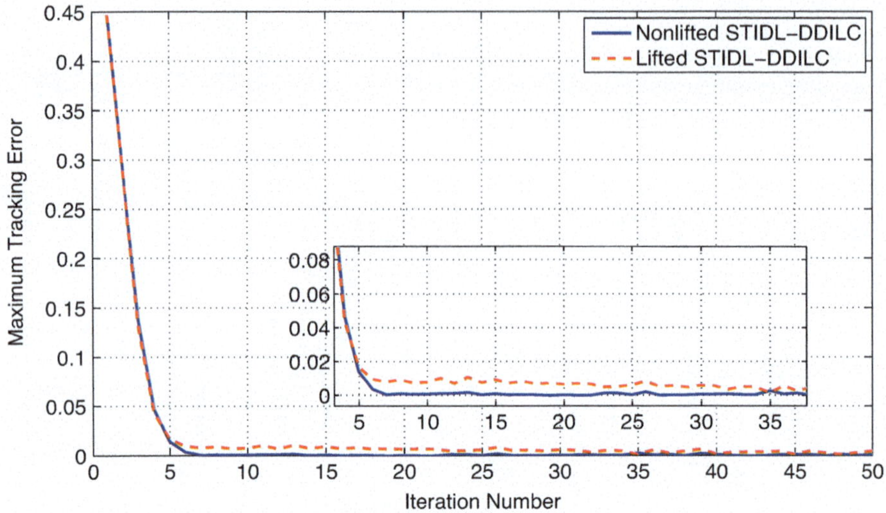


**Fig. 4.7** The convergence of tracking error in Example 2

The controller parameters are selected as $u_k(0) = 0.92$, $\hat{\varphi}_0(t) = 0.1$, $\rho_i = 1, i = 1, \cdots, N$, $\eta = 0.1$, $\lambda = 0.5$, $\mu = 1$. Applying the nonlifted STIDL-DDILC along with the lifted STIDL-DDILC for a comparison, the simulation result is shown in Fig. 4.7.

From Figs. 4.6 and 4.7, the applicability and effectiveness of the proposed non-lifted STIDL-DDILC approach is obvious: a satisfactory performance is achieved in the presence of various random disturbances. Further, the nonlifted STIDL-DDILC performs better than the lifted STIDL-DDILC by taking advantage of more control information from previous time instants of the same iteration.

## 4.5 Summary

This chapter presents a STIDL method for relaxing some critical conditions in the CFIDL, PFIDL, and FFIDL methods. By using the lifting technique, a lifted STIDL-DDILC is first proposed in a optimization design framework. Meanwhile, a nonlifted STIDL-DDILC is further presented as an alternative for reducing the potential computational complexity if the controlled plant has a large operation length, or a small sampling period in real implementation. Both theoretical analysis and simulation study are provided to verify the effectiveness of the two proposed STIDL-DDILC methods.

## References


Amann N, Owens DH, Rogers E (1994) 2D systems theory applied to learning control systems. In: The 33rd conference on decision and control. IEEE. https://ieeexplore.ieee.org/document/411268

Amann N, Owens H, Rogers E (1996) Iterative learning control for discrete-time systems with exponential rate of convergence. IEE-Control Theory Appl 143:217–224

Barton KL, Bristow DA, Alleyne AG (2010) A numerical method for determining monotonicity and convergence rate in iterative learning control. Int J Control 83:219–226

Boyd S, Vandenberghe L (2004) Convex optimization. Cambridge University Press

Bristow DA (2008) Weighting matrix design for robust monotonic convergence in norm optimal iterative learning control. In: American control conference. IEEE. https://ieeexplore.ieee.org/document/4587213

Dijkstra B, Bosgra OH (2002) Extrapolation of optimal lifted system ILC solution, with application to a waferstage. In: American control conference. IEEE. https://ieeexplore.ieee.org/document/1025176

Freeman CT (2012) Constrained point-to-point iterative learning control with experimental verification. Control Eng Pract 20(5):489–498

Gunnarsson S, Norrlof M (2001) On the design of ILC algorithms using optimization. Automatica 37:2011–2016

Haber A, Fraanje R, Verhaegen M (2012) Linear computational complexity robust ILC for lifted systems. Automatica 48:1102–1110

Hakvoort W, Aarts R, van Dijk J, Jonker J (2008) Lifted system iterative learning control applied to an industrial robot. Control Eng Pract 16:377–391

Hakvoort W, Aarts R, van Dijk J, Jonker J (2009) A computationally efficient algorithm of iterative learning control for discrete-time linear time-varying systems. Automatica 45:2925–2929

Harte TJ, Hatonen J, Owens DH (2005) Discrete-time inverse model-based iterative learning control: stability, monotonicity and robustness. Int J Control 78(8):577–586

Hätönen J (2004) Issues of algebra and optimality in iterative learning control. University of Oulu Press

Horn RA, Johnson CR (2012) Matrix analysis. Cambridge University Press

Jury EI (1964) Theory and application of the z-Transform method. Wiley, New York

Lee JH, Lee KS, Kim WC (2000) Model-based iterative learning control with a quadratic criterion for time-varying linear systems. Automatica 36:641–657

Logsdon JS, Biegler LT (1989) Accurate solution of differential-algebraic optimization problems. Ind Eng Chem Res 28:1628–1639

Mishra S, Topcu U, Tomizuka M (2011) Optimization-based constrained iterative learning control. IEEE Trans Control Syst Technol 19:1613–1621

Owens DH, Freeman CT, Chu B (2013) Norm optimal iterative learning control for planar tracking tasks. In: The 11th IFAC international workshop on adaptation and learning in control and signal processing. SinceDirect. https://www.sciencedirect.com/science/article/pii/S1474667016330233

Rice JK, Verhaegen M (2010) A structured matrix approach to efficient calculation of LQG repetitive learning controllers in the lifted setting. Int J Control 83:1265–1276

Rogers E, Owens DH, Werner H, Freeman CT, Lewin PL, Kichhoff S, Schmidt C, Lichtenberg G (2010) Norm-optimal iterative learning control with application to problems in accelerator-based free electron lasers and rehabilitation robotics. Eur J Control 5:497–522

Sun HQ, Alleyne AG (2014) A computationally efficient norm optimal iterative learning control approach for LTV systems. Automatica 50:141–148

van de Wijdeven J, Donkers T, Bosgra O (2009) Iterative learning control for uncertain systems: robust monotonic convergence analysis. Automatica 45(10):2383–2391

Yang S, Qu Z, Fan X, Nian X (2008) Novel iterative learning controls for linear discrete-time systems based on a performance index over iterations. Automatica 44:1366–1372

# Chapter 5
# Data-Driven ILC for Systems with Package Dropouts

## 5.1 Introduction

Recently, network control system (NCS) (Ge et al. 2017) has been extensively applied in various fields owing to its low cost, simple installation, less wiring, easy maintenance, and high reliability. However, the implementation of ILC in NCSs which execute repeatedly is still an open problem (Huang et al. 2014). The performance of the network control system will be deteriorated when the system suffers from time delays, data loss, communication noise, etc. Therefore, the robustness of ILC with data noise is critical in the successful implementation of ILC for NCSs.

In Bu et al. (2011), Bu and Hou (2011), both first-order and high-order learning control algorithms are designed for linear systems with measurement output losses, and the convergence is guaranteed under mathematical expectation. Further, two ILC methods are designed with strict almost-sure convergence for stochastic linear systems (Shen et al. 2017). By introducing a Markov chain, a P-type ILC scheme allowing successive data dropouts (Shen and Xu 2017; Shen et al. 2017) is proposed for linear systems. Aiming for nonlinear affine systems, an ILC method with both intermittent and successive updating algorithms (Shen et al. 2018) is proposed in the presence of input and output data losses. Further, a dropped data compensation by the latest previous data is explored in the networked ILC scheme for a nonlinear affine system (Liu and Ruan 2018; Bu et al. 2013). In the work of Liu and Ruan (2017), an extended networked ILC is designed for nonlinear affine systems where only the dropped input information is compensated.

However, most of the existing ILC methods for data noise are limited to linear systems (Bu et al. 2011; Bu and Hou 2011; Shen et al. 2017; Shen and Xu 2017; Shen et al. 2017) or affined nonlinear systems (Shen et al. 2018; Liu and Ruan 2018; Bu et al. 2013; Liu and Ruan 2017). In other words, they are model-based depending on an exact linear model or an affined model structure of the controlled plant. Consequently, many difficulties may be encountered when applying these methods (Bu et al. 2011; Bu and Hou 2011; Shen et al. 2017; Shen and Xu 2017;

 
R. Chi et al., *Data-Driven Iterative Learning Control for Discrete-Time Systems*, Intelligent Control and Learning Systems 2,
https://doi.org/10.1007/978-981-19-5950-9_5

Shen et al. 2017, 2018; Liu and Ruan 2018; Bu et al. 2013; Liu and Ruan 2017) into practical systems that are difficult to model. Further, the existing networked ILC schemes (Bu et al. 2011; Bu and Hou 2011; Shen et al. 2017; Shen and Xu 2017; Shen et al. 2017, 2018; Liu and Ruan 2018; Bu et al. 2013; Liu and Ruan 2017) mainly adopt P-type or D-type learning laws where the learning gains remain unchanged once being selected properly which causes a poor ability in dealing with large uncertainty and disturbance. In addition, for the data dropout problem, the lost data is mostly compensated by using the latest previous data. But up to now, no work has been done to compensate for the lost data using a predicted one at the same time instant through a predictive model. Although several data-driven ILC methods (Chi et al. 2012, 2017, 2018; Lin et al. 2019) have been developed more recently for nonlinear nonaffine repetitive systems, the data package losses at both input and output sides have not been considered wherein.

This chapter reconsiders the ILC problem with both input and output data dropouts under a data-driven framework. A linear data model (LDM) is developed at first to describe the input and output dynamics of the controlled plant in the iteration domain. Consequently, an iterative predictive model is built by virtue of the gained LDM which virtually exists in the computer for the purpose of predicting and compensating the output data dropouts. Then, a compensation-based data-driven ILC (cDDILC) is proposed by introducing two Bernoulli random variables to describe the random input and output data losses. The lost output data is compensated using the predictive one and the lost input data is replaced by the one in the latest previous iteration of the same time instant.

Although this work starts with an unknown linear time-varying system, the proposed method is data-driven in nature because no model information is needed for the controller design, analysis, and implementation. Therefore, the proposed cDDILC method is all-purposed and can be directly applied without any modification to practical nonlinear nonaffine systems whose models are unavailable. The effectiveness of the proposed cDDILC is guaranteed rigorously in theory and verified through simulations.

This chapter is organized as follows. Section 5.2 formulates the ILC problem with data dropout. Section 5.3 presents the design of a compensation-based DDILC. Section 5.4 shows the convergence analysis in mathematics. Section 5.5 provides simulation results to verify the effectiveness of the proposed method. Section 5.6 gives some conclusions.

## 5.2 Problem Formulation

Consider a class of MIMO linear time-varying systems

$$\begin{cases} \boldsymbol{x}_k(t+1) = \boldsymbol{A}(t)\boldsymbol{x}_k(t) + \boldsymbol{B}(t)\boldsymbol{u}_k(t) \\ \boldsymbol{y}_k(t) = \boldsymbol{C}(t)\boldsymbol{x}_k(t) \end{cases} \tag{5.1}$$

where $\boldsymbol{y}_k(t) \in R^n$ and $\boldsymbol{u}_k(t) \in R^n$ are system outputs and inputs, respectively; $\boldsymbol{x}_k(t) \in R^m$ is the system state; $t \in \{0, \ldots, N\}$, where $N \in Z^+$ represents the endpoint; $k \in \{0, 1, \ldots\}$ denotes the iteration index. $\boldsymbol{A}(t) \in R^{m\times m}$, $\boldsymbol{B}(t) \in R^{m\times n}$, and $\boldsymbol{C}(t) \in R^{n\times m}$ are unknown coefficient matrices.

**Assumption 5.1** The initial states of system (5.1) satisfy the identical initial condition, i.e., $\boldsymbol{x}_k(0) = \boldsymbol{x}_0$, $\forall k$, where $\boldsymbol{x}_0$ is a constant vector.

Similar to Chi et al. (2012), one can obtain the following equation from (5.1),

$$\begin{aligned}\boldsymbol{y}_k(t+1) &= \boldsymbol{C}(t+1)\prod_{i=0}^{t}\boldsymbol{A}(t-i)\boldsymbol{x}_k(0)\\&\quad+\boldsymbol{C}(t+1)\sum_{i=0}^{t}\prod_{j=0}^{i-1}\boldsymbol{A}(t-j)\boldsymbol{B}(t-i)\boldsymbol{u}_k(t-i)\\&=\boldsymbol{C}(t+1)\prod_{i=0}^{t}\boldsymbol{A}(t-i)\boldsymbol{x}_k(0)+\boldsymbol{\Phi}(t)\boldsymbol{U}_k(t)\end{aligned}\tag{5.2}$$

where $\boldsymbol{U}_k(t) = \left[\boldsymbol{u}_k^{\mathrm{T}}(0), \boldsymbol{u}_k^{\mathrm{T}}(1), \ldots, \boldsymbol{u}_k^{\mathrm{T}}(t)\right]^{\mathrm{T}} \in R^{(n*(t+1))\times 1}$, $\boldsymbol{u}_k(t) = \left[u_{1,k}(t), u_{2,k}(t), \ldots, u_{n,k}(t)\right]^{\mathrm{T}} \in R^{n\times 1}$, $\boldsymbol{\Phi}(t) = [\varphi(0), \varphi(1), \ldots \varphi(t)] \in R^{n\times(n^*(t+1))}$ and $\varphi(i) = \boldsymbol{C}(t+1)\prod_{j=i+1}^{t}\boldsymbol{A}(j)\boldsymbol{B}(i) \in R^{n\times n}$, $i \in \{0, \ldots, t\}$.

Then, by using Assumption 5.1, one can build a virtual LDM to describe the I/O behavior of the system (5.1) between two consecutive iterations

$$\begin{aligned}\Delta\boldsymbol{y}_k(t+1) &= \boldsymbol{y}_k(t+1) - \boldsymbol{y}_{k-1}(t+1)\\&=\boldsymbol{C}(t+1)\prod_{i=0}^{t}\boldsymbol{A}(t-i)\boldsymbol{x}_k(0)+\boldsymbol{\Phi}(t)\boldsymbol{U}_k(t)\\&\quad-\boldsymbol{C}(t+1)\prod_{i=0}^{t}\boldsymbol{A}(t-i)\boldsymbol{x}_{k-1}(0)+\boldsymbol{\Phi}(t)\boldsymbol{U}_{k-1}(t)\\&=\boldsymbol{\Phi}(t)\Delta\boldsymbol{U}_k(t)\end{aligned}\tag{5.3}$$

where $\Delta\boldsymbol{U}_k(t) = \boldsymbol{U}_k(t) - \boldsymbol{U}_{k-1}(t)$.

Because $\boldsymbol{A}(t)$, $\boldsymbol{B}(t)$, and $\boldsymbol{C}(t)$ are all real matrices with appropriate dimensions and $t \in \{0, \ldots, N\}$ is finite, it is clear that the matrices $\boldsymbol{A}(t)$, $\boldsymbol{B}(t)$, and $\boldsymbol{C}(t)$ are bounded. As a direct result, $\boldsymbol{\Phi}(t)$ is also bounded for $t \in \{0, \ldots, N\}$. Without loss of generality, assume $0 < \|\boldsymbol{\Phi}(t)\| \le b_\Phi$ in subsequent analysis where $b_\Phi > 0$ is a constant.

**Remark 5.1** Since all the coefficient matrices are unknown in (5.1), the traditional model-based control schemes cannot be used directly for this system. Meanwhile, from Eq. (5.3), one can see that all the unknown system matrices are compressed into a single parameter matrix $\boldsymbol{\Phi}(t)$, which can be estimated using the obtained LDM (5.3).

**Remark 5.2** The LDM (5.3) exists virtually in the computer for the only purpose of controller design and analysis. And it has no special physical meaning but only describes the data relationship of the original controlled plant between two consecutive iterations. In addition, it can be also employed as an iterative predictive model in the subsequent controller design.

## 5.3 Controller Design

When a control system is conducted via network transmission where the calculated input and the measurement output of the plant are transmitted through the network, the delays and losses often occur in the transmission of I/O data due to the network congestion and bandwidth limitation. In this section, therefore, the ILC problem is focused on the repetitive system (5.1) where the data dropout occurs on both the controller side and the actuator side. The control goal is to seek a control input $\boldsymbol{u}_k(t)$ such that the tracking error $\boldsymbol{e}_k(t+1) = \boldsymbol{y}_d(t+1) - \boldsymbol{y}_k(t+1)$ converges iteratively even though there exists data dropout, where $\boldsymbol{y}_d(t+1) \in R^n$ represents a desired trajectory and is bounded for all $t \in \{0, \ldots, N\}$.

By using the LDM (5.3), a compensation-based DDILC (cDDILC) with package dropout is presented as follows:

$$\hat{\boldsymbol{y}}_k(t+1) = \hat{\boldsymbol{y}}_{k-1}(t+1) + \hat{\boldsymbol{\Phi}}_k(t)\Delta\boldsymbol{U}_k(t) \tag{5.4}$$

$$\tilde{\boldsymbol{y}}_k(t+1) = \alpha_k(t+1)\boldsymbol{y}_k(t+1) + (1-\alpha_k(t+1))\,\hat{\boldsymbol{y}}_k(t+1) \tag{5.5}$$

$$\tilde{\boldsymbol{u}}_k(t) = \boldsymbol{u}_{k-1}(t) + \frac{\rho\hat{\boldsymbol{\varphi}}_k^{\mathrm{T}}(t)}{\lambda + \left\|\hat{\boldsymbol{\varphi}}_k(t)\right\|^2}\left(\tilde{\boldsymbol{e}}_{k-1}(t+1) - \sum_{i=0}^{t-1}\hat{\boldsymbol{\varphi}}_k(i)\Delta\boldsymbol{u}_k(i)\right) \tag{5.6}$$

$$\boldsymbol{u}_k(t) = \beta_k(t)\tilde{\boldsymbol{u}}_k(t) + (1-\beta_k(t))\,\boldsymbol{u}_{k-1}(t) \tag{5.7}$$

$$\begin{aligned}\hat{\boldsymbol{\Phi}}_k(t) &= \hat{\boldsymbol{\Phi}}_{k-1}(t)\\ &+ \frac{\eta\alpha_k(t+1)\left(\Delta\boldsymbol{y}_{k-1}(t+1) - \hat{\boldsymbol{\Phi}}_{k-1}(t)\Delta\boldsymbol{U}_{k-1}(t)\right)}{\mu + \|\Delta\boldsymbol{U}_{k-1}(t)\|^2}\Delta\boldsymbol{U}_{k-1}^{\mathrm{T}}(t)\end{aligned} \tag{5.8}$$

$$\hat{\boldsymbol{\Phi}}_k(t) = \hat{\boldsymbol{\Phi}}_0(t),\ \text{if}\ \hat{\boldsymbol{\varphi}}_k(i)\hat{\boldsymbol{\varphi}}_0(i) < 0\ \text{or}\ \left\|\hat{\boldsymbol{\Phi}}_k(t)\right\| \le \varepsilon,\quad i \in \{0, 1, \ldots, t\} \tag{5.9}$$

where $\hat{\boldsymbol{y}}_k(t+1) \in R^n$ represents the predictive output; $\hat{\boldsymbol{\Phi}}_k(t)$ is used to estimate $\boldsymbol{\Phi}(t)$; $\alpha_k(t+1)$ and $\beta_k(t)$ are two Bernoulli random variables describing the random data loss, in which the value of 1 or 0 indicates that the data transmission is normal or lost, respectively; $\lambda > 0$ and $\mu > 0$ are two weights; $\rho > 0$ and $0 < \eta < 2$ are two step-size factors; $\tilde{\boldsymbol{y}}_k(t+1) \in R^n$ represents the compensated output; $\boldsymbol{u}_k(t)$ is the actual control input and $\tilde{\boldsymbol{u}}_k(t) \in R^n$ represents the calculated control input;

$\tilde{\boldsymbol{e}}_{k-1}(t+1) = \boldsymbol{y}_d(t+1) - \tilde{\boldsymbol{y}}_{k-1}(t+1)$; $\Delta\boldsymbol{u}_k(t) = \boldsymbol{u}_k(t) - \boldsymbol{u}_{k-1}(t)$; $\varepsilon$ is a small positive constant; $\hat{\boldsymbol{\Phi}}_0(t)$ is the initial estimation of $\hat{\boldsymbol{\Phi}}_k(t)$ at the first iteration.

**Remark 5.3** One can see from the proposed cDDILC method (5.4)–(5.9) that controller (5.6) and estimator (5.8) are located in the same network side, they both use the I/O data transmitted from the controlled plant that is reflected by using the Bernoulli random variable $\alpha_k(t+1)$. Meanwhile, in the output compensator (5.5) and the input compensator (5.7), two Bernoulli random variables $\alpha_k(t+1)$ and $\beta_k(t)$ are used to address the data dropouts in the input and output signals, respectively. Equation (5.4) is a locally iterative predictive model used to predict the system output in a real-time wise, and (5.9) is a reset algorithm to enhance the estimated performance of (5.8).

The control system block diagram of the proposed cDDILC is shown in Fig. 5.1, where $\delta^{-1}$ represents an iteration backward operator, i.e., $\delta^{-1}\boldsymbol{x}_k(t) = \boldsymbol{x}_{k-1}(t)$; $z^{-1}$ represents an time backward operator, i.e., $z^{-1}\boldsymbol{x}_k(t) = \boldsymbol{x}_k(t-1)$. $\beta_k(t)$ and $\alpha_k(t+1)$ are used to denote whether the data transmission is a success or failure.

**Remark 5.4** There are two networks and two memories in Fig. 5.1. The I/O data produced by the controlled plant is transmitted through Network 1 and stored in Memory 1, which is used in controller (5.6) and estimator (5.8), as well as the output compensator (5.5). The control input signals calculated by the proposed learning control law (5.6) are transmitted through Network 2 and stored in Memory 2, which acts on the controlled plant to compensate the data dropout of the input signals in the input compensator (5.7).

**Remark 5.5** From Fig. 5.1, one can see that the real control input $\boldsymbol{u}_k(t)$ is available for the iterative learning controller and the feasibility of the proposed cDDILC is clear. Both the input and output data dropouts are considered and compensated in the proposed control method (5.4)–(5.9). For the missing measurement output $\boldsymbol{y}_k(t+1)$, it is replaced and compensated by the predictive output value $\hat{\boldsymbol{y}}_k(t+1)$; the control input in the latest previous iteration, $\boldsymbol{u}_{k-1}(t)$, is used to compensate for the missing $\boldsymbol{u}_k(t)$.

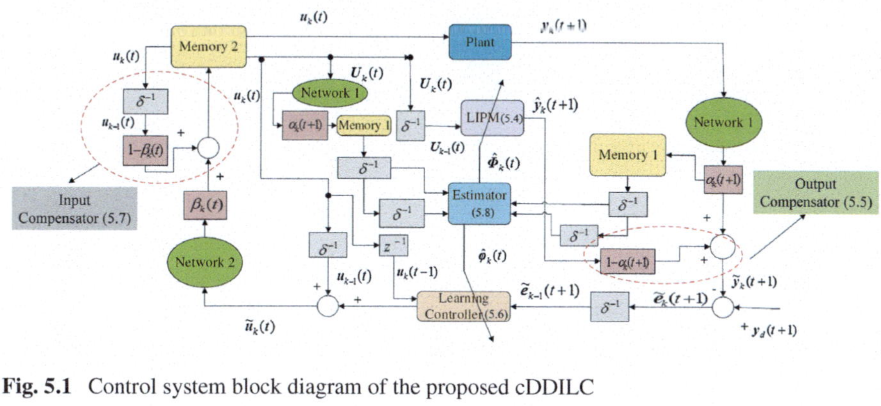


**Fig. 5.1** Control system block diagram of the proposed cDDILC

## 5.4 Convergence Analysis

The convergence property of the proposed cDDILC scheme is shown as the following theorem.

**Theorem 5.1** *For a class of linear time-varying systems (5.1) with the presence of random data loss on both input and output sides, if Assumption 5.1 holds, then the proposed cDDILC scheme (5.4)–(5.9) guarantees that:*

*(i) the parameter estimation $\hat{\boldsymbol{\Phi}}_k(t)$ is bounded, $\forall t \in \{0, 1, \ldots, N\}$, $\forall k$;*
*(ii) the tracking error $\boldsymbol{e}_k(t+1)$ converges to a small bound over the finite time interval $t \in \{0, 1, \ldots, N\}$ as $k$ approaches to infinity.*

***Proof*** *Part 1. The boundedness of* $\hat{\boldsymbol{\Phi}}_k(t)$

Define a parameter estimation error as $\tilde{\boldsymbol{\Phi}}_k(t) = \boldsymbol{\Phi}(t) - \hat{\boldsymbol{\Phi}}_k(t)$. Subtracting $\boldsymbol{\Phi}(t)$ from both sides of (5.8), it is easy to get

$$\tilde{\boldsymbol{\Phi}}_k(t) = \tilde{\boldsymbol{\Phi}}_{k-1}(t)\left(\boldsymbol{I}_1 - \frac{\alpha_k(t+1)\eta\Delta\boldsymbol{U}_{k-1}(t)\Delta\boldsymbol{U}_{k-1}^{\mathrm{T}}(t)}{\mu + \|\Delta\boldsymbol{U}_{k-1}(t)\|^2}\right) \tag{5.10}$$

where $\boldsymbol{I}_1 \in R^{(n*(t+1))\times(n*(t+1))}$ is a unit matrix.

One has,

$$\left\|\tilde{\boldsymbol{\Phi}}_{k-1}(t)\left(\boldsymbol{I}_1 - \frac{\alpha_k(t+1)\eta\Delta\boldsymbol{U}_{k-1}(t)\Delta\boldsymbol{U}_{k-1}^{\mathrm{T}}(t)}{\mu + \left\|\Delta\boldsymbol{U}_{k-1}(t)\right\|^2}\right)\right\|^2 = \left\|\tilde{\boldsymbol{\Phi}}_{k-1}(t)\right\|^2 + \left(-2 + \frac{\alpha_k(t+1)\eta\left\|\Delta\boldsymbol{U}_{k-1}(t)\right\|^2}{\mu + \left\|\Delta\boldsymbol{U}_{k-1}(t)\right\|^2}\right)\frac{\alpha_k(t+1)\eta\left\|\tilde{\boldsymbol{\Phi}}_{k-1}(t)\Delta\boldsymbol{U}_{k-1}(t)\right\|^2}{\mu + \left\|\Delta\boldsymbol{U}_{k-1}(t)\right\|^2}. \tag{5.11}$$

Since $0 < \eta < 2$, $\mu > 0$, and $\alpha_k(t+1) \in \{0, 1\}$, one has

$$\left(-2 + \frac{\alpha_k(t+1)\eta\|\Delta\boldsymbol{U}_{k-1}(t)\|^2}{\mu + \|\Delta\boldsymbol{U}_{k-1}(t)\|^2}\right) < 0. \tag{5.12}$$

Then, it can be found a constant $0 < d_1 < 1$, such that

$$\left\|\tilde{\boldsymbol{\Phi}}_{k-1}(t)\left(\boldsymbol{I}_1 - \frac{\alpha_k(t+1)\eta\Delta\boldsymbol{U}_{k-1}(t)\Delta\boldsymbol{U}_{k-1}^{\mathrm{T}}(t)}{\mu + \left\|\Delta\boldsymbol{U}_{k-1}(t)\right\|^2}\right)\right\| \le d_1\left\|\tilde{\boldsymbol{\Phi}}_{k-1}(t)\right\|^2. \tag{5.13}$$

In terms of (5.13), one has

$$\left\|\tilde{\boldsymbol{\Phi}}_k(t)\right\| \le d_1\left\|\tilde{\boldsymbol{\Phi}}_{k-1}(t)\right\| \le \cdots \le d_1^k\left\|\tilde{\boldsymbol{\Phi}}_0(t)\right\|. \tag{5.14}$$

Hence, $\tilde{\boldsymbol{\Phi}}_k(t)$ is bounded because $\hat{\boldsymbol{\Phi}}_0(t)$ is chosen bounded and $\boldsymbol{\Phi}_k(t)$ is bounded too. Therefore, the boundedness of $\hat{\boldsymbol{\Phi}}_k(t)$ can be guaranteed, $\forall t \in \{0, \ldots, N\}$ and $\forall k$.

*Part 2. The bounded convergence of tracking error*

Define an output estimation error as $\boldsymbol{\sigma}_k(t+1) = \boldsymbol{y}_k(t+1) - \hat{\boldsymbol{y}}_k(t+1)$. According to (5.3) and (5.4), we have

$$\begin{aligned}\sigma_k(t+1) &= \boldsymbol{y}_k(t+1) - \hat{\boldsymbol{y}}_k(t+1)\\ &= \boldsymbol{y}_{k-1}(t+1) + \boldsymbol{\Phi}(t)\Delta\boldsymbol{U}_k(t) - \hat{\boldsymbol{y}}_{k-1}(t+1) - \hat{\boldsymbol{\Phi}}_k(t)\Delta\boldsymbol{U}_k(t)\\ &= \boldsymbol{\sigma}_{k-1}(t+1) + \tilde{\boldsymbol{\Phi}}_k(t)\Delta\boldsymbol{U}_k(t)\end{aligned} \tag{5.15}$$

From (5.15), it yields

$$\boldsymbol{\sigma}_k(t+1) = \sum_{j=0}^{k} \tilde{\boldsymbol{\Phi}}_j(t)\Delta\boldsymbol{U}_j(t) \tag{5.16}$$

where the output estimation error $\boldsymbol{\sigma}_k(t+1) \in R^n$ is set as zero for $k < 0$.

According to (5.14), we know that $\left\|\tilde{\boldsymbol{\Phi}}_k(t)\right\|$ can converge to zero when $k \to \infty$. Apparently, we have $\lim_{k\to\infty}\left\|\tilde{\boldsymbol{\Phi}}_k(t)\Delta\boldsymbol{U}_k(t)\right\| = 0$. Therefore, $\lim_{k\to\infty}\left\|\sum_{j=0}^{k}\tilde{\boldsymbol{\Phi}}_j(t)\Delta\boldsymbol{U}_j(t)\right\|$ is bounded. Consequently, one can obtain that $\boldsymbol{\sigma}_k(t+1)$ is bounded. In the subsequent analysis, it is assumed $\|\boldsymbol{\sigma}_k(t+1)\| \le b_\sigma$ where $b_\sigma$ is a positive constant.

Since $\hat{\boldsymbol{\Phi}}_k(t)$ is bounded and $\beta_k(t) \in \{0, 1\}$, there exists a suitable $\lambda$ such that

$$\left\|\frac{\beta_k(t)\hat{\boldsymbol{\varphi}}_k(t)}{\lambda + \left\|\hat{\boldsymbol{\varphi}}_k(t)\right\|^2}\right\| \le \frac{\beta_k(t)\left\|\hat{\boldsymbol{\varphi}}_k(t)\right\|}{2\sqrt{\lambda}\left\|\hat{\boldsymbol{\varphi}}_k(t)\right\|} \le \frac{1}{2\sqrt{\lambda}} = d_2 < \frac{0.5}{b_\Phi} \tag{5.17}$$

$$0 < \left\|\frac{\beta_k(t)\hat{\boldsymbol{\varphi}}_k(t)\hat{\boldsymbol{\varphi}}_k(i)}{\lambda + \left\|\hat{\boldsymbol{\varphi}}_k(t)\right\|^2}\right\| \le b_\Phi\frac{\beta_k(t)\left\|\hat{\boldsymbol{\varphi}}_k(t)\right\|}{2\sqrt{\lambda}\left\|\hat{\boldsymbol{\varphi}}_k(t)\right\|} < \frac{b_\Phi}{2\sqrt{\lambda}} = d_3 < 0.5 \tag{5.18}$$

$$\left(\sum_{i=0}^{t-1}\left\|\frac{\beta_k(t)\hat{\boldsymbol{\varphi}}_k^{\mathrm{T}}(t)\hat{\boldsymbol{\varphi}}_k(i)}{\lambda + \left\|\hat{\boldsymbol{\varphi}}_k(t)\right\|^2}\right\|\right)^{\frac{1}{t}} \le d_4 \tag{5.19}$$

where $d_2$, $d_3$, and $d_4$ are properly constants.

According to (5.6) and (5.7), one has

$$
\begin{aligned}
\boldsymbol{u}_k(t) &= \beta_k(t)\left(\boldsymbol{u}_{k-1}(t) + \frac{\rho\hat{\boldsymbol{\varphi}}_k^{\mathrm{T}}(t)}{\lambda + \left\|\hat{\boldsymbol{\varphi}}_k(t)\right\|^2}\left(\tilde{\boldsymbol{e}}_{k-1}(t+1) - \sum_{i=0}^{t-1}\hat{\boldsymbol{\varphi}}_k(i)\Delta\boldsymbol{u}_k(i)\right)\right) \\
&\quad + (1-\beta_k(t))\,\boldsymbol{u}_{k-1}(t) \\
&= \boldsymbol{u}_{k-1}(t) + \frac{\rho\beta_k(t)\hat{\boldsymbol{\varphi}}_k^{\mathrm{T}}(t)}{\lambda + \left\|\hat{\boldsymbol{\varphi}}_k(t)\right\|^2}\left(\tilde{\boldsymbol{e}}_{k-1}(t+1) - \sum_{i=0}^{t-1}\hat{\boldsymbol{\varphi}}_k(i)\Delta\boldsymbol{u}_k(i)\right) \qquad (5.20)
\end{aligned}
$$

Let $\boldsymbol{A}_k(t) = \begin{bmatrix} \mathbf{0} & \boldsymbol{I}_2 & \mathbf{0} & \cdots & \mathbf{0} \\ \mathbf{0} & \mathbf{0} & \boldsymbol{I}_2 & \cdots & \mathbf{0} \\ \vdots & \vdots & \vdots & \ddots & \vdots \\ \mathbf{0} & \mathbf{0} & \mathbf{0} & \cdots & \boldsymbol{I}_2 \\ \mathbf{0} & -\frac{\rho\beta_k(t)\hat{\boldsymbol{\varphi}}_k^{\mathrm{T}}(t)\hat{\boldsymbol{\varphi}}_k(0)}{\lambda+\left\|\hat{\boldsymbol{\varphi}}_k(t)\right\|^2} & -\frac{\rho\beta_k(t)\hat{\boldsymbol{\varphi}}_k^{\mathrm{T}}(t)\hat{\boldsymbol{\varphi}}_k(1)}{\lambda+\left\|\hat{\boldsymbol{\varphi}}_k(t)\right\|^2} & \cdots & -\frac{\rho\beta_k(t)\hat{\boldsymbol{\varphi}}_k^{\mathrm{T}}(t)\hat{\boldsymbol{\varphi}}_k(t-1)}{\lambda+\left\|\hat{\boldsymbol{\varphi}}_k(t)\right\|^2} \end{bmatrix} \in$ $R^{(n*(t+1))\times(n*(t+1))}$, and $\boldsymbol{C} = [\mathbf{0}, \ldots, \mathbf{0}, \boldsymbol{I}_2]^{\mathrm{T}} \in R^{(n*(t+1))\times n}$, where $\mathbf{0} \in R^{n\times n}$ is a zero matrix and $\boldsymbol{I}_2 \in R^{n\times n}$ is an unit matrix.

One has

$$
\Delta\overline{\boldsymbol{U}}_k(t) = \boldsymbol{A}_k(t)\Delta\overline{\boldsymbol{U}}_k(t-1) + \frac{\rho\beta_k(t)\boldsymbol{C}\hat{\boldsymbol{\varphi}}_k^{\mathrm{T}}(t)}{\lambda + \left\|\hat{\boldsymbol{\varphi}}_k(t)\right\|^2}\tilde{\boldsymbol{e}}_{k-1}(t+1) \qquad (5.21)
$$

where $\Delta\overline{\boldsymbol{U}}_k(t) = \left[\Delta\boldsymbol{u}_k^{\mathrm{T}}(0), \ldots, \Delta\boldsymbol{u}_k^{\mathrm{T}}(t)\right]^{\mathrm{T}} \in R^{(n*(t+1))\times 1}$ denotes an iterative difference vector of control input with fixed dimension. For example, $\Delta\overline{\boldsymbol{U}}_k(t-1) = \left[\mathbf{0}^{\mathrm{T}}, \Delta\boldsymbol{u}_k^{\mathrm{T}}(0), \ldots, \Delta\boldsymbol{u}_k^{\mathrm{T}}(t-1)\right]^{\mathrm{T}} \in R^{(n*(t+1))\times 1}$, $\mathbf{0} = [0, \ldots, 0]^{\mathrm{T}} \in R^n$, and $\Delta\boldsymbol{u}_k(t) = \mathbf{0}$ when $t$ is less than zero. The characteristic equation of $\boldsymbol{A}_k(t)$ is

$$
\det\left(z^{t+1}\boldsymbol{I}_2 + \frac{\rho\beta_k(t)\hat{\boldsymbol{\varphi}}_k^{\mathrm{T}}(t)\hat{\boldsymbol{\varphi}}_k(t-1)}{\lambda + \left\|\hat{\boldsymbol{\varphi}}_k(t)\right\|^2}z^t + \cdots + \frac{\rho\beta_k(t)\hat{\boldsymbol{\varphi}}_k^{\mathrm{T}}(t)\hat{\boldsymbol{\varphi}}_k(0)}{\lambda + \left\|\hat{\boldsymbol{\varphi}}_k(t)\right\|^2}z\right) = 0 \qquad (5.22)
$$

According to (5.19), by selecting a suitable $\rho$, one has

$$
\rho\sum_{i=0}^{t-1}\left\|\frac{\beta_k(t)\hat{\boldsymbol{\varphi}}_k^{\mathrm{T}}(t)\hat{\boldsymbol{\varphi}}_k(i)}{\lambda + \left\|\hat{\boldsymbol{\varphi}}_k(t)\right\|^2}\right\| \le \rho d_4^t < 1 \qquad (5.23)
$$

Then, by virtue of Lemma 4.2, it yields

$$
\begin{aligned}
|z|^t &\le \rho \sum_{i=0}^{t-1} \left\| \frac{\beta_k(t)\hat{\boldsymbol{\varphi}}_k^{\mathrm{T}}(t)\hat{\boldsymbol{\varphi}}_k(i)}{\lambda + \left\|\hat{\boldsymbol{\varphi}}_k(t)\right\|^2} \right\| |z|^{t-i} \\
&\le \rho \sum_{i=0}^{t-1} \left\| \frac{\beta_k(t)\hat{\boldsymbol{\varphi}}_k^{\mathrm{T}}(t)\hat{\boldsymbol{\varphi}}_k(i)}{\lambda + \left\|\hat{\boldsymbol{\varphi}}_k(t)\right\|^2} \right\| \\
&\le \rho d_4^t
\end{aligned}
\tag{5.24}
$$

which means $|z|^t \le \rho d_4^t < 1$.

To further select suitable $\rho$ and $\lambda$ such that $|z| \le \sqrt[t]{\rho} d_4 < 0.5$. Then, by virtue of Lemma 3.2, we have

$$
\|\boldsymbol{A}_k(t)\|_v \le \sqrt[t]{\rho} d_4 + \delta \le d_5 < 0.5 \tag{5.25}
$$

where $\|\cdot\|_v$ denotes a proper matrix norm, and $0 < d_5 < 0.5$ is a constant.

From Eq. (5.21), one has

$$
\begin{aligned}
\left\|\Delta\overline{\boldsymbol{U}}_k(t)\right\|_v \le& \|\boldsymbol{A}_k(t)\|_v \left\|\Delta\overline{\boldsymbol{U}}_k(t-1)\right\|_v \\
&+\rho \left\| \frac{\beta_k(t)\hat{\boldsymbol{\varphi}}_k(t)}{\lambda + \left\|\hat{\boldsymbol{\varphi}}_k(t)\right\|^2} \right\| \|\tilde{\boldsymbol{e}}_{k-1}(t+1)\|.
\end{aligned}
\tag{5.26}
$$

Then, according to (5.17), (5.25), and (5.26), one can derive that

$$
\left\|\Delta\overline{\boldsymbol{U}}_k(t)\right\|_v \le \rho d_2 \sum_{i=0}^{t} d_5^{t-i} \|\tilde{\boldsymbol{e}}_{k-1}(i+1)\| \tag{5.27}
$$

where $\overline{\boldsymbol{U}}_k(t)$ is zero for $t < 0$.

In the light of the definition of $\tilde{\boldsymbol{e}}_k(t+1)$, it follows

$$
\begin{aligned}
\tilde{\boldsymbol{e}}_k(t+1) &= \boldsymbol{y}_d(t+1) - \alpha_k(t+1)\boldsymbol{y}_k(t+1) - (1-\alpha_k(t+1))\,\hat{\boldsymbol{y}}_k(t+1) \\
&= \boldsymbol{y}_d(t+1) - \alpha_{k-1}(t+1)\boldsymbol{y}_{k-1}(t+1) - (1-\alpha_{k-1}(t+1))\,\hat{\boldsymbol{y}}_{k-1}(t+1) \\
&\quad + \alpha_{k-1}(t+1)\boldsymbol{y}_{k-1}(t+1) + (1-\alpha_{k-1}(t+1))\,\hat{\boldsymbol{y}}_{k-1}(t+1) \\
&\quad - \alpha_k(t+1)\boldsymbol{y}_k(t+1) - (1-\alpha_k(t+1))\,\hat{\boldsymbol{y}}_k(t+1) \\
&= \tilde{\boldsymbol{e}}_{k-1}(t+1) + \alpha_{k-1}(t+1)\boldsymbol{\sigma}_{k-1}(t+1) - \alpha_k(t+1)\boldsymbol{\sigma}_k(t+1) \\
&\quad - \hat{\boldsymbol{\Phi}}_k(t)\Delta\boldsymbol{U}_k(t).
\end{aligned}
\tag{5.28}
$$

According to (5.21), we can rewrite (5.28) as

$$
\begin{aligned}
\tilde{\boldsymbol{e}}_k(t+1) =& \left(\boldsymbol{I}_2 - \frac{\rho\beta_k(t)\hat{\boldsymbol{\varphi}}_k^{\mathrm{T}}(t)\hat{\boldsymbol{\varphi}}_k(t)}{\lambda + \left\|\hat{\boldsymbol{\varphi}}_k(t)\right\|^2}\right) \tilde{\boldsymbol{e}}_{k-1}(t+1) + \alpha_{k-1}(t+1)\boldsymbol{\sigma}_{k-1}(t+1) \\
&- \alpha_k(t+1)\boldsymbol{\sigma}_k(t+1) - \hat{\boldsymbol{\Phi}}_k(t)\boldsymbol{A}_k(t)\Delta\overline{\boldsymbol{U}}_k(t-1).
\end{aligned}
\tag{5.29}
$$

From Eq. (5.29), we have

$$\begin{aligned}\|\tilde{\boldsymbol{e}}_k(t+1)\| \leq & \left\| \left( \boldsymbol{I}_2 - \frac{\rho\beta_k(t)\hat{\boldsymbol{\varphi}}_k^{\mathrm{T}}(t)\hat{\boldsymbol{\varphi}}_k(t)}{\lambda + \left\|\hat{\boldsymbol{\varphi}}_k(t)\right\|^2} \right) \right\| \|\tilde{\boldsymbol{e}}_{k-1}(t+1)\| \\ & + \|\alpha_{k-1}(t+1)\boldsymbol{\sigma}_{k-1}(t+1)\| \\ & + \|\alpha_k(t+1)\boldsymbol{\sigma}_k(t+1)\| \\ & + \left\|\hat{\boldsymbol{\Phi}}_k(t)\right\|_v \|\boldsymbol{A}_k(t)\|_v \left\|\Delta\overline{\boldsymbol{U}}_k(t-1)\right\|_v. \end{aligned} \tag{5.30}$$

Similarly, one can find a suitable $\rho$ and $\lambda$ such that

$$0 < \left\| \left( \boldsymbol{I}_2 - \frac{\rho\beta_k(t)\hat{\boldsymbol{\varphi}}_k(t)\hat{\boldsymbol{\varphi}}_k(t)}{\lambda + \left\|\hat{\boldsymbol{\varphi}}_k(t)\right\|^2} \right) \right\| = d_6 < 1 \tag{5.31}$$

where $0 < d_6 = 1 - \rho d_3 < 1$ is a constant.

We have obtained that $\forall t$ and $\forall k$, $\|\sigma_k(t+1)\| \leq b_\sigma$ are bounded. Further, $\alpha_k(t+1) \in \{0, 1\}$, so $\|\alpha_{k-1}(t+1)\sigma_{k-1}(t+1)\| + \|\alpha_k(t+1)\sigma_k(t+1)\| \leq 2b_\sigma$ is bounded. Then, from (5.30), we can obtain

$$\begin{aligned}\|\tilde{\boldsymbol{e}}_k(t+1)\| \leq & \, d_6 \|\tilde{\boldsymbol{e}}_{k-1}(t+1)\| + \rho b_\Phi d_2 \sum_{i=0}^{t-1} d_5^{t-i} \|\tilde{\boldsymbol{e}}_{k-1}(i+1)\| \\ & + 2b_\sigma. \end{aligned} \tag{5.32}$$

Define $\left\|\boldsymbol{e}_{\max,k}\right\| = \max_{t\in\{0,1,\ldots,N-1\}} \{\|\boldsymbol{e}_k(t+1)\|\}$. Assume that $\left\|\boldsymbol{e}_{\max,k}\right\| = \|\boldsymbol{e}_k(\tau)\|$, where $\tau$ is some instant over $\{1, \ldots, N\}$.

By virtue of (5.32), we have

$$\begin{aligned}\left\|\tilde{\boldsymbol{e}}_{\max,k}\right\| = \|\tilde{\boldsymbol{e}}_k(\tau)\| \leq & \, d_6 \|\tilde{\boldsymbol{e}}_{k-1}(\tau)\| + \rho d_3 \sum_{i=1}^{\tau} d_5^{\tau-i} \|\tilde{\boldsymbol{e}}_{k-1}(i)\| \\ & + 2b_\sigma. \end{aligned} \tag{5.33}$$

Define $d_7 = \rho d_3$, one can rewrite (5.33) as

$$\begin{aligned}\left\|\tilde{\boldsymbol{e}}_{\max,k}\right\| & = \|\tilde{\boldsymbol{e}}_k(\tau)\| \\ & \leq d_6 \|\tilde{\boldsymbol{e}}_{k-1}(\tau)\| + d_7 \sum_{i=1}^{\tau} d_5^{\tau-i} \|\tilde{\boldsymbol{e}}_{k-1}(i)\| + 2b_\sigma \\ & \leq \left( d_6 + d_7 \frac{d_5}{1-d_5} \right) \left\|\tilde{\boldsymbol{e}}_{\max,k-1}\right\| + 2b_\sigma. \end{aligned} \tag{5.34}$$

From (5.31), we know that $0 < d_7 = 1 - d_6 < 1$, then one can obtain

$$
\begin{aligned}
d_6 + d_7 \frac{d_5}{1-d_5} &= 1 - d_7 + d_7 \frac{d_5}{1-d_5} \\
&= 1 + \frac{(2d_5 - 1)\, d_7}{1-d_5}.
\end{aligned} \tag{5.35}
$$

According to (5.25), it is clear that $0 < d_5 < 0.5$ and $d_7$ is located in the region of (0, 1). So, it is obvious that

$$
0 < 1 + \frac{(2d_5 - 1)\, d_7}{1-d_5} = d_8 < 1. \tag{5.36}
$$

Therefore, (5.34) can be rewritten as

$$
\begin{aligned}
\left\| \tilde{\boldsymbol{e}}_{\max,k} \right\| &\le d_8 \left\| \tilde{\boldsymbol{e}}_{\max,k-1} \right\| + 2b_\sigma \le \cdots \\
&\le d_8^k \left\| \tilde{\boldsymbol{e}}_{\max,0} \right\| + \frac{2b_\sigma}{1-d_8}.
\end{aligned} \tag{5.37}
$$

Since $\left\| \tilde{\boldsymbol{e}}_{\max,0} \right\|$ is bounded, it is obvious that $\lim_{k\to\infty} \left\| \tilde{\boldsymbol{e}}_{\max,k} \right\| \le \frac{2b_\sigma}{1-d_8}$ converges to a small bound. In addition, from the definition of $\tilde{\boldsymbol{e}}_k(t+1)$, we have

$$
\begin{aligned}
\tilde{\boldsymbol{e}}_k(t+1) &= \boldsymbol{y}_d(t+1) - \alpha_k(t+1)\boldsymbol{y}_k(t+1) - (1-\alpha_k(t+1))\, \hat{\boldsymbol{y}}_k(t+1) \\
&= \alpha_k(t+1)\boldsymbol{y}_d(t+1) - \alpha_k(t+1)\boldsymbol{y}_k(t+1) + (1-\alpha_k(t+1))\, \boldsymbol{y}_d(t+1) \\
&\quad - (1-\alpha_k(t+1))\, \hat{\boldsymbol{y}}_k(t+1) \\
&= \alpha_k(t+1)\boldsymbol{e}_k(t+1) + (1-\alpha_k(t+1)) \left( \boldsymbol{y}_d(t+1) - \hat{\boldsymbol{y}}_k(t+1) \right) \\
&= \alpha_k(t+1)\boldsymbol{e}_k(t+1) + (1-\alpha_k(t+1)) \\
&\quad \times \left( \boldsymbol{y}_d(t+1) - \boldsymbol{y}_k(t+1) + \boldsymbol{y}_k(t+1) - \hat{\boldsymbol{y}}_k(t+1) \right).
\end{aligned} \tag{5.38}
$$

Rearranging (5.38) yields

$$
\boldsymbol{e}_k(t+1) = \tilde{\boldsymbol{e}}_k(t+1) - (1-\alpha_k(t+1))\, \boldsymbol{\sigma}_k(t+1). \tag{5.39}
$$

Taking norms on both side of (5.39), one can obtain

$$
\|\boldsymbol{e}_k(t+1)\| \le \|\tilde{\boldsymbol{e}}_k(t+1)\| + |(1-\alpha_k(t+1))|\, \|\boldsymbol{\sigma}_k(t+1)\|. \tag{5.40}
$$

Since $\lim_{k\to\infty} \|\tilde{\boldsymbol{e}}_k(t+1)\|$ and $\lim_{k\to\infty} \|\boldsymbol{\sigma}_k(t+1)\|$are bounded, one can conclude that $\|\boldsymbol{e}_k(t+1)\|$ converges to a bounded range when $k \to \infty$, i.e., $\lim_{k\to\infty} \|\boldsymbol{e}_k(t+1)\| \le \frac{2b_\sigma}{1-d_8} + |(1-\alpha_k(t+1))|\, b_\sigma$. The bounded convergence of $\boldsymbol{e}_k(t+1)$ is evident. □

## 5.5 Simulation Study

***Example 1***. Consider a linear system

$$\begin{cases} \dot{x}_{1,k} = x_{2,k} \\ \dot{x}_{2,k} = x_{3,k} \\ \dot{x}_{3,k} = x_{4,k} - 149.36u_k \\ \dot{x}_{4,k} = -663.82x_{1,k} - 368.52x_{2,k} - 170.28x_{3,k} - 21.5x_{4,k} + 3808.68u_k \end{cases} \tag{5.41}$$

The simulation time is $2s$. The sampled period is $h = 0.1s$.
The desired trajectory is

$$y_d(t) = 10^{-6}(t-1)^3(4 - 0.03(t-1)). \tag{5.42}$$

*Case 1: The data loss rates for both measurement output and calculated input are* 10%. The initial values are set as $y_k(0) = 0$, $\hat{y}_k(0) = 0$, $u_0(t) = 0$, and $\hat{\varphi}_0(t) = 126$, and the controller parameters are set as $\lambda = 0.5$, $\rho = 1.5$, $\mu = 0.1$, $\eta = 2$, $\varepsilon = 0.00001$. Then, applying the cDDILC scheme (5.4)–(5.9), the convergence of tracking error is shown in Fig. 5.2, where Y-axis represents the maximum absolute tracking error $e_{\max}(k) = \max_{t\in\{1,2,\ldots,20\}}\{|e_k(t)|\}$ and the X-axis denotes the iteration numbers. Figure 5.3 shows the output tracking performance at the 30th iteration, where the vertical axis denotes the system output and the horizontal axis denotes the time instant. It is seen that the proposed cDDILC achieves the asymptotic convergence iteratively under the influence of data dropout.

*Case 2: The data loss rates for both measurement output and calculated input are 0%, i.e., no data loss occurs in both input and output sides.* Under the same simulation conditions, applying the proposed cDDILC (5.4)–(5.9), the simulation results are shown in Figs. 5.2 and 5.3. One can see that $e_{\max}(k)$ can converge to zero and the system output can perfectly track the desired trajectory. So, it has a good convergence performance when data transmission is normal.

By comparison, one can conclude that the cDDILC (5.4)–(5.9) can compensate the adverse effect of packet losses to achieve a well tracking performance with a trade-off of slightly reduced convergence performance.

***Example 2***. Consider a nonlinear discrete-time system

$$\begin{aligned} y_k(t+1) = & \frac{2.5y_k(t)y_k(t-1)}{1 - y_k^2(t) + y_k^2(t-1)} + 0.7\sin\left(0.5\left(y_k(t) + y_k(t-1)\right)\right) \\ & + 1.4u_k(t-1) + 1.2u_k(t), \quad 0 < t \le 40. \end{aligned} \tag{5.43}$$

The desired trajectory is

$$y_d(t+1) = 5 \times (-1)^{\mathrm{round}(t/80)}, \ 0 < t \le 40. \tag{5.44}$$

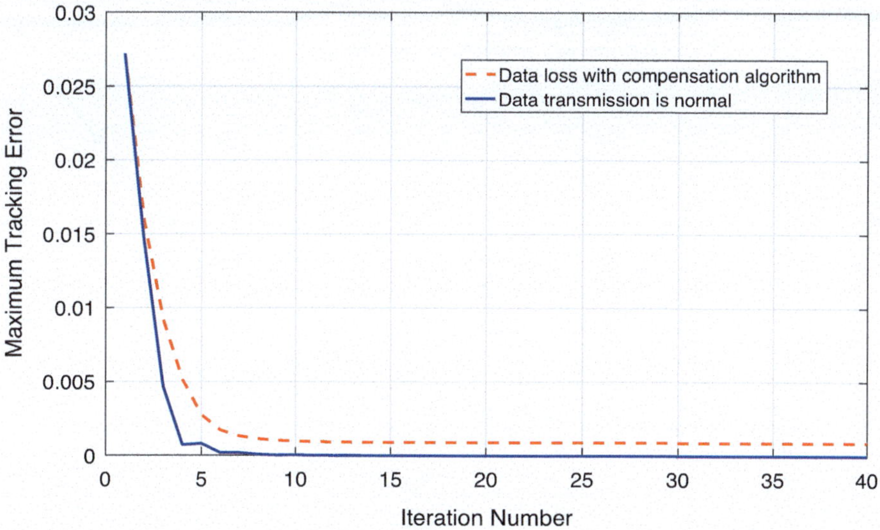


**Fig. 5.2** Comparison of tracking error under different cases in Example 1

The data loss rates for both measurement output and calculated input are 10%. The initial values are set as $y_k(0) = 0.1$, $\hat{y}_k(0) = 0.5$, $u_0(t) = 1$, $\hat{\varphi}_0(t) = 11$, and the controller parameters are selected as $\lambda = 2$, $\rho = 4$, $\mu = 1$, $\eta = 1$, $\varepsilon = 0.00001$. Then, applying the cDDILC method (5.4)–(5.9), the profiles of maximum tracking error and system output at the 100th iteration are shown in Figs. 5.4 and 5.5, respectively. One can see that the proposed cDDILC achieves an asymptotic convergence of tracking error along iteration direction to make the system output track the desired trajectory satisfactorily.

## 5.6 Summary

The problem of random data loss at both input and output sides is considered and a compensation-based data-driven ILC method is presented for repetitive systems. A virtual linear data model is developed in iteration domain to equivalently reformulate the input-output dynamic relationship of the repetitive control system. Two Bernoulli random variables are used to describe the random data loss at both input and output sides. The lost output data is compensated by the predictive one which is predicted using the virtual linear data model. The lost input data is replaced by the one in the latest previous iteration of the same time instant. The proposed cDDILC does not depend on any prior knowledge of the mathematical model, system orders, or delays of the controlled plant.

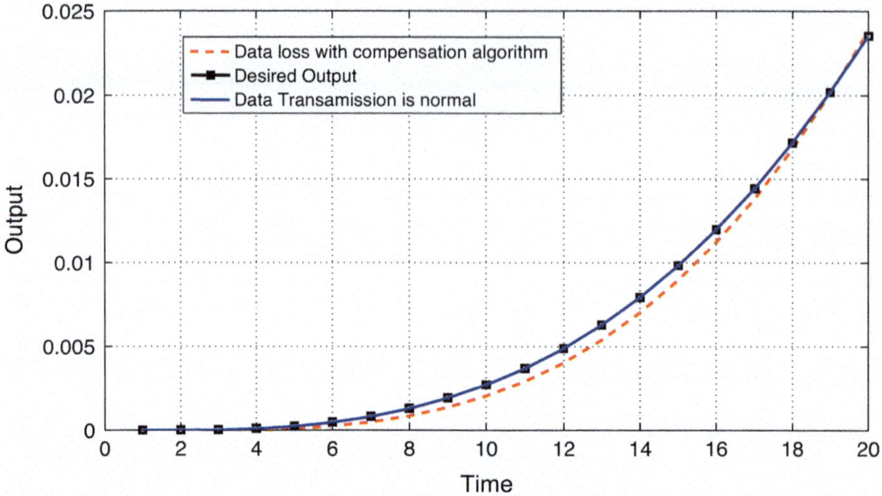


**Fig. 5.3** Comparison of system output under different cases in Example 1

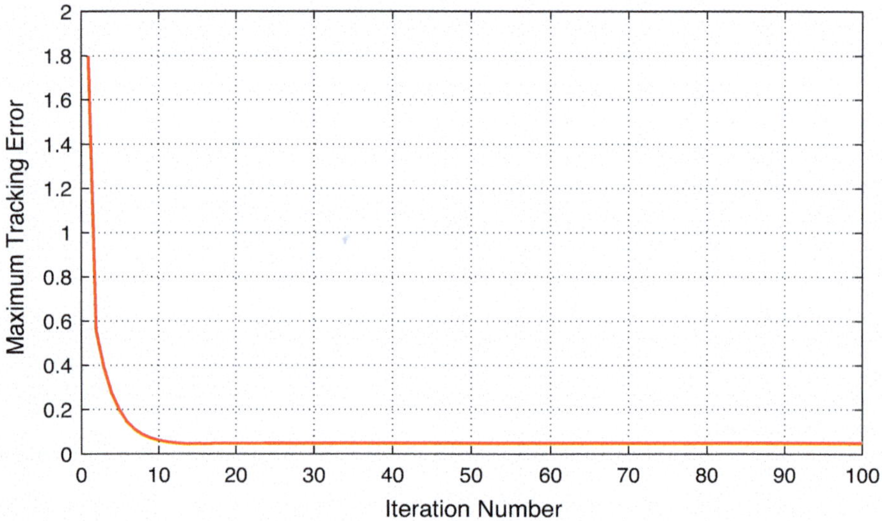


**Fig. 5.4** Tracking error in Example 2

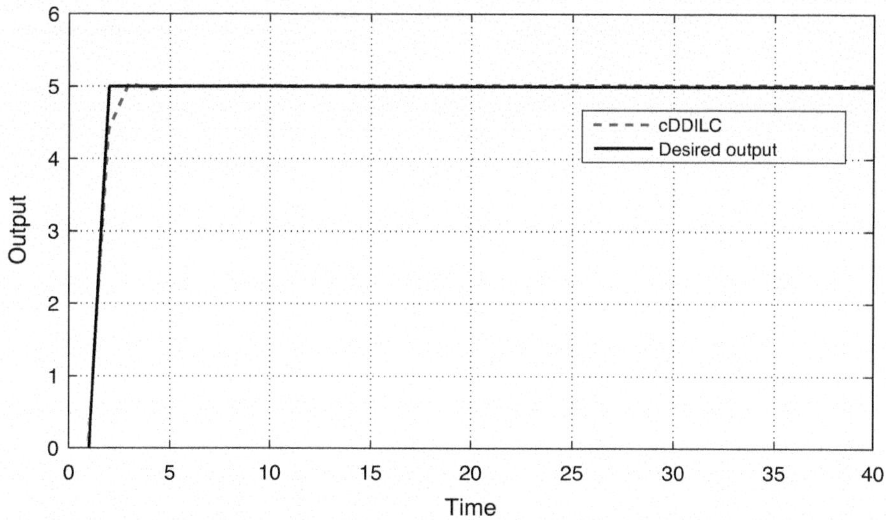


**Fig. 5.5** System output in Example 2

## References

Bu XH, Hou ZS (2011) Stability of iterative learning control with data dropouts via asynchronous dynamical system. Int J Autom Comput 8(1):29–36

Bu XH, Hou ZS, Yu FS (2011) Stability of first and high order iterative learning control with data dropouts. Int J Control Autom Syst 9(5):843–849

Bu XH, Yu F, Hou ZS, Wang F (2013) Iterative learning control for a class of nonlinear systems with random packet losses. Nonlinear Anal Real World Appl 14(1):567–580

Chi RH, Wang DW, Hou ZS, Jin ST (2012) Data-driven optimal terminal iterative learning control. J Process Control 22:2026–2037

Chi RH, Liu XH, Zhang RK, Hou ZS, Huang B (2017) Constrained data-driven optimal iterative learning control. J Process Control 55:10–29

Chi RH, Hou ZS, Jin ST, Huang B (2018) Computationally efficient data-driven higher order optimal iterative learning control. IEEE Trans Neural Netw Learn Syst 29(12):5971–5980

Ge X, Yang F, Han Q (2017) Distributed networked control systems: a brief overview. Inf Sci 380:117–131

Huang LX, Fang Y, Wang T (2014) Method to improve convergence performance of iterative learning control systems over wireless networks in presence of channel noise. IET Control Theory A 8(3):175–182

Lin N, Chi RH, Huang B, Hou ZS (2019) Multi-lag input iterative dynamic linearization based DDAILC. J Franklin Inst 356(1):457–473

Liu J, Ruan X (2017) Networked iterative learning control for discrete-time systems with stochastic packet dropouts in input and output channels. Adv Differ Equ 1:1–21

Liu J, Ruan X (2018) Networked iterative learning control design for nonlinear systems with stochastic output packet dropouts. Asian J Control 20(3):1077–1087

Shen D, Xu JX (2017) A novel Markov chain based ILC analysis for linear stochastic systems under general data dropouts environments. IEEE Trans Automat Control 62(11):5850–5857

Shen D, Zhang C, Xu Y (2017) Two updating schemes of iterative learning control for networked control systems with random data dropouts. Inf Sci 381:352–370

Shen D, Jin Y, Xu Y (2017) Learning control for linear systems under general data dropouts at both measurement and actuator sides: a Markov chain approach. Asian J Control 354(13):5091–5109
Shen D, Zhang C, Xu Y (2018) Intermittent and successive ILC for stochastic nonlinear systems with random data dropouts. Asian J Control 20(3):1102–1114

# Chapter 6
# Data-Driven ILC for Systems with Varying Trial Lengths

## 6.1 Introduction

Many practical systems that run repeatedly have varying ranges or lengths of run time in each iteration (Nahrstaedt et al. 2008; Longman and Mombaur 2006; Seel et al. 2016; Freeman et al. 2010; Freeman 2014; Zhu and Wang 2018; Guth et al. 2013). For instance, in treating the drop foot patients (Nahrstaedt et al. 2008; Longman and Mombaur 2006; Seel et al. 2016; Freeman et al. 2010; Freeman 2014; Zhu and Wang 2018) with functional electrical stimulation, the electrical stimulation must be stopped when the patient feels uncomfortable. In a crane control system (Guth et al. 2013) for trajectory tracking, when the load is approaching the obstacle, the rope is wound up to ensure that the load does not touch the obstacle. Because the position of the obstacle is different, the operation time is also different in different runs.

It is worth pointing out that the systems (Nahrstaedt et al. 2008; Longman and Mombaur 2006; Seel et al. 2016; Freeman et al. 2010; Freeman 2014; Zhu and Wang 2018; Guth et al. 2013) operate repetitively by nature although they have inconsistent running time in each iteration. For example, repeated functional electrical stimulation is applied to the diseased upper limb to enable the arms to move repetitively for the purpose of rehabilitation (Freeman et al. 2010; Freeman 2014; Zhu and Wang 2018). Iterative learning control (ILC) (Meng and Moore 2020; Yu and Li 2017, Bu et al. 2019) is most effective for such repetitive processes by learning historical operation experience to improve control performance. However, most of the existing ILCs require that the interval length of each run be completely repeated. When the running length of one batch cannot reach the desired one, the output information is not available at the time span from the terminated point to the desired endpoint and thus the control input at this time span cannot be updated commonly. Therefore, traditional ILCs are not suitable for the tracking control problem when the trial lengths change.

Aiming at solving the above problem, some works (Seel et al. 2011, 2017; Meng and Zhang 2018; Shen et al. 2016a, b; Li et al. 2014, 2015) have been conducted. In



R. Chi et al., *Data-Driven Iterative Learning Control for Discrete-Time Systems*, Intelligent Control and Learning Systems 2,
https://doi.org/10.1007/978-981-19-5950-9_6

Seel et al. (2011, 2017), a P-type ILC with Q-filter is proposed where the problem of variable batch lengths is addressed by defining a maximum pass length error. Similarly, assuming that there is at least one batch run length equal to the maximum batch process length for a fixed iteration interval, a modified P-type learning method (Meng and Zhang 2018) has been designed subject to variable running lengths. In addition, the random change of the pass lengths is considered in (Shen et al. 2016a), and a P-type ILC is developed by introducing a random variable. Shen et al. (2016b) take the traditional P-type ILC with Arimoto-like gain to deal with the random change of batch length. Also, iteration-average operator-based ILC methods (Li et al. 2014, 2015) are proposed for variable trial lengths to collect and utilize the control information of previous iterations to improve the performance.

However, a common assumption in the above methods (Seel et al. 2011, 2017; Meng and Zhang 2018; Shen et al. 2016a, b; Li et al. 2014, 2015) is that the tracking error be zero for the time span beyond the batch length if the batch length does not reach the expected maximum one. Obviously, it is not practicable to force the output to be the same as the desired one due to noises and disturbances. Further, the control input signals over the time span beyond the batch length cannot be updated due to the zero tracking errors which may deteriorate the control performance with a reduced convergence rate. The second problem existing in the above works (Seel et al. 2011, 2017; Meng and Zhang 2018; Shen et al. 2016a, b; Li et al. 2014, 2015) is that most of the ILC methods adopt P-type and PD-type algorithms. Although these algorithms are easy to implement with light computation burden, they are weak in dealing with uncertainties due to the fixed learning gain which is also hard to select properly.

In this chapter, the ILC problem for variable trial lengths is reconsidered from a viewpoint that an auxiliary predictive model is built to compensate the outputs beyond the batch length. And an auxiliary predictive compensation-based data-driven ILC (APC-DDILC) scheme is proposed consequently. By introducing a random variable, an expanded output is defined via argumentation of the actual output and the predictive one, and an expanded tracking error is also defined consequently. Then, we propose an APC-DDILC method by means of the expanded tracking error and output. Both rigorous analysis and extensive simulations are provided in this chapter to verify the proposed method and demonstrate its effectiveness.

The main features of the proposed method are summarized as (i) An auxiliary predictive model is established to predict the system output at the time span beyond the actual operation interval of the controlled plant, which is then used in the learning control law as a compensation to accelerate convergence rate and to improve performance. (ii) The learning gain is nonlinear and time-iteration-varying, and it can be updated through the parameter estimation law by the use of I/O data. Obviously, the iteratively variant learning gain can enhance the adaptability and robustness against the internal and external uncertainties. (iii) The proposed method only uses I/O data for the controller design rather than a mechanistic model where the linear system model is used only to show how the controller is derived and analyzed. Therefore, the proposed method is data-driven and can be directly applied to practical complex plants that have no first-principle model information available.

This chapter is organized as follows. Section 6.2 formulates the control problem. The controller design and analysis are provided in Sects. 6.3 and 6.4, respectively. Section 6.5 gives some simulation results. Section 6.6 concludes this chapter.

## 6.2 Problem Formulation

Consider a class of linear discrete-time systems

$$\begin{cases} \boldsymbol{x}_k(t+1) = \boldsymbol{A}(t)\boldsymbol{x}_k(t) + \boldsymbol{B}(t)u_k(t) \\ y_k(t) = \boldsymbol{C}(t)\boldsymbol{x}_k(t) \end{cases} \tag{6.1}$$

where $y_k(t) \in R$ represents system output; $u_k(t) \in R$ represents control input; $\boldsymbol{x}_k(t) \in R^n$ represents the system state; $t \in \{0, 1, \ldots, N_k\}$ denotes time instant; $k \in \{0, 1, \ldots\}$ is the iteration index. Note that the operation length, $N_k$, of system (6.1) varies iteratively which is different from (5.1) where the operation length is fixed. $\boldsymbol{A}(t)$, $\boldsymbol{B}(t)$, and $\boldsymbol{C}(t)$ are unknown matrices with appropriate dimensions.

Consider system (6.1) satisfying Assumption 5.1. Then, a LDM of (6.1) can be built by following the similar steps as (5.2)–(5.3),

$$y_k(t+1) - y_{k-1}(t+1) + \boldsymbol{\varphi}(t)\Delta\boldsymbol{u}_k(t) \tag{6.2}$$

where $\boldsymbol{u}_k(t) = [u_k(0), \ldots, u_k(t)]^{\mathrm{T}} \in R^{t+1}$, $\Delta\boldsymbol{u}_k(t) = \boldsymbol{u}_k(t) - \boldsymbol{u}_{k-1}(t)$, and $\boldsymbol{\varphi}(t) = [\phi(0), \phi(1), \ldots, \phi(t)] \in R^{1\times(t+1)}$.

Because $\boldsymbol{A}(t)$, $\boldsymbol{B}(t)$, and $\boldsymbol{C}(t)$ are all real matrices with appropriate dimensions and $t \in \{0, \ldots, N\}$ is finite, it is clear that $\boldsymbol{\varphi}(t)$ is bounded. Without loss of generality, one can assume that $\|\boldsymbol{\varphi}(t)\| \le b_\varphi$, where $b_\varphi$ is a positive constant.

The control task of this section is to design a learning controller to drive system output $y_k(t+1)$ to track the target trajectory $y_d(t+1)$ over the time interval $\{0, 1, \ldots, N_k\}$, that is, the tracking error $e_k(t+1) = y_d(t+1) - y_k(t+1)$, $t \in \{0, 1, \ldots, N_k\}$, converges to zero along iteration axis.

## 6.3 Controller Design

Define an expanded output as

$$\bar{y}_k(t+1) = \beta_k(t)y_k(t+1) + (1-\beta_k(t))\hat{y}_k(t+1) \tag{6.3}$$

where $\beta_k(t) = \begin{cases} 1, & 0 \le t \le N_k \\ 0, & N_k < t \le N_d \end{cases}$ is a random variable, and $\hat{y}_k(t+1)$ is a predicted output used to compensate for the output of the absent part.

According to (6.3), Eq. (6.2) can be rewritten as

$$y_k(t+1) = \bar{y}_{k-1}(t+1) + \boldsymbol{\varphi}(t)\,\Delta \boldsymbol{u}_k(t). \tag{6.4}$$

Then, we take (6.4) as an auxiliary model to predict the system output, i.e.,

$$\hat{y}_k(t+1) = \bar{y}_{k-1}(t+1) + \hat{\boldsymbol{\varphi}}_k(t)\,\Delta \boldsymbol{u}_k(t) \tag{6.5}$$

where $\hat{\boldsymbol{\varphi}}_k(t) = \left[\hat{\phi}_k(0), \ldots, \hat{\phi}_k(t)\right] \in R^{t+1}$ is the estimation of $\boldsymbol{\varphi}(t)$.

Correspondingly, define an expanded tracking error as $\bar{e}_k(t+1) = y_d(t+1) - \bar{y}_k(t+1)$. Then, an auxiliary predictive compensation-based data-driven ILC (APC-DDILC) is presented as

$$u_k(t) = u_{k-1}(t) + \frac{\rho_t \hat{\phi}_k(t)\,\bar{e}_{k-1}(t+1)}{\lambda + \left|\hat{\phi}_k(t)\right|^2} - \frac{\hat{\phi}_k(t)\sum_{i=0}^{t-1}\rho_i \hat{\phi}_k(i)\Delta u_k(i)}{\lambda + \left|\hat{\phi}_k(t)\right|^2} \tag{6.6}$$

$$\hat{\boldsymbol{\varphi}}_k(t) = \hat{\boldsymbol{\varphi}}_{k-1}(t) + \frac{\eta\left(\bar{y}_{k-1}(t+1) - \bar{y}_{k-2}(t+1) - \hat{\boldsymbol{\varphi}}_{k-1}(t)\,\Delta \boldsymbol{u}_{k-1}(t)\right)\Delta \boldsymbol{u}_{k-1}^{\mathrm{T}}(t)}{\mu + \|\Delta \boldsymbol{u}_{k-1}(t)\|^2} \tag{6.7}$$

$$\hat{\boldsymbol{\varphi}}_k(t) = \hat{\boldsymbol{\varphi}}_0(t), \text{ if } \left\|\hat{\boldsymbol{\varphi}}_k(t)\right\| \le \varepsilon \text{ or } sgn\left(\hat{\phi}_k(i)\right) \ne sgn\left(\hat{\phi}_0(i)\right) \tag{6.8}$$

where $0 < \eta < 2$, $\mu > 0$, $\rho_i > 0$, $i \in \{0, 1, \ldots, t\}$, $\lambda > 0$ and $\varepsilon > 0$ are some proper positive constants; $\hat{\boldsymbol{\varphi}}_0(t)$ is the initial value of $\hat{\boldsymbol{\varphi}}_k(t)$, and $\hat{\phi}_k(i)$ is the $(i+1)$th element of $\hat{\boldsymbol{\varphi}}_k(t)$.

**Remark 6.1** Note that auxiliary model (6.5) is used to predict the system output between $N_k + 1$ and $N_d$ which is then compensated in the proposed learning control law and estimation algorithm through the defined expanded output and tracking error. It will accelerate convergence speed and improve the control performance because the control input over this time span can be updated uninterruptedly by using the predicted nonzero tracking error. However, many existing works (Seel et al. 2011, 2017; Meng and Zhang 2018; Shen et al. 2016a, b; Li et al. 2014, 2015) just simply set the system outputs over this time interval to be the desired outputs, which causes a problem in the sense that the corresponding control inputs cannot be updated due to the zero tracking errors.

**Remark 6.2** The learning gain in (6.6) is time-iteration-varying and is updated utilizing I/O data in real time. Therefore, the proposed APC-DDILC is superior to the PID-type ILC methods (Meng and Zhang 2018; Shen et al. 2016a, b) with constant learning gains. In addition, more control inputs at the previous time instants are used in the control law (6.6) for further improving control performance.

**Remark 6.3** Note that (6.8) is a reset algorithm which is added to improve the tracking performance of the estimation algorithm (6.7) by ensuring $\hat{\boldsymbol{\varphi}}_k(t)$ not zero, similar to a persistent excitation condition (Hou and Xiong 2019). Another purpose of adding (6.8) is to make all the elements of $\hat{\boldsymbol{\varphi}}_k(t)$ have the same sign as that of $\boldsymbol{\varphi}(t)$ correspondingly, which is used in the subsequent convergence analysis. To guarantee this condition, the initial estimation, $\hat{\boldsymbol{\varphi}}_0(t)$, should be selected carefully to have its elements signs be the same as that of the corresponding element of $\boldsymbol{\varphi}(t)$. For an unknown system, the signs of the elements of $\boldsymbol{\varphi}(t)$ can be estimated by using the I/O data obtained from a closed-loop experiment.

## 6.4 Convergence Analysis

The convergence property of the proposed APC-DDILC scheme is shown as the following theorem.

**Theorem 6.1** *For a class of linear systems (6.1), if Assumption 5.1 holds, by selecting the controller parameters properly, the proposed APC-DDILC scheme (6.6)–(6.8) guarantees that:*

*(i) the parameter estimation $\hat{\boldsymbol{\varphi}}_k(t)$ is bounded, $\forall t \in \{0, 1, \ldots, N_k\}$, $\forall k$.*
*(ii) both $\bar{e}_k(t+1)$ and $e_k(t+1)$, $t \in \{0, 1, \ldots, N_k\}$, converge to zero in the sense of expectation as $k$ approaches to infinity, i.e., $\lim_{k\to\infty} E\{\bar{e}_k(t+1)\} = 0$ and $\lim_{k\to\infty} E\{e_k(t+1)\} = 0$.*

***Proof*** *Part 1. The boundedness of $\hat{\boldsymbol{\varphi}}_k(t)$*

Define a parameter estimation error as $\tilde{\boldsymbol{\varphi}}_k(t) = \boldsymbol{\varphi}(t) - \hat{\boldsymbol{\varphi}}_k(t)$. Subtracting $\boldsymbol{\varphi}(t)$ from both sides of (6.7), one has

$$\tilde{\boldsymbol{\varphi}}_k(t) = \tilde{\boldsymbol{\varphi}}_{k-1}(t) - \frac{\eta\left(\bar{y}_{k-1}(t+1) - \bar{y}_{k-2}(t+1) - \hat{\boldsymbol{\varphi}}_{k-1}(t)\Delta \boldsymbol{u}_{k-1}(t)\right)\Delta \boldsymbol{u}_{k-1}^{\mathrm{T}}(t)}{\mu + \left\|\Delta \boldsymbol{u}_{k-1}(t)\right\|^2}. \tag{6.9}$$

In terms of Eqs. (6.3)–(6.5), we have

$$\begin{aligned} &\bar{y}_{k-1}(t+1) - \bar{y}_{k-2}(t+1) - \hat{\boldsymbol{\varphi}}_{k-1}(t)\Delta \boldsymbol{u}_{k-1}(t) \\ &\quad = \beta_{k-1}(t)\, y_{k-1}(t+1) + (1 - \beta_{k-1}(t))\,\hat{y}_{k-1}(t+1) - \hat{y}_{k-1}(t+1) \\ &\quad = \beta_{k-1}(t)\,\tilde{\boldsymbol{\varphi}}_{k-1}(t)\,\Delta \boldsymbol{u}_{k-1}(t). \end{aligned} \tag{6.10}$$

Substituting (6.10) into (6.9), one can derive that

$$\begin{aligned} \tilde{\boldsymbol{\varphi}}_k(t) &= \tilde{\boldsymbol{\varphi}}_{k-1}(t) - \frac{\eta\beta_{k-1}(t)\tilde{\boldsymbol{\varphi}}_{k-1}(t)\Delta \boldsymbol{u}_{k-1}(t)\Delta \boldsymbol{u}_{k-1}^{\mathrm{T}}(t)}{\mu + \|\Delta \boldsymbol{u}_{k-1}(t)\|^2} \\ &= \tilde{\boldsymbol{\varphi}}_{k-1}(t)\left(\boldsymbol{I} - \frac{\eta\beta_{k-1}(t)\Delta \boldsymbol{u}_{k-1}(t)\Delta \boldsymbol{u}_{k-1}^{\mathrm{T}}(t)}{\mu + \|\Delta \boldsymbol{u}_{k-1}(t)\|^2}\right). \end{aligned} \tag{6.11}$$

Squaring the right-hand side of (6.11), one obtains

$$\begin{aligned}&\left\|\tilde{\boldsymbol{\varphi}}_{k-1}(t)\left(\boldsymbol{I}-\frac{\eta\beta_{k-1}(t)\Delta\boldsymbol{u}_{k-1}(t)\Delta\boldsymbol{u}_{k-1}^{\mathrm{T}}(t)}{\mu+\|\Delta\boldsymbol{u}_{k-1}(t)\|^2}\right)\right\|^2\\&=\left\|\tilde{\boldsymbol{\varphi}}_{k-1}(t)\right\|^2+\left(-2+\frac{\eta\beta_{k-1}(t)\|\Delta\boldsymbol{u}_{k-1}(t)\|^2}{\mu+\|\Delta\boldsymbol{u}_{k-1}(t)\|^2}\right)\\&\times\frac{\eta\beta_{k-1}(t)\left\|\tilde{\boldsymbol{\varphi}}_{k-1}(t)\Delta\boldsymbol{u}_{k-1}^{\mathrm{T}}(t)\right\|^2}{\mu+\|\Delta\boldsymbol{u}_{k-1}(t)\|^2}.\end{aligned}\tag{6.12}$$

Since $0<\eta<2$, and $\mu>0$, then

$$-2+\frac{\eta\beta_{k-1}(t)\|\Delta\boldsymbol{u}_{k-1}(t)\|^2}{\mu+\|\Delta\boldsymbol{u}_{k-1}(t)\|^2}<0.\tag{6.13}$$

From (6.12) and (6.13), it is implied that

$$\left\|\tilde{\boldsymbol{\varphi}}_{k-1}(t)\left(\boldsymbol{I}-\frac{\eta\beta_{k-1}(t)\Delta\boldsymbol{u}_{k-1}(t)\Delta\boldsymbol{u}_{k-1}^{\mathrm{T}}(t)}{\mu+\|\Delta\boldsymbol{u}_{k-1}(t)\|^2}\right)\right\|^2\le\left\|\tilde{\boldsymbol{\varphi}}_{k-1}(t)\right\|^2.\tag{6.14}$$

Thus, there must exist a coefficient $d_k$, $0<d_k\le 1$, such that

$$\left\|\tilde{\boldsymbol{\varphi}}_k(t)\right\|\le d_k\left\|\tilde{\boldsymbol{\varphi}}_{k-1}(t)\right\|\le d_kd_{k-1}\left\|\tilde{\boldsymbol{\varphi}}_{k-2}(t)\right\|\le\cdots\le\bar{d}_k\left\|\tilde{\boldsymbol{\varphi}}_0(t)\right\|\tag{6.15}$$

where $\bar{d}_k=\prod_{i=1}^{k}d_i$, and $d_k=1$ only when $\beta_{k-1}(t)=0$. Even if $d_k=1$, as long as $0<d_k<1$ occurs in infinite iterations, $\bar{d}_k$ still approaches to zero. Therefore $\lim_{k\to\infty}\bar{d}_k=0$ and $\lim_{k\to\infty}\left\|\tilde{\boldsymbol{\varphi}}_k(t)\right\|=0$ can be derived from (6.15).

Since $\boldsymbol{\varphi}_k(t)$ is bounded, the boundedness of $\hat{\boldsymbol{\varphi}}_k(t)$ can be obtained directly $\forall t\in\{0,\ldots,N_k\}$, and $\forall k$.

*Part 2. The convergence of output tracking error*

To show that $e_k(t+1)$ converges along the iteration direction, we present the dynamics of $\bar{e}_k(t+1)$ as follows,

$$\begin{aligned}\bar{e}_k(t+1)&=\bar{e}_{k-1}(t+1)+\bar{y}_{k-1}(t+1)-\bar{y}_k(t+1)\\&=\bar{e}_{k-1}(t+1)+\bar{y}_{k-1}(t+1)-\beta_k(t)y_k(t+1)\\&-(1-\beta_k(t))\hat{y}_k(t+1)\\&=\bar{e}_{k-1}(t+1)-(1-\beta_k(t))\hat{\boldsymbol{\varphi}}_k(t)\Delta\boldsymbol{u}_k(t)\\&-\beta_k(t)\boldsymbol{\varphi}(t)\Delta\boldsymbol{u}_k(t).\end{aligned}\tag{6.16}$$

Since the ILC problem of variable time interval has been converted to the standard ILC problem over a fixed time interval $[0,N_d]$ by introducing predictive output over $[N_k+1,N_d]$ as the compensation of the lacking actual output, the following expression can be obtained from Eq. (6.16),

$$\begin{aligned}\bar{e}_k(N_d)&=\bar{e}_{k-1}(N_d)-((1-\beta_k(N_d-1))\hat{\boldsymbol{\psi}}_k(N_d-1)\\&+\beta_k(N_d-1)\boldsymbol{\psi}(N_d-1))\Delta\boldsymbol{u}_k(N_d-1)\end{aligned}\tag{6.17}$$

where $\hat{\boldsymbol{\psi}}_k(N_d-1) = [\hat{\phi}_k(0)\ \hat{\phi}_k(1)\ \cdots\ \hat{\phi}_k(N_d-1)] \in R^{1\times N_d}$ and $\boldsymbol{\psi}(N_d-1) = [\phi(0)\ \phi(1)\ \cdots\ \phi(N_d-1)] \in R^{1\times N_d}$.

Denote $b_k(i) = \frac{\hat{\phi}_k(i)}{\lambda+\left|\hat{\phi}_k(i)\right|^2}$, $\bar{\boldsymbol{B}}_k(i) = \left[0\ 0 \cdots 0\ \rho_i b_k(i)\right]^{\mathrm{T}} \in R^{N_d}$, $\bar{\boldsymbol{u}}_k(i) = [u_k(i-N_d+1)\ u_k(i-N_d+2)\ \cdots\ u_k(i)]^{\mathrm{T}} \in R^{N_d}$, and

$$\bar{\boldsymbol{A}}_k(i) = \begin{bmatrix} 0 & 1 & 0 & \cdots & 0 \\ 0 & 0 & 1 & \cdots & 0 \\ \vdots & \vdots & \vdots & \ddots & \vdots \\ 0 & 0 & 0 & \cdots & 1 \\ 0 & l_k(N_d-1)b_k(i) & l_k(N_d-2)b_k(i) & \cdots & l_k(1)b_k(i) \end{bmatrix} \in R^{N_d\times N_d}$$

where $l_k(j_o) = -\rho_{i-j_o}\hat{\phi}_k(i-j_o)$, $j_o = 1, \ldots, N_d-1; i \in \{0, 1, \ldots, N_d-1\}$; $\bar{\boldsymbol{u}}_k(i)$ has a fixed dimension, which is different from $\boldsymbol{u}_k(t)$ whose dimension varies with time instant; $\hat{\phi}_k(t) = 0$, if $t < 0$.

Then, according to control law (6.6), we can get

$$\Delta\bar{\boldsymbol{u}}_k(N_d-1) = \bar{\boldsymbol{A}}_k(N_d-1)\Delta\bar{\boldsymbol{u}}_k(N_d-2) + \bar{\boldsymbol{B}}_k(N_d-1)\bar{e}_{k-1}(N_d). \tag{6.18}$$

Further, (6.18) can be rewritten as

$$\begin{aligned}\Delta\bar{\boldsymbol{u}}_k(N_d-1) &= \prod_{i=1}^{N_d-1}\bar{\boldsymbol{A}}_k(i)\bar{\boldsymbol{B}}_k(0)\bar{e}_{k-1}(1) + \prod_{i=2}^{N_d-1}\bar{\boldsymbol{A}}_k(i)\bar{\boldsymbol{B}}_k(1)\bar{e}_{k-1}(2) \\ &\quad + \cdots + \bar{\boldsymbol{A}}_k(N_d-1)\bar{\boldsymbol{B}}_k(N_d-2)\bar{e}_{k-1}(N_d-1) \\ &\quad + \bar{\boldsymbol{B}}_k(N_d-1)\bar{e}_{k-1}(N_d) \\ &= \sum_{j=1}^{N_d}\prod_{i=j}^{N_d-1}\bar{\boldsymbol{A}}_k(i)\bar{\boldsymbol{B}}_k(j-1)\bar{e}_{k-1}(j)\end{aligned} \tag{6.19}$$

where $\prod_{i=N_d}^{N_d-1}\bar{\boldsymbol{A}}_k(i) = 1$.

According to the definition of $\bar{\boldsymbol{u}}_k(N_d-1)$ and $\boldsymbol{u}_k(N_d-1)$, one has $\bar{\boldsymbol{u}}_k(N_d-1) = \boldsymbol{u}_k(N_d-1)$. Then one can derive from (6.19) that

$$\Delta\boldsymbol{u}_k(N_d-1) = \sum_{j=1}^{N_d}\prod_{i=j}^{N_d-1}\bar{\boldsymbol{A}}_k(i)\bar{\boldsymbol{B}}_k(j-1)\bar{e}_{k-1}(j). \tag{6.20}$$

Substituting (6.20) into (6.17), yields

$$\bar{e}_k(N_d) = \bar{e}_{k-1}(N_d) - ((1-\beta_k(N_d-1))\hat{\boldsymbol{\psi}}_k(N_d-1) \\ +\beta_k(N_d-1)\boldsymbol{\psi}(N_d-1))\sum_{j=1}^{N_d}\prod_{i=j}^{N_d-1}\bar{\boldsymbol{A}}_k(i)\bar{\boldsymbol{B}}_k(j-1)\bar{e}_{k-1}(j). \tag{6.21}$$

Denote $\boldsymbol{\theta}_k(i) = (1-\beta_k(i))\hat{\boldsymbol{\psi}}_k(i) + \beta_k(i)\boldsymbol{\psi}(i),\quad i \in \{0, 1, \ldots, N_d - 1\}$ and $\boldsymbol{H}_k(N_d-1)$ is a matrix as shown in Eq. (6.22),

$$\boldsymbol{H}_k(N_d-1) = \begin{bmatrix} q_k^o(0) & 0 & \cdots & 0 \\ -\boldsymbol{\theta}_k(1)\bar{\boldsymbol{A}}_k(1)\bar{\boldsymbol{B}}_k(0) & q_k^o(1) & \cdots & 0 \\ \vdots & \vdots & \ddots & \vdots \\ p_k^o(1) & p_k^o(2) & \cdots & q_k^o(N_d-1) \end{bmatrix} \tag{6.22}$$

where $p_k^o(j) = -\boldsymbol{\theta}_k(N_d-1)\prod_{i=j}^{N_d-1}\bar{\boldsymbol{A}}_k(i)\bar{\boldsymbol{B}}_k(j-1);\ q_k^o(j) = 1-\boldsymbol{\theta}_k(j)\bar{\boldsymbol{B}}_k(j)$.

Then, Eq. (6.21) can be written in a compact form as

$$\boldsymbol{M}_k(N_d) = \boldsymbol{H}_k(N_d-1)\boldsymbol{M}_{k-1}(N_d) \tag{6.23}$$

where $\boldsymbol{M}_k(N_d) = \left[\bar{e}_k(1) \cdots \bar{e}_k(N_d)\right]^{\mathrm{T}} \in R^{N_d}$.

Applying the expectation operator on both sides of (6.23), yields

$$E\{\boldsymbol{M}_k(N_d)\} = E\{\boldsymbol{H}_k(N_d-1)\}E\{\boldsymbol{M}_{k-1}(N_d)\} \\ = \bar{\boldsymbol{H}}_k(N_d-1)E\{\boldsymbol{M}_{k-1}(N_d)\} \tag{6.24}$$

where $\bar{\boldsymbol{H}}_k(N_d-1) = E\{\boldsymbol{H}_k(N_d-1)\}$. Then, taking norms on both sides of Eq. (6.24), one derives

$$\|E\{\boldsymbol{M}_k(N_d)\}\| \le \left\|\bar{\boldsymbol{H}}_k(N_d-1)\right\| \|E\{\boldsymbol{M}_{k-1}(N_d)\}\|. \tag{6.25}$$

Since $\boldsymbol{H}_k(N_d-1)$ is a lower triangular square matrix, $\bar{\boldsymbol{H}}_k(N_d-1)$ is also a lower triangular square matrix, and the spectral radius of $\bar{\boldsymbol{H}}_k(N_d-1)$ is shown in the following formula,

$$s\left(\bar{\boldsymbol{H}}_k(N_d-1)\right) = \max_{i\in\{0,1,\ldots,N_d-1\}}\left|E\left\{1-\boldsymbol{\theta}_k(i)\bar{\boldsymbol{B}}_k(i)\right\}\right| \\ = \max_{i\in\{0,1,\ldots,N_d-1\}}\left|1-E\left\{\frac{\rho_i(1-\beta_k(i))\left|\hat{\phi}_k(i)\right|^2+\rho_i\beta_k(i)\phi(i)\hat{\phi}_k(i)}{\lambda+\left|\hat{\phi}_k(i)\right|^2}\right\}\right|. \tag{6.26}$$

According to the reset algorithm, $\boldsymbol{\varphi}_k(t)$ has the same sign as $\hat{\boldsymbol{\varphi}}_k(t)$. So, it follows that

$$0 < \frac{\left|\hat{\phi}_k(i)\right|^2}{\lambda+\left|\hat{\phi}_k(i)\right|^2} \le \frac{\left|\hat{\phi}_k(i)\right|^2}{2\sqrt{\lambda}\left|\hat{\phi}_k(i)\right|} \le \frac{\left|\hat{\phi}_k(i)\right|}{2\sqrt{\lambda}} \tag{6.27}$$

$$0 < \frac{\hat{\phi}_k(i)\phi(i)}{\lambda + \left|\hat{\phi}_k(i)\right|^2} \leq \frac{\left|\hat{\phi}_k(i)\right| |\phi(i)|}{2\sqrt{\lambda}\left|\hat{\phi}_k(i)\right|} \leq \frac{|\phi(i)|}{2\sqrt{\lambda}}. \tag{6.28}$$

Taking mathematical expectations of (6.27) and (6.28), yields

$$0 < E\left\{\frac{\left|\hat{\phi}_k(i)\right|^2}{\lambda + \left|\hat{\phi}_k(i)\right|^2}\right\} \leq E\left\{\frac{\left|\hat{\phi}_k(i)\right|^2}{2\sqrt{\lambda}\left|\hat{\phi}_k(i)\right|}\right\} \leq \frac{E\left\{\left|\hat{\phi}_k(i)\right|\right\}}{2\sqrt{\lambda}} \tag{6.29}$$

$$0 < E\left\{\frac{\hat{\phi}_k(i)\phi(i)}{\lambda + \left|\hat{\phi}_k(i)\right|^2}\right\} \leq E\left\{\frac{\left|\hat{\phi}_k(i)\right| |\phi(i)|}{2\sqrt{\lambda}\left|\hat{\phi}_k(i)\right|}\right\} \leq \frac{E\{|\phi(i)|\}}{2\sqrt{\lambda}}. \tag{6.30}$$

Since we have shown that $\hat{\boldsymbol{\varphi}}_k(t)$ is bounded, it is easy to get that $\hat{\phi}_k(i)$, $i \in \{0, 1, \ldots, t\}$, the $(i+1)$th element of $\hat{\boldsymbol{\varphi}}_k(t)$, is bounded. Further, $E\left\{\left|\hat{\phi}_k(i)\right|\right\}$ is bounded. Similarly, $E\{|\phi(i)|\}$ is bounded. Without loss of generality, we assume $E\left\{\left|\hat{\phi}_k(i)\right|\right\} \leq b_{\hat{\phi}}$ and $E\{|\phi(i)|\} \leq b_{\phi}$.

Thus, from (6.29) and (6.30), we can further get

$$\begin{aligned} &E\left\{\frac{\rho_i\left(1-\beta_k(i)\right)\left|\hat{\phi}_k(i)\right|^2+\rho_i\beta_k(i)\phi(i)\hat{\phi}_k(i)}{\lambda+\left|\hat{\phi}_k(i)\right|^2}\right\} \\ &= \rho_i\left(1-p_k\right)E\left\{\frac{\left|\hat{\phi}_k(i)\right|^2}{\lambda+\left|\hat{\phi}_k(i)\right|^2}\right\} + \rho_i p_k E\left\{\frac{\phi(i)\hat{\phi}_k(i)}{\lambda+\left|\hat{\phi}_k(i)\right|^2}\right\} \\ &\leq \rho_i\left(1-p_k\right)\frac{b_{\hat{\phi}}}{2\sqrt{\lambda}} + \rho_i p_k \frac{b_{\phi}}{2\sqrt{\lambda}}. \end{aligned} \tag{6.31}$$

By selecting $\lambda$ and $\rho_i$ properly such that $\lambda > \left(\rho_i^2\left((1-p_k)\, b_{\hat{\phi}} + p_k b_{\phi}\right)^2\right)\Big/ 16$, where $p_k = E\{\beta_k(i)\}$, and combined with (6.26) and (6.31), one has

$$s\left(\bar{\boldsymbol{H}}_k(N_d-1)\right) < 1. \tag{6.32}$$

In terms of Lemma 3.2, one can find a constant $0 < \tilde{d} < 1$ such that

$$\left\|\bar{\boldsymbol{H}}_k(N_d-1)\right\| < s\left(\bar{\boldsymbol{H}}_k(N_d-1)\right) + \delta \leq \tilde{d} < 1. \tag{6.33}$$

From (6.25) and (6.33), we have

$$\|E\{\boldsymbol{M}_k(N_d)\}\| \leq \tilde{d}\,\|E\{\boldsymbol{M}_{k-1}(N_d)\}\| \leq \cdots \leq \tilde{d}^k\,\|E\{\boldsymbol{M}_0(N_d)\}\| \tag{6.34}$$

which implies

$$\lim_{k\to\infty} \|E\{\boldsymbol{M}_k(N_d)\}\| = 0. \tag{6.35}$$

Accordingly, $E\{\bar{e}_k(t+1)\}$, for $t \in \{0, 1, \ldots, N_d - 1\}$, also converges to zero as the iteration increases because $\bar{e}_k(t+1)$ is the $(t+1)$th element of $E\{\boldsymbol{M}_k(N_d)\}$, that is $\lim_{k\to\infty} E\{\bar{e}_k(t+1)\} = 0$.

From the definition of $\bar{e}_k(t+1)$ and $e_k(t+1)$, as well as Eqs. (6.4) and (6.5), we can derive that

$$\begin{aligned}\bar{e}_k(t+1) &= y_d(t+1) - \beta_k(t)\, y_k(t+1) - (1-\beta_k(t))\, \hat{y}_k(t+1) \\ &= e_k(t+1) + (1-\beta_k(t))\, \tilde{\boldsymbol{\varphi}}_k(t)\, \Delta \boldsymbol{u}_k(t).\end{aligned} \tag{6.36}$$

Rearranging (6.36) yields

$$e_k(t+1) = \bar{e}_k(t+1) - (1-\beta_k(t))\, \tilde{\boldsymbol{\varphi}}_k(t)\, \Delta \boldsymbol{u}_k(t). \tag{6.37}$$

Denote $g_k(t+1) = \tilde{\boldsymbol{\varphi}}_k(t)\, \Delta \boldsymbol{u}_k(t)$. Because $\lim_{k\to\infty} \left\|\tilde{\boldsymbol{\varphi}}_k(t)\right\| = 0$, one obtains

$$\lim_{k\to\infty} |g_k(t+1)| = 0 \tag{6.38}$$

i.e.,

$$\lim_{k\to\infty} |E\{g_k(t+1)\}| = 0. \tag{6.39}$$

Taking expectations and norms on both sides of (6.37), one has

$$0 \le |E\{e_k(t+1)\}| \le |E\{\bar{e}_k(t+1)\}| + |E\{(1-\beta_k(t))\, g_k(t+1)\}|. \tag{6.40}$$

Taking limits on both sides of Eq. (6.40), and according to (6.39) and $\lim_{k\to\infty} E\{\bar{e}_k(t+1)\} = 0$, we can get

$$0 \le \|\lim_{k\to\infty} E\{e_k(t+1)\}\| \le 0. \tag{6.41}$$

Therefore, we can derive $\lim_{k\to\infty} E\{e_k(t+1)\} = 0$. □.

## 6.5 Simulation Study

Consider the linear time-varying discrete-time system (6.1), where $\boldsymbol{A}(t)$, $\boldsymbol{B}(t)$ and $\boldsymbol{C}(t)$ are shown below

$$\boldsymbol{A}(t) = \begin{bmatrix} 0.2\exp(-t/100) & -0.6 & 0 \\ 0 & 0.5 & \sin t \\ 0 & 0 & 0.7 \end{bmatrix},$$

$$\boldsymbol{B}(t) = \left[0\ 0.3 \sin t\ 1\right]^{\mathrm{T}},$$

and

$$\boldsymbol{C}(t) = \left[0\ 0.1\ 1 + 0.1 \cos t\right].$$

The control task is to track the desired trajectory

$$y_d\,(t+1) = \sin\left(2t\pi/50\right) + \sin\left(2t\pi/5\right),\ t \in \{0, 1, \ldots, 50\}.$$

In this simulation, the expected running length is $N_d = 50$, and $N_{\min}$ is selected as 40. The iteration length $N_k$ varies over the discrete set $\{40, 41, \ldots, 50\}$ randomly, which is shown in Fig. 6.1. The probability distribution of each batch length is shown in Fig. 6.2.

Set $x_{k,i}(0) = 0, i = 1, 2, 3, u_0(t) = 0, \hat{\boldsymbol{\varphi}}_0(t) = 1, \eta = 0.8, \mu = 1, \lambda = 1$, and $\rho = 1.5$. The time-dynamics and iteration-dynamics of $\hat{\boldsymbol{\varphi}}_k\,(t)$ are illustrated in Figs. 6.3 and 6.4, respectively, by applying the proposed APC-DDILC scheme (6.6)–(6.8). One can see that the estimated parameter $\hat{\boldsymbol{\varphi}}_k\,(t)$ varies very smoothly along the iteration direction though its time-dynamics is evident.

Meanwhile, the output tracking profiles at the 5th, 15th, 30th, and 50th iterations are shown in Fig. 6.5, where the vertical axis denotes the system output and the horizontal axis denotes the time instant. The tracking error is shown in Fig. 6.6, where the X-axis denotes the iteration numbers and Y-axis represents the mean value of absolute tracking error, $e_{m,k} = \sum_{t=3}^{N_k} |e_k(t)| \Big/ N_d$. From Fig. 6.5 one can see

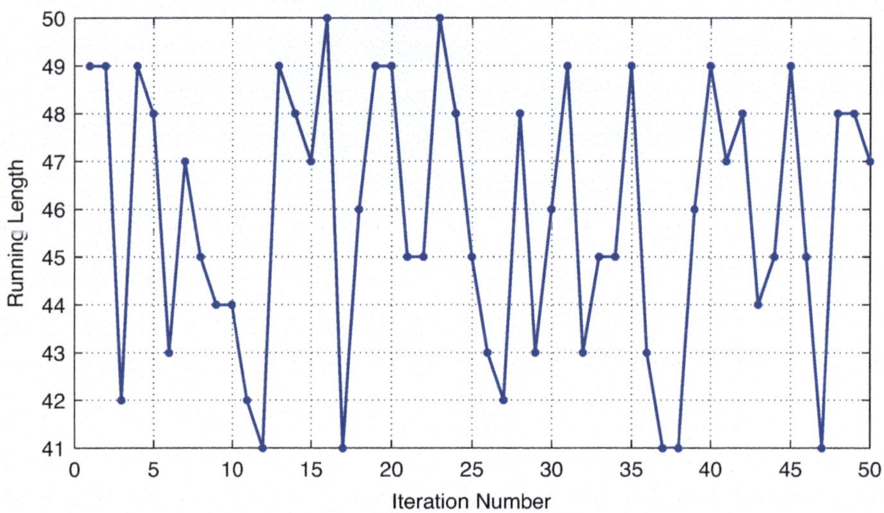


**Fig. 6.1** Running length of each iteration

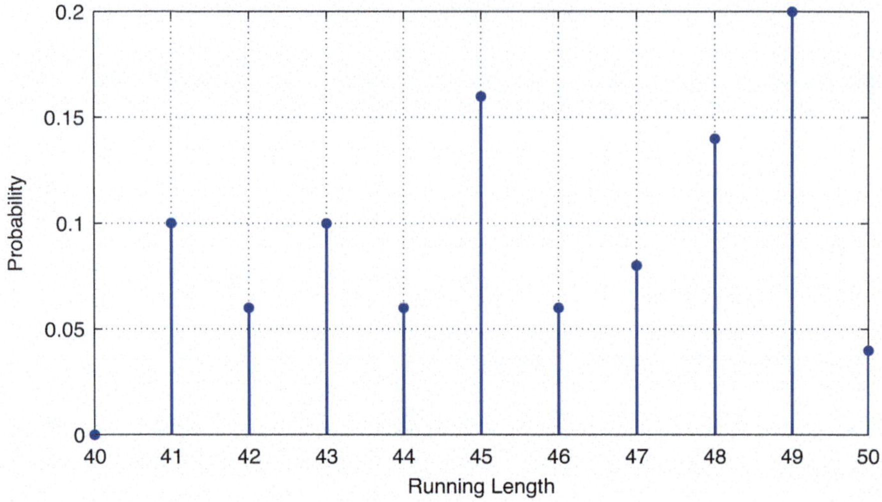


**Fig. 6.2** Probability of the varying iteration length

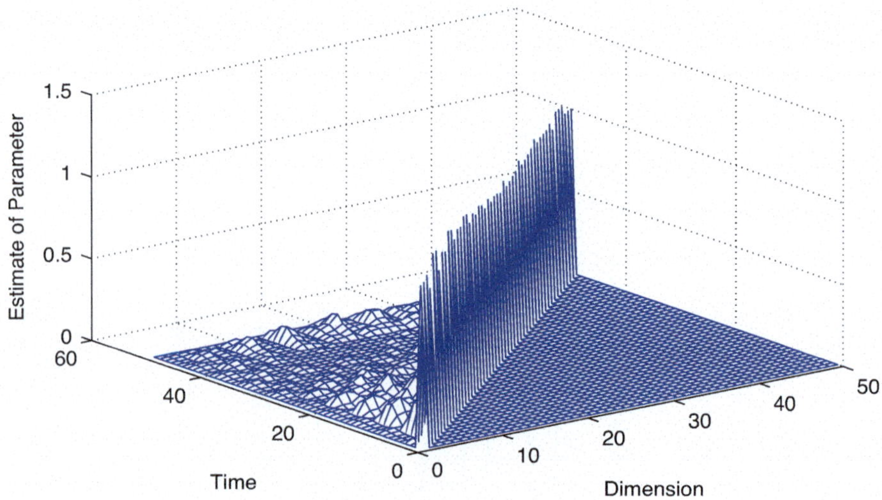


**Fig. 6.3** Parameter estimation $\hat{\varphi}_{10}(t)$

that iteration lengths are different at each iteration, and as the number of the iteration increases, the tracking performance is improved. From Fig. 6.6, it is clear that the tracking error gradually converges as the iteration number increases.

To make a comparison with the existing methods, a P-type ILC method proposed in Shen et al. (2016a) is also applied,

$$u_{k+1}(t) = u_k(t) + Le_k^*(t+1)$$

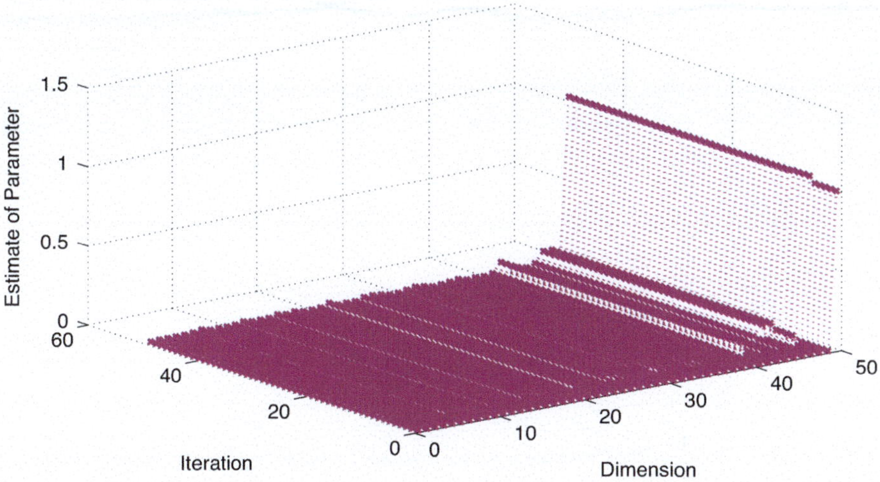


**Fig. 6.4** Parameter estimation $\hat{\varphi}_k$ (50)

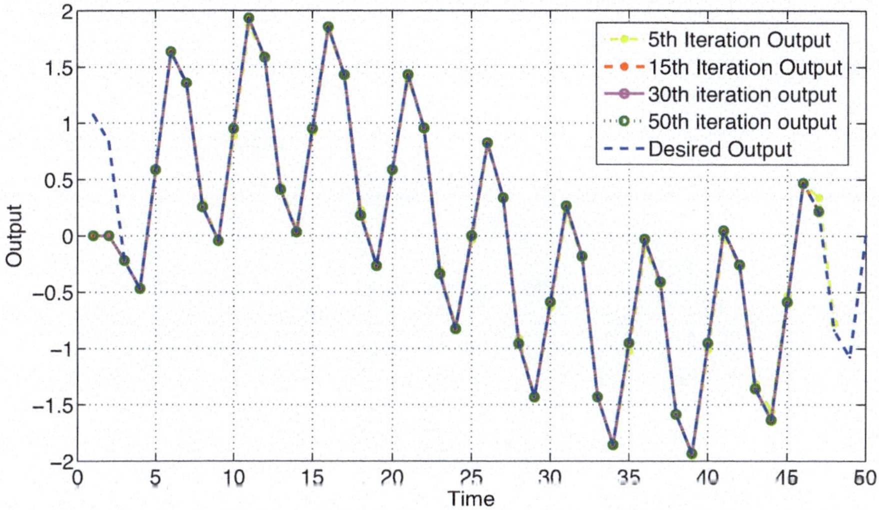


**Fig. 6.5** Output tracking performance

where $e_k^*(t) = \begin{cases} e_k(t)\ ,\ 0 \le t \le N_k \\ 0\ ,\ N_k < t \le N_d \end{cases}$, and $L$ is a learning gain.

It is clear in this method that the errors in the time interval beyond the batch length are set as zero and thus the corresponding control input cannot be updated any longer.

The simulation settings are the same as that for the APC-DDILC method, that is $x_{k,i}(0) = 0, i = 1, 2, 3$, and $u_k(0) = 0$. Setting $L = 0.5$ and applying the P-type ILC method (Shen et al. 2016a), the simulation result is given in Fig. 6.6. Comparatively,

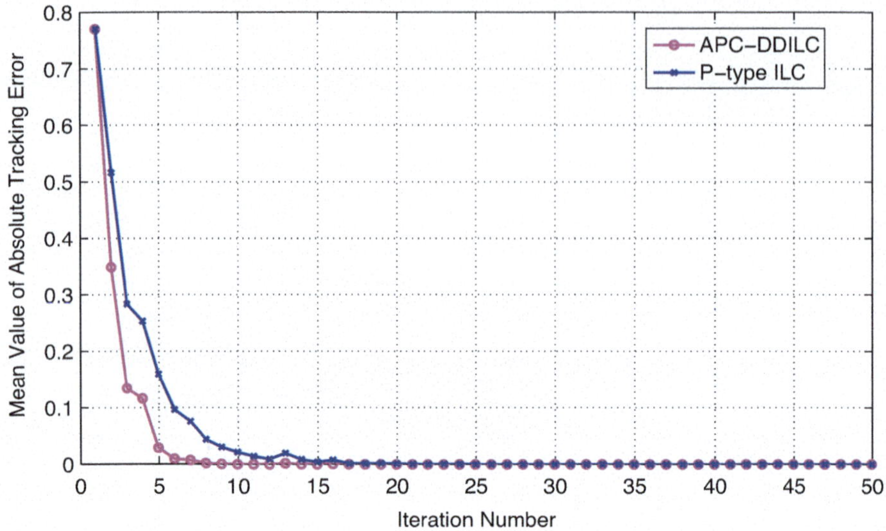


**Fig. 6.6** Convergence of tracking error

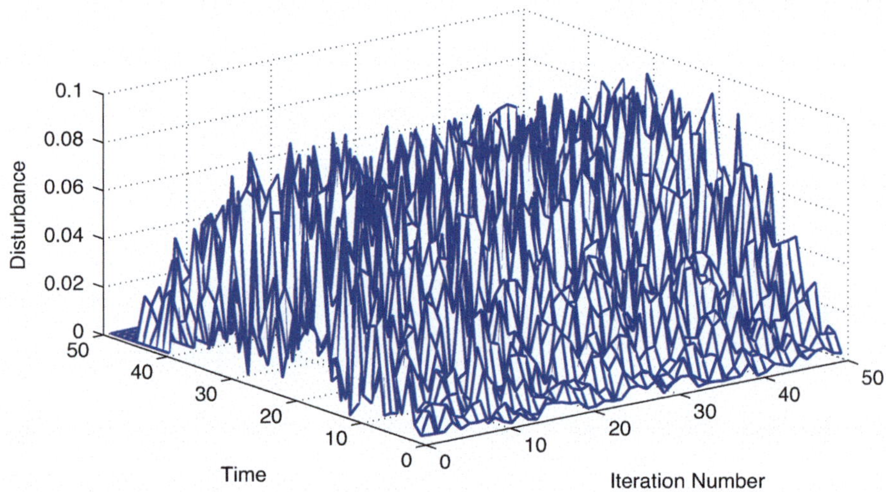


**Fig. 6.7** Stochastic disturbance

the presented APC-DDILC method outperforms the P-type ILC because the control action can be tuned all the time instead of keeping unchanged when the running time beyond the trial length is not available as that in Shen et al. (2016a).

Further, to show the robustness of the APC-DDILC approach to external disturbances, a stochastic disturbance $v_k(t) = 0.1\text{rand} \times \sin(t\pi / 50)$ is added into the system, which is time-iteration-varying, as shown in Fig. 6.7.

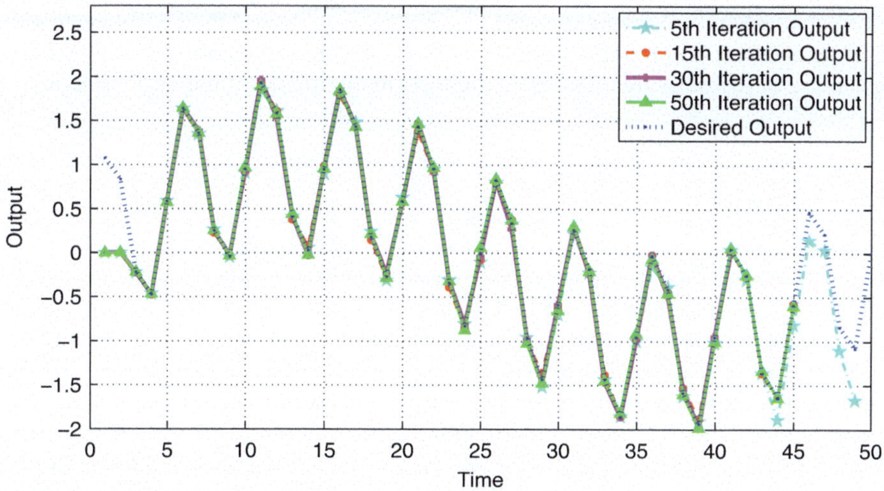


**Fig. 6.8** Robust output tracking performance with disturbances

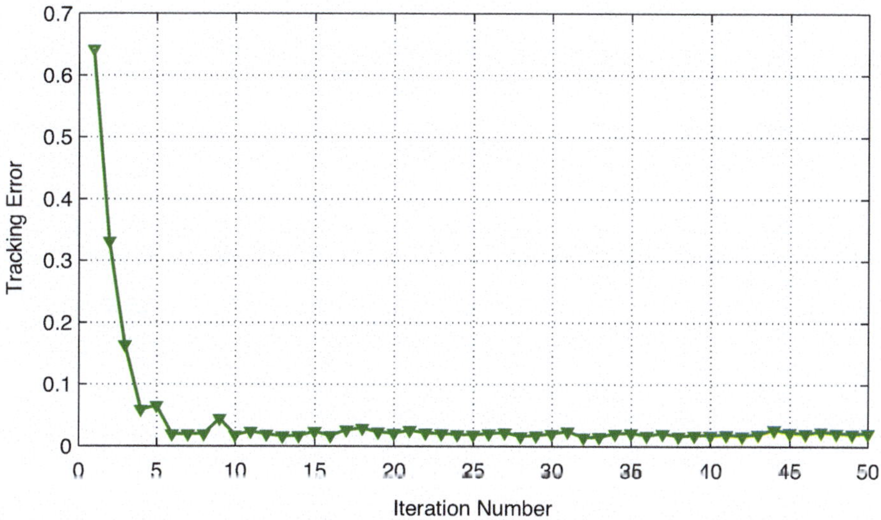


**Fig. 6.9** Robust convergence of tracking error with disturbances

Select the same parameters of the APC-DDILC as that used in the previous cases, i.e., $x_{k,i}(0) = 0$, $i = 1, 2, 3$, $u_0(t) = 0$, $\hat{\varphi}_0(t) = 1$, $\eta = 0.8$, $\mu = 1$, $\lambda = 1$, and $\rho = 1.5$. The output tracking performance is shown in Fig. 6.8, and the tracking error convergence is shown in Fig. 6.9. It can be seen from Figs. 6.8 and 6.9, although the tracking error converges to a bounded range, the control performance is still acceptable. Therefore, one can conclude that the proposed APC-DDILC can well regulate external disturbances because its learning gain is capable of being updated by the use of the measured data which reflects the influence of the disturbances.

## 6.6 Summary

An auxiliary predictive compensation-based DDILC is presented to deal with iteration-varying batch lengths. The auxiliary predictive model is built through a state transformation equation of the original system and a consequent iteration-difference of the output. The expanded output and tracking error are defined by introducing a random variable to contain both the predictive output and the actual output. The developed APC-DDILC consists of a learning control law and a parameter estimation algorithm whose learning gain is variable and estimated via measured data in real time. The boundedness of parameter estimation and the convergence of tracking error are proved. Simulation results illustrate the effectiveness of the APC-DDILC as well as its superior control performance to the other methods that have no compensation for the unequal trial length.

## References

Bu X, Yu Q, Hou Z, Qian W (2019) Model free adaptive iterative learning consensus tracking control for a class of nonlinear multiagent systems. IEEE Trans Syst Man Cybern Syst 49(4):677–686

Freeman CT (2014) Newton-method based iterative learning control for robot-assisted rehabilitation using FES. Mechatron 24:934–943

Freeman CT, Hughes AM, Burridge JH, Chappell PH, Lewin PL, Rogers E (2010) Iterative learning control of FES applied to the upper extremity for rehabilitation. Control Eng Pract 17(3):368–381

Guth M, Seel T, Raisch J (2013) Iterative learning control with variable pass length applied to trajectory tracking on a crane with output constraints. In: Proceedings of the 52nd IEEE conference on decision and control. IEEE. https://ieeexplore.ieee.org/document/6760946

Hou Z, Xiong S (2019) On model-free adaptive control and its stability analysis. IEEE Trans Autom Control 64(11):4555–4569

Li X, Xu J, Huang D (2014) An iterative learning control approach for linear systems with randomly varying trial lengths. IEEE Trans Autom Control 59(7):1954–1960

Li X, Xu J, Huang D (2015) Iterative learning control for nonlinear dynamic systems with randomly varying trial lengths. Int J Adapt Control Signal Process 29(11):1341–1353

Longman RW, Mombaur KD (2006) Investigating the use of iterative learning control and repetitive control to implement periodic gaits. In: Fast motions in biomechanics and robotics. Springer, Berlin

Meng D, Moore KL (2020) Contraction mapping-based robust convergence of iterative learning control with uncertain, locally-Lipschitz nonlinearity. IEEE Trans Syst Man Cybern Syst 50(2):442–454

Meng D, Zhang J (2018) Deterministic convergence for learning control systems over iteration-dependent tracking intervals. IEEE Trans Neural Networks Learn Syst 29(8):3885–3892

Nahrstaedt H, Schauer T, Shalaby R, Hesse S, Raisch J (2008) Automatic control of a drop-foot stimulator based on angle measurement using bioimpedance. Artif Organs 32(8):649–654

Seel T, Schauer T, Raisch J (2011) Iterative learning control for variable pass length systems. IFAC Proc Volumes 44(1):4880–4885

Seel T, Werner C, Raisch J, Schauer T (2016) Iterative learning control of a drop foot neuroprosthesis Generating physiological foot motion in paretic gait by automatic feedback control. Control Eng Pract 48:87–97

Seel T, Schauer T, Raisch J (2017) Monotonic convergence of iterative learning control systems with variable pass length. Int J Control 90(3):393–406

Shen D, Zhang W, Wang YQ, Chien CJ (2016b) On almost sure and mean square convergence of P-type ILC under randomly varying iteration lengths. Automatica 63:359–365

Shen D, Zhang W, Xu JX (2016a) Iterative learning control for discrete nonlinear systems with randomly iteration varying lengths. Syst Control Lett 96:81–87

Yu M, Li C (2017) Robust adaptive iterative learning control for discrete-time nonlinear systems with time-iteration-varying parameters. IEEE Trans Syst Man Cybern Syst 47(7):1737–1745

Zhu X, Wang J (2018) Double iterative compensation learning control for active training of upper limb rehabilitation robot. Int J Control Autom Syst 2:1–11

# Chapter 7
# Data-Driven ILC for Systems with Quantized Data

## 7.1 Introduction

Data transmission is a common characteristic of network control systems (NCSs) and multiagent systems (MASs), and it is also a critical issue due to the limited transmission bandwidth of networks. Aiming for saving the limited resource, it is desirable to quantize signals before data transmissions. As an earlier work, a logarithmic quantizer (Bu et al. 2015, 2017) was used to quantize the output measurements when designing the ILC law. Xu et al. (2017) have extended the results of Bu et al. (2015) by using quantized error information. In Yin et al. (2018) and Jin (2018), the input signals are quantized before they are fed to the controlled plant. An ILC method is proposed in Shen and Xu (2017) employing a probabilistic quantizer. Both quantization and event-triggered transmission methods are considered in Xiong et al. (2016) for the learning consensus protocol of a discrete-time network. A quantized ILC method using finite-leveled sigma-delta quantization is proposed for MASs (Zhang and Li 2017). The authors in Gao et al. (2017) have discussed both the quantization and the multi-rate sampling effects on the iterative learning fault-tolerant control for networked processes. In Huo and Shen (2020, 2019), the zero-convergence problem of ILC is addressed by incorporating a finite uniform quantizer with an encoding-decoding method. However, one can observe in these studies that they often adopt a P-type learning algorithm with fixed learning gain and the theoretical results are most limited to linear systems. From a practical viewpoint, the fixed learning gain may hinder wide applications of these ILC methods because it greatly reduces the adaptability in presence of uncertainties and exogenous disturbances.

Although many adaptive ILC methods with iteratively tunable learning gains have been developed to deal with the nonrepetitive uncertainties by introducing an adaptation mechanism into the iteration domain (Hui et al. 2020; Li and Li 2014; Jin 2016; Shen et al. 2019), little work has been devoted to the quantized systems. Besides, model-dependence is also a critical drawback of the existing quantized ILC methods, preventing them from extensive applications. Several data-driven ILC

 
R. Chi et al., *Data-Driven Iterative Learning Control for Discrete-Time Systems*, Intelligent Control and Learning Systems 2,
https://doi.org/10.1007/978-981-19-5950-9_7

methods (Janssens et al. 2013; Chi et al. 2018a, b, 2015, 2022) have been presented under an optimal ILC framework. However, no work has been found on quantized data-driven ILC except for the work (Bu et al. 2018) proposed by Bu et al., where, nevertheless, the introduced linear data model (LDM) contains only one parameter which is too simple to represent a complex nonlinear system.

Inspired by the above observations, this chapter proposes a quantized data-driven iterative learning control (QDDILC) for nonlinear nonaffine discrete-time systems whose iterative input-output (I/O) dynamics is reformulated as a new linear increment form by using the state transition technique and the iterative dynamic linearization method. By using a logarithmic quantizer to quantify the tracking error, the QDDILC scheme is developed bypassing the mechanistic modeling. The convergence of the proposed QDDILC is proved mathematically in the presence of data quantization. The main contributions of this chapter are summarized as:

(i) The proposed QDDILC method includes both a quantized learning control law and a quantized parameter updating law. The latter makes it possible to automatically tune the learning gains in the learning control law by using the real-time I/O data, thereby enhancing the robustness against uncertainties.

(ii) Different from Bu et al. (2018), the LDM used in this chapter contains much more additional parameters and input signals, thus enhancing the control performance.

(iii) Besides the contraction mapping principle, a new double-dynamics analysis method using mathematical induction is introduced to address the coupled dynamics between inputs and tracking errors in the convergence analysis.

This chapter is organized as follows. Section 7.2 shows the problem statement. The QDDILC is proposed in Sect. 7.3. Section 7.4 presents the convergence analysis. Simulation study is conducted in Sect. 7.5. Section 7.6 gives some conclusions.

## 7.2 Problem Formulation

Consider a class of nonlinear nonaffine discrete-time systems,

$$y_k(t+1) = f\left(y_k(t), \ldots, y_k(t-n_y), u_k(t), \ldots, u_k(t-n_u)\right) \tag{7.1}$$

where $y_k(t)$ and $u_k(t)$ are system output and input, respectively, where $y_k(t) = 0$ and $u_k(t) = 0$, $\forall t < 0$; $n_y$ and $n_u$ are two unknown orders; $t \in \{0, 1, \ldots, N\}$ is time instant where $N \in Z^+$ represents the endpoint; $k \in \{0, 1, \ldots\}$ indicates the iteration number; $f(\cdots) : R^{n_u+n_y+2} \mapsto R$ is an unknown nonlinear scalar function, $f(0, 0) = 0$.

Consider nonlinear system (7.1) satisfying assumptions 4.1, 4.2, and 4.3.

From Theorem 4.1, we can transform system (7.1) into the following STIDL-based LDM,

$$\Delta y_k(t+1) = \boldsymbol{\varphi}_k(t)\Delta \boldsymbol{u}_k(t) \tag{7.2}$$

where $\Delta y_k(t) = y_k(t) - y_{k-1}(t)$, $\Delta \boldsymbol{u}_k(t) = \boldsymbol{u}_k(t) - \boldsymbol{u}_{k-1}(t)$; $\boldsymbol{\varphi}_k(t) = [\phi_k(0), \ldots, \phi_k(t)]$, $\boldsymbol{u}_k(t) = [u_k(0), \ldots, u_k(t)]^{\mathrm{T}}$; $\phi_k(i)$ and $u_k(i)$ are the $(i+1)$th element of $\boldsymbol{\varphi}_k(t)$ and $\boldsymbol{u}_k(t)$, respectively, $i \in \{0, \ldots, t\}$.

In terms of Assumption 4.2, it follows that $\left\|\boldsymbol{\varphi}_k(t)\right\| \leq b_\varphi$, where $b_\varphi > 0$ is a constant.

## 7.3 Controller Design

This section aims at addressing the design of learning control law using quantized measurements to make system output $y_k(t+1)$ track all desired discrete-time points of the trajectory $y_d(t+1)$ exactly, where $y_d(t+1)$ is bounded, $\forall t \in \{0, \ldots, N\}$.

Adopt the following logarithmic set of quantization levels,

$$H = \left\{\pm h_i : h_i = r^i h_0, i = 0, \pm 1, \pm 2, \ldots\right\} \cup \{0\}, h_0 > 0$$

where $0 < r < 1$ is a constant related to quantization density; $h_i$ is a quantitative level; and $h_0$ is the initial value of the quantization level.

The associated quantizer $q(\cdot)$ is given as

$$q(v) = \begin{cases} 0 & if \quad v = 0 \\ h_i & if \quad \underline{h}_i < h_i < \bar{h}_i \\ -q(-v) & if \quad v < 0 \end{cases}$$

where $\underline{h}_i = h_i/(1+\vartheta)$ and $\bar{h}_i = h_i/(1-\vartheta)$ with $\vartheta = (1-r)/(1+r)$ and $0 < \vartheta < 1$; $q(v)$ is the quantized data with $v$ being the actual data. It is clear that $q(v)$ is symmetry and time-invariant, that is, $q(-v) = -q(v)$.

By virtue of the sector bound method (Fu and Xie 2005), the quantization error is defined as $q(v) - v = \delta(v)v$, where $|\delta(v)| < \vartheta_v$ with $\vartheta_v = (1-r_v)/(1+r_v)$, where $r_v$ is a constant related to the quantization density of the quantized data $q(v)$.

Define a quantized tracking error as $q\left(e_k(t+1)\right) = w_{e,k} e_k(t+1)$, where $e_k(t+1) = y_d(t+1) - y_k(t+1)$, $w_{e,k} = \left(1 + \delta\left(e_k(t+1)\right)\right)$, $\left|\delta\left(e_k(t+1)\right)\right| < \vartheta_e$ and $\vartheta_e = (1-r_e)/(1+r_e)$ with $0 < \vartheta_e < 1$.

Design an objective function with quantized tracking error,

$$J\left(u_k(t)\right) = |q(e_k(t+1))|^2 + \lambda |u_k(t) - u_{k-1}(t)|^2$$

Substituting the STDL-based LDM (7.2) into the above equation, one has

$$J\left(u_k(t)\right) = \left|q(e_{k-1}(t+1)) - \boldsymbol{\varphi}_k(t) \Delta \boldsymbol{u}_k(t)\right|^2 + \lambda |u_k(t) - u_{k-1}(t)|^2 \tag{7.3}$$

where $\lambda > 0$ denotes a weighting factor.

Using the definition of quantized error, (7.3) becomes,

$$J\left(u_k(t)\right)=\left|w_{e,k-1}e_{k-1}(t+1)-\sum_{i=0}^{t-1}\phi_k(i)\Delta u_k(i)-\phi_k(t)\Delta u_k(t)\right|^2 \\ +\lambda|u_k(t)-u_{k-1}(t)|^2 \tag{7.4}$$

Solving the optimization problem of (7.4) with respect to $u_k(t)$, a quantized learning control law is obtained as,

$$u_k(t)=u_{k-1}(t)+\frac{\rho w_{e,k-1}\phi_k(t)e_{k-1}(t+1)}{\lambda+|\phi_k(t)|^2}-\frac{\rho\phi_k(t)\sum_{i=0}^{t-1}\phi_k(i)\Delta u_k(i)}{\lambda+|\phi_k(t)|^2} \tag{7.5}$$

where $\rho>0$ is a step-size constant.

Since $\phi_k(i), i\in\{0,\ldots,t\}$, is unavailable in (7.5), a quantized parameter updating law is presented by defining an objective function using quantized measurements,

$$J\left(\hat{\boldsymbol{\varphi}}_k(t)\right)=\left\|q(\Delta y_{k-1}(t+1))-\hat{\boldsymbol{\varphi}}_k(t)\Delta\boldsymbol{u}_{k-1}(t)\right\|^2+\mu\left\|\hat{\boldsymbol{\varphi}}_k(t)-\hat{\boldsymbol{\varphi}}_{k-1}(t)\right\|^2 \tag{7.6}$$

where $q(\Delta y_k(t))=(1+\delta(\Delta y_k(t)))\,\Delta y_k(t)$; $\mu>0$ is a weighting constant; and $\hat{\boldsymbol{\varphi}}_k(t)$ is used to estimate $\boldsymbol{\varphi}_k(t)$.

According to (7.2), we can rewrite (7.6) as

$$\begin{aligned}&J\left(\hat{\boldsymbol{\varphi}}_k(t)\right)\\&=\left\|w_{\Delta,k-1}\Delta y_{k-1}(t+1)-\hat{\boldsymbol{\varphi}}_{k-1}(t)\Delta\boldsymbol{u}_{k-1}(t)-\left(\hat{\boldsymbol{\varphi}}_k(t)-\hat{\boldsymbol{\varphi}}_{k-1}(t)\right)\Delta\boldsymbol{u}_{k-1}(t)\right\|^2\\&+\mu\left\|\hat{\boldsymbol{\varphi}}_k(t)-\hat{\boldsymbol{\varphi}}_{k-1}(t)\right\|^2\end{aligned} \tag{7.7}$$

Solving the optimization problem of (7.7) with respect to $\hat{\boldsymbol{\varphi}}_k(t)$, a quantized parameter estimation updating law is obtained as,

$$\hat{\boldsymbol{\varphi}}_k(t)=\hat{\boldsymbol{\varphi}}_{k-1}(t)+\eta\frac{\left(w_{\Delta,k-1}\Delta y_{k-1}(t+1)-\hat{\boldsymbol{\varphi}}_{k-1}(t)\Delta\boldsymbol{u}_{k-1}(t)\right)\Delta\boldsymbol{u}_{k-1}^{\mathrm{T}}(t)}{\mu+\|\Delta\boldsymbol{u}_{k-1}(t)\|^2} \tag{7.8}$$

where $\eta>0$ is a step-size constant; $w_{\Delta,k-1}=1+\delta\left(\Delta y_{k-1}(t+1)\right)$, $|\delta\left(\Delta y_{k-1}(t+1)\right)|<\vartheta_\Delta$ and $\vartheta_\Delta=(1-r_\Delta)/(1+r_\Delta)$ with $0<\vartheta_\Delta<1$.

In summary, the proposed quantized DDILC (QDDILC) method is constructed as follows,

$$u_k(t)=u_{k-1}(t)+\rho\frac{\hat{\phi}_k(t)\left(w_{e,k-1}e_{k-1}(t+1)-\sum_{i=0}^{t-1}\hat{\phi}_k(i)\Delta u_k(i)\right)}{\lambda+\left|\hat{\phi}_k(t)\right|^2} \tag{7.9}$$

$$\hat{\boldsymbol{\varphi}}_k(t)=\hat{\boldsymbol{\varphi}}_{k-1}(t)+\eta\frac{\left(w_{\Delta,k-1}\Delta y_{k-1}(t+1)-\hat{\boldsymbol{\varphi}}_{k-1}(t)\Delta\boldsymbol{u}_{k-1}(t)\right)\Delta\boldsymbol{u}_{k-1}^{\mathrm{T}}(t)}{\mu+\|\Delta\boldsymbol{u}_{k-1}(t)\|^2} \tag{7.10}$$

$$\hat{\varphi}_k(t) = \hat{\varphi}_0(t), \text{ if } \operatorname{sgn}(\hat{\phi}_k(i)) \neq \operatorname{sgn}(\hat{\phi}_0(i)), \text{ or } \left\|\hat{\varphi}_k(t)\right\| \leq \varepsilon \tag{7.11}$$

where (7.11) is a reset algorithm used to strengthen the ability of parameter updating law (7.10) in tracking time-iteration-varying estimation; $\hat{\varphi}_0(t)$ is an initial estimation of $\hat{\varphi}_k(t)$ ; $\varepsilon > 0$ is small sufficiently.

**Remark 7.1** Although the STDL-based LDM (7.2) is employed to attain the control method, the proposed QDDILC is data-driven because no mechanistic model information is required in (7.9)–(7.11). Therefore, it is a good alternative of model-based methods for systems whose models are unavailable or have large uncertainty and strong nonlinearity.

**Remark 7.2** Different from the existing DDILC methods without considering quantization problems (Janssens et al. 2013; Chi et al. 2018a, b, 2015, 2022), it is the quantized measurements that are used in both the learning control algorithm (7.9) and the parameter updating law (7.10), and the subsequent analysis is based on the data quantization.

**Remark 7.3** Different from the quantized DDILC (Bu et al. 2018), the learning control law (7.9) employs more input information obtained from previous instants of the current operation with corresponding parameter estimations as a feedback to improve tracking performance.

**Remark 7.4** Different from the quantized non-adaptive ILC (Bu et al. 2015, 2017; Xu et al. 2017; Yin et al. 2018; Jin 2018; Shen and Xu 2017; Xiong et al. 2016; Zhang and Li 2017; Gao et al. 2017; Huo and Shen 2020, 2019), the learning gain of control law (7.9) is tunable in real time by using parameter estimations derived from the quantized adaptive updating law (7.10), therefore improving the adaptability of the proposed QDDILC to the uncertainties.

**Remark 7.5** The controller design is conducted directly for a discrete-time system. As a result, the control performance of the proposed QDDILC is not influenced when being applied in realistic systems.

## 7.4 Convergence Analysis

Before conducting the convergence analysis, a lemma is given as follows.

**Lemma 7.1** (Huo and Shen 2019) *Consider the following general system*

$$\boldsymbol{\kappa}_{k+1}(t) = \boldsymbol{A}_k(t)\boldsymbol{\kappa}_k(t) + \boldsymbol{P}_k(t)$$

*where* $\boldsymbol{\kappa}_k(t) \in R^m$, $\boldsymbol{P}_k(t) \in R^m$ *and* $\boldsymbol{A}_k(t) \in R^{m\times m}$. *For any* $t \in \{0, 1, \ldots, N-1\}$, *if there exists an iterative sequence* $\{\varpi_s(t)\}_{s=0}^{s=\infty}$, *where* $\varpi_0(t) = 1$ *and* $0 < \varpi_{s+1}(t) - \varpi_s(t) \leq \chi(t)$ *with* $1 \leq \chi(t) < \infty$, *such that*

$$\left\| \prod_{\varpi=\kappa_s(t)}^{\varpi_{s+1}(t)-1} A_k(t) \right\|_\infty \leq \varsigma_1 < 1$$

*holds,* $\forall k \in \{1, 2, \ldots\}$, *where* $0 < \varsigma_1 < 1$ *is a constant, then we have the following two results: (1)* $\boldsymbol{\kappa}_k(t)$ *is bounded, provided that* $\boldsymbol{P}_k(t)$ *is bounded; (2)* $\lim_{k\to\infty} \boldsymbol{\kappa}_k(t) = 0$, *provided that* $\lim_{k\to\infty} \boldsymbol{P}_k(t) = 0$.

The convergence of the proposed QDDILC is shown in the following theorem.

**Theorem 7.1** *For a class of nonlinear systems (7.1), if Assumptions 4.1, 4.2 and 4.3 hold and* $\lambda > \rho^2 b_\varphi^2/4$, *then the proposed QDDILC method (7.9)–(7.11) guarantees that:*

*(i) the parameter estimation* $\hat{\boldsymbol{\varphi}}_k(t)$ *is bounded over the finite time interval* $t \in \{0, 1, \ldots, N\}$, $\forall k$;
*(ii) the tracking error* $e_k(t+1)$ *converges to zero over the finite time interval* $t \in \{0, 1, \ldots, N\}$ *as* $k$ *approaches to infinity, i.e.,* $\lim_{k\to\infty} e_k(t+1) = 0$;
*(iii) the system output* $y_k(t)$ *and the control input signal* $u_k(t)$ *are bounded,* $\forall t \in \{0, 1, \ldots, N\}$, $\forall k$.

***Proof*** The Lyapunov-like method used in traditional adaptive ILC analysis cannot be directly employed here because the systems considered is nonlinear nonaffine in nature, where no parametric structure is available. In contrast, both the contraction mapping principle and the double dynamics analysis (DDA) method with mathematical induction are introduced to conduct the analysis of convergence and stability. The analysis consists of three parts shown as follows.

*Part 1. The boundedness of* $\hat{\boldsymbol{\varphi}}_k(t)$

Define $\tilde{\boldsymbol{\varphi}}_k(t) = \boldsymbol{\varphi}_k(t) - \hat{\boldsymbol{\varphi}}_k(t)$ as a parameter estimation error.

Subtracting $\boldsymbol{\varphi}_k(t)$ from both sides of (7.10), yields,

$$\begin{aligned} \tilde{\boldsymbol{\varphi}}_k(t) = {} & \tilde{\boldsymbol{\varphi}}_{k-1}(t)\left(\boldsymbol{I} - \boldsymbol{R}_{k-1}(t)\right) + \Delta\boldsymbol{\varphi}_k(t) \\ & - \delta(\Delta y_{k-1}(t+1))\boldsymbol{\varphi}_{k-1}(t)\boldsymbol{R}_{k-1}(t) \end{aligned} \tag{7.12}$$

where $\boldsymbol{R}_{k-1}(t) = \eta\frac{\Delta\boldsymbol{u}_{k-1}(t)\Delta\boldsymbol{u}_{k-1}^{\mathrm{T}}(t)}{\mu+\|\Delta\boldsymbol{u}_{k-1}(t)\|^2}$.

Since $\left\|\boldsymbol{\varphi}_k(t)\right\| \leq b_\varphi$ and $|\delta(\Delta y_{k-1}(t+1))| < \vartheta_\Delta$, taking norms on both sides of Eq. (7.12), one has

$$\left\|\tilde{\boldsymbol{\varphi}}_k(t)\right\| \leq \left\|\tilde{\boldsymbol{\varphi}}_{k-1}(t)\left(\boldsymbol{I} - \boldsymbol{R}_{k-1}(t)\right)\right\| + \vartheta_\Delta b_\varphi \left\|\boldsymbol{R}_{k-1}(t)\right\| + 2b_\varphi \tag{7.13}$$

Considering the following equation,

$$\left\|\tilde{\boldsymbol{\varphi}}_{k-1}(t)\left(\mathbf{I}-\boldsymbol{R}_{k-1}(t)\right)\right\|^{2}=\left\|\tilde{\boldsymbol{\varphi}}_{k-1}(t)\right\|^{2}+\left(-2+\frac{\eta\|\Delta\boldsymbol{u}_{k-1}(t)\|^{2}}{\mu+\|\Delta\boldsymbol{u}_{k-1}(t)\|^{2}}\right)\times\frac{\eta\left\|\Delta\boldsymbol{u}_{k-1}^{\mathrm{T}}(t)\tilde{\boldsymbol{\varphi}}_{k-1}(t)\right\|^{2}}{\mu+\|\Delta\boldsymbol{u}_{k-1}(t)\|^{2}} \tag{7.14}$$

Owing to $0<\eta<2$ and $\mu>0$, it is clear that

$$\left(-2+\frac{\eta\|\Delta\boldsymbol{u}_{k-1}(t)\|^{2}}{\mu+\|\Delta\boldsymbol{u}_{k-1}(t)\|^{2}}\right)<0 \tag{7.15}$$

Equations (7.14) and (7.15) together imply that

$$\left\|\tilde{\boldsymbol{\varphi}}_{k-1}(t)\left(\mathbf{I}-\boldsymbol{R}_{k-1}(t)\right)\right\|^{2}<\left\|\tilde{\boldsymbol{\varphi}}_{k-1}(t)\right\|^{2}$$

So, one can find a proper constant $0<d_1<1$ such that

$$\left\|\tilde{\boldsymbol{\varphi}}_{k-1}(t)\left(\mathbf{I}-\boldsymbol{R}_{k-1}(t)\right)\right\|^{2}\leq d_1\left\|\tilde{\boldsymbol{\varphi}}_{k-1}(t)\right\|^{2} \tag{7.16}$$

Further, owing to $0<\eta<2$ and $\mu>0$, one has

$$0<\|\boldsymbol{R}_{k-1}(t)\|\leq\frac{\eta\,\|\Delta\boldsymbol{u}_{k-1}(t)\|\,\left\|\Delta\boldsymbol{u}_{k-1}^{\mathrm{T}}(t)\right\|}{\mu+\|\Delta\boldsymbol{u}_{k-1}(t)\|^{2}}\leq C_1<2 \tag{7.17}$$

where $C_1$ is a positive constant.

In views of (7.13), (7.16) and (7.17), one obtains

$$\begin{aligned}\left\|\tilde{\boldsymbol{\varphi}}_{k}(t)\right\| &\leq d_1\left\|\tilde{\boldsymbol{\varphi}}_{k-1}(t)\right\|+(C_1\vartheta_{\Delta}+2)\,b_{\varphi}\\ &\leq\cdots\leq d_1^{k}\left\|\tilde{\boldsymbol{\varphi}}_{0}(t)\right\|+\frac{(C_1\vartheta_{\Delta}+2)\,b_{\varphi}}{1-d_1}\end{aligned} \tag{7.18}$$

which implies $\lim\limits_{k\to\infty}\tilde{\boldsymbol{\varphi}}_{k}(t)\leq(C_1\vartheta_{\Delta}+2)\,b_{\varphi}\big/(1-d_1)$ because the initial estimation error $\tilde{\boldsymbol{\varphi}}_{0}(t)$ is bounded due to the boundedness of $\boldsymbol{\varphi}_{k}(t)$ and $\hat{\boldsymbol{\varphi}}_{0}(t)$.

Then, $\hat{\boldsymbol{\varphi}}_{k}(t)$ is bounded too, and we can assume $\left\|\hat{\boldsymbol{\varphi}}_{k}(t)\right\|\leq b_{\hat{\varphi}}$ in the following analysis where $b_{\hat{\varphi}}>0$ is a proper constant. Further, it is obviously that $\left|\hat{\phi}_{k}(i)\right|\leq b_{\hat{\phi}}$. So, conclusion (i) of Theorem 7.1 is proved.

*Part 2. The convergence of tracking error*

According to the definition of tracking error and (7.2), we get

$$e_k(t+1)=e_{k-1}(t+1)-\phi_k(t)\Delta u_k(t)-\sum_{i=0}^{t-1}\phi_k(i)\Delta u_k(i) \tag{7.19}$$

Substituting the learning control law (7.9) into (7.19), yields,

$$e_k(t+1) = a_{1,k}(t)e_{k-1}(t+1) + p_{1,k}(t) \tag{7.20}$$

where $a_{1,k}(t) = 1 - \frac{\rho w_{e,k-1}\phi_k(t)\hat{\phi}_k(t)}{\lambda+\left|\hat{\phi}_k(t)\right|^2}$ and

$$p_{1,k}(t) = \frac{\rho\phi_k(t)\hat{\phi}_k(t)\sum_{i=0}^{t-1}\hat{\phi}_k(i)\Delta u_k(i)}{\lambda+\left|\hat{\phi}_k(t)\right|^2} - \sum_{i=0}^{t-1}\phi_k(i)\Delta u_k(i) \tag{7.21}$$

Rewrite (7.5) as

$$\begin{aligned} u_k(t) = u_{k-1}(t) &- \frac{\rho\hat{\phi}_k(t)\sum_{i=0}^{t-1}\hat{\phi}_k(i)\Delta u_k(i)}{\lambda+\left|\hat{\phi}_k(t)\right|^2} \\ &+ \frac{\rho\hat{\phi}_k(t)w_{e,k-1}}{\lambda+\left|\hat{\phi}_k(t)\right|^2}(y_d(t+1) - y_{k-1}(t+1)) \end{aligned} \tag{7.22}$$

Similar to Eq. (7.2), we can obtain the dynamic relationship between the $i$th iteration and the $j$th iteration as follows,

$$y_i(t+1) - y_j(t+1) = \boldsymbol{\varphi}_{i,j}(t)(\boldsymbol{u}_i(t) - \boldsymbol{u}_j(t)) \tag{7.23}$$

where $\boldsymbol{u}_i(t) = [u_i(0), u_i(1), \ldots, u_i(t)]^{\mathrm{T}}$ and $\boldsymbol{u}_j(t) = [u_j(0), u_j(1), \ldots, u_j(t)]^{\mathrm{T}}$; $\boldsymbol{\varphi}_{i,j}(t) = [\phi_{i,j}(0), \phi_{i,j}(1), \ldots, \phi_{i,j}(t)]$ is the partial derivative of $g_t(\cdots)$ with respect to $\boldsymbol{u}_i(t)$ at some points in the interval $[\boldsymbol{u}_i(t), \boldsymbol{u}_j(t)]$.

In terms of (7.23), for $i = k-1$, $j = 0$, one has

$$\begin{aligned} y_{k-1}(t+1) = y_0(t+1) &- \sum_{i=0}^{t}\phi_{k-1,0}(i)u_0(i) \\ &+ \phi_{k-1,0}(t)u_{k-1}(t) + \sum_{i=0}^{t-1}\phi_{k-1,0}(i)u_{k-1}(i) \end{aligned} \tag{7.24}$$

where the initial states $u_0(i)$, $i \in \{0, \ldots, t\}$, is assumed to be bounded.

Substituting (7.24) into (7.22), yields

$$u_k(t) = a_{2,k}(t)u_{k-1}(t) + p_{2,k}(t) \tag{7.25}$$

where $a_{2,k}(t) = 1 - \frac{\rho w_{e,k-1}\hat{\phi}_k(t)\phi_{k-1,0}(t)}{\lambda+\left|\hat{\phi}_k(t)\right|^2}$ and

$$
\begin{aligned}
p_{2,k}(t) = & \frac{\rho\hat{\phi}_k(t)}{\lambda + \left|\hat{\phi}_k(t)\right|^2}\Bigg(w_{e,k-1}e_0(t+1) - \sum_{i=0}^{t-1}\hat{\phi}_k(i)\Delta u_k(i) \\
& -w_{e,k-1}\left(\sum_{i=0}^{t-1}\phi_{k-1,0}(i)u_{k-1}(i) - \sum_{i=0}^{t}\phi_{k-1,0}(i)u_0(i)\right)\Bigg)
\end{aligned} \tag{7.26}
$$

Due to the double-dynamics in (7.20) and (7.25), we now introduce the DDA method and mathematical induction as the main tools to conduct the convergence analysis in two steps.

Step (1): Let $t = 0$. Note that there is an implicit assumption that the sum of the inputs is zero when $t < 0$ from the constraint on the system that $u_k(t) = 0$, $\forall t < 0$. In addition, $\left|\phi_{i,j}(t)\right| \le b_\phi$ and the initial input $u_0(t)$ are bounded, $\forall i, j \in \{0, 1, \ldots\}$, $\forall t \in \{0, \ldots, N\}$, and $\forall k$.

According to Eq. (7.21), one has

$$
p_{1,k}(0)=0, \ \forall k \tag{7.27}
$$

that is, $p_{1,k}(0)$ is bounded, and further we have $\lim_{k\to\infty} p_{1,k}(0) = 0$.

Since $w_{e,k-1} = 1 + \delta\left(e_{k-1}(t+1)\right)$ and $0 < \left|\delta\left(e_{k-1}(t+1)\right)\right| < \vartheta_e < 1$, we have

$$
0 < w_{e,k-1} < 2 \quad \text{and} \quad 0 < \left|w_{e,k-1}\right| < 1 + \vartheta_e < 2 \tag{7.28}
$$

which means that $w_{e,k-1}$ is bounded.

We know that $\phi_k(t)$ is of fixed sign in accordance with Assumption 4.3. Moreover, in the implementation of the proposed method, the initial estimation of $\hat{\phi}_k(t)$, i.e., $\hat{\phi}_0(t)$, is set to have the same sign as that of $\phi_k(t)$. So, based on the reset algorithm (7.11), we have $\phi_k(t)\hat{\phi}_k(t)$ is positive.

By selecting $\rho$ and $\lambda$ properly such that $\lambda > \rho^2 b_\phi^2/4$, according to (7.28), we have,

$$
\begin{aligned}
0 &< \frac{\rho w_{e,k-1}\phi_k(t)\hat{\phi}_k(t)}{\lambda + \left|\hat{\phi}_k(t)\right|^2} \\
&= \left|\frac{\rho w_{e,k-1}\phi_k(t)\hat{\phi}_k(t)}{\lambda + \left|\hat{\phi}_k(t)\right|^2}\right| \\
&\le \frac{\rho\left(1+\vartheta_e\right)\left|\phi_k(t)\right|}{2\sqrt{\lambda}} \\
&\le \frac{\rho b_\phi}{\sqrt{\lambda}} < 2
\end{aligned} \tag{7.29}
$$

which leads to

$$0 \leq \left|a_{1,k}(t)\right| \leq d_2 < 1 \tag{7.30}$$

where $d_2$ is a positive constant.

Then, according to (7.20), (7.27), (7.30) and Lemma 7.1, we have

$$|e_k(1)| \leq b_e(0) \quad \text{and} \quad \lim_{k\to\infty} e_k(1) = 0, \quad \forall k \in \{0, 1, \ldots\} \tag{7.31}$$

which means $e_k(1)$ is bounded, where $b_e(0) > 0$ is a constant.

Using learning control law (7.9), we get

$$\Delta u_k(0) = \rho \frac{\hat{\phi}_k(0) w_{e,k-1} e_{k-1}(1)}{\lambda + \left|\hat{\phi}_k(0)\right|^2}$$

then together with (7.31), we have $\lim_{k\to\infty} \Delta u_k(0) = 0$. Besides, we know from Eq. (7.26) that

$$p_{2,k}(0) = \frac{\rho \hat{\phi}_k(0) w_{e,k-1}}{\lambda + \left|\hat{\phi}_k(0)\right|^2} \left(e_0(1) + \phi_{k-1,0}(0) u_0(0)\right) \tag{7.32}$$

Using inequality $a^2 + b^2 \geq 2ab$, one has

$$\left|\frac{\rho \hat{\phi}_k(t)}{\lambda + \left|\hat{\phi}_k(t)\right|^2}\right| \leq \frac{\rho}{2\sqrt{\lambda}} \triangleq C_2 \tag{7.33}$$

where $C_2$ is a proper positive constant.

Then, together with (7.28), Eq. (7.32) becomes

$$\begin{aligned} \left|p_{2,k}(0)\right| &\leq \left|\frac{\rho \hat{\phi}_k(0)}{\lambda + \left|\hat{\phi}_k(0)\right|^2}\right| \left|w_{e,k-1}\right| \left(\left|e_0(1)\right| + \left|\phi_{k-1,0}(0)\right| \left|u_0(0)\right|\right) \\ &< C_2 \left(1 + \vartheta_e\right) \left(b_e(0) + b_\phi \left|u_0(0)\right|\right) \end{aligned} \tag{7.34}$$

that is, $p_{2,k}(0)$ is bounded.

Similarly, $\hat{\phi}_k(t)\phi_{k-1,0}(t)$ is positive. In the light of (7.28) together with the condition of $\lambda > \rho^2 b_\phi^2 / 4$, we have

$$
\begin{aligned}
0 &< \frac{\rho w_{e,k-1}\hat{\phi}_k(t)\phi_{k-1,0}(t)}{\lambda+\left|\hat{\phi}_k(t)\right|^2} \\
&= \left|\frac{\rho w_{e,k-1}\hat{\phi}_k(t)\phi_{k-1,0}(t)}{\lambda+\left|\hat{\phi}_k(t)\right|^2}\right| \\
&\leq \frac{\rho\left(1+\vartheta_e\right)\left|\phi_{k-1,0}(t)\right|}{2\sqrt{\lambda}} \\
&\leq \frac{\rho b_\phi}{\sqrt{\lambda}} \\
&< 2
\end{aligned}
\tag{7.35}
$$

which implies $0 \leq \left|a_{2,k}(t)\right| < 1$.

So, from (7.25), (7.34), (7.35) and Lemma 7.1, it is derived that $|u_k(0)| \leq b_u(0)$ is bounded, where $b_u(0) > 0$ is a constant.

Step (2): For $t \in \{0, \ldots, N-1\}$, we assume $\lim\limits_{k\to\infty} e_k(t+1) = 0$, $\lim\limits_{k\to\infty} \Delta u_k(t) = 0$, and $|u_k(t)| \leq b_u(t)$ is bounded, where $b_u(t)$ denotes a positive constant. Note that it has been proved that $u_0(t)$ is bounded, $\forall t \in \{0, \ldots, N\}$, and $\left|\phi_{i,j}(t)\right| \leq b_\phi, \forall i, j, t$, and $|\hat{\phi}_k(i)| \leq b_{\hat{\phi}}$ are also bounded, and $\forall i \in \{0, \ldots, t\}$.

For $t = N$, (7.20) becomes

$$
e_k(N+1) = a_{1,k}(N)e_{k-1}(N+1) + p_{1,k}(N) \tag{7.36}
$$

where $a_{1,k}(N) = 1 - \frac{\rho w_{e,k-1}\phi_k(N)\hat{\phi}_k(N)}{\lambda+\left|\hat{\phi}_k(N)\right|^2}$ and

$$
p_{1,k}(N) = \frac{\rho\phi_k(N)\hat{\phi}_k(N)\sum\limits_{i=0}^{N-1}\hat{\phi}_k(i)\Delta u_k(i)}{\lambda+\left|\hat{\phi}_k(N)\right|^2} - \sum_{i=0}^{N-1}\phi_k(i)\Delta u_k(i) \tag{7.37}
$$

Similar to (7.33), we have the relationship

$$
\left|\frac{\rho\phi_k(N)\hat{\phi}_k(N)}{\lambda+\left|\hat{\phi}_k(N)\right|^2}\right| \leq \rho\frac{|\phi_k(N)|}{2\sqrt{\lambda}} \leq C_2 b_\phi
$$

Then, according to the boundedness of $\hat{\phi}_k(i)$, one has

$$\begin{aligned}\left|p_{1,k}(N)\right| &\leq \left|\frac{\rho\phi_k(N)\hat{\phi}_k(N)}{\lambda+\left|\hat{\phi}_k(N)\right|^2}\right| \sum_{i=0}^{N-1}\left|\hat{\phi}_k(i)\right| |\Delta u_k(i)| + \sum_{i=0}^{N-1}|\phi_k(i)|\,|\Delta u_k(i)| \\ &\leq C_2 b_\phi b_{\hat{\phi}} \sum_{i=0}^{N-1}|\Delta u_k(i)| + b_\phi \sum_{i=0}^{N-1}|\Delta u_k(i)| \end{aligned} \tag{7.38}$$

Further, we can get that $\left|p_{1,k}(N)\right| \leq 2\left(C_2 b_{\hat{\phi}}+1\right) b_\phi \sum_{i=0}^{N-1} b_u(i)$ is bounded owing to $|u_k(t)| \leq b_u(t)$ for all $t \in \{0, \ldots, N-1\}$.

Since $\lim_{k\to\infty} \Delta u_k(t) = 0, \forall t \in \{0, \ldots, N-1\}$, by virtue of (7.38), we have

$$\lim_{k\to\infty} p_{1,k}(N) = 0 \tag{7.39}$$

Following (7.28)–(7.30), we have $\phi_k(N)\hat{\phi}_k(N)$ is positive, and $0 \leq \left|a_{1,k}(N)\right| < 1$ because $\lambda > \rho^2 b_{\hat{\phi}}^2/4$. So, together with Lemma 7.1, one can conclude from (7.38) and (7.39) that,

$$|e_k(N+1)| \leq b_e(N) \quad \text{and} \quad \lim_{k\to\infty} e_k(N+1) = 0, \quad \forall k \tag{7.40}$$

that is, $e_k(N+1)$ is bounded where $b_e(N) > 0$ is a constant.

Then, for $t = N$, combining (7.28) and (7.33), Eq. (7.9) becomes

$$\begin{aligned}|\Delta u_k(N)| &\leq \left|\frac{\rho\hat{\phi}_k(N)}{\lambda+\left|\hat{\phi}_k(N)\right|^2}\right| \left(\left|w_{e,k-1}\right| |e_{k-1}(N+1)| + \sum_{i=0}^{N-1}\left|\hat{\phi}_k(i)\right| |\Delta u_k(i)|\right) \\ &< C_2\left((1+\vartheta_e)\,|e_{k-1}(N+1)| + \sum_{i=0}^{N-1} b_{\hat{\phi}}\,|\Delta u_k(i)|\right)\end{aligned}$$

Because $\lim_{k\to\infty} \Delta u_k(t) = 0, \forall t \in \{0, \ldots, N-1\}$, according to (7.40), we can get that $\lim_{k\to\infty} \Delta u_k(N) = 0$.

Again, in terms of (7.26), (7.28) and (7.33), one obtains

$$\begin{aligned}\left|p_{2,k}(N)\right| &\leq \left|\frac{\rho\hat{\phi}_k(N)}{\lambda+\left|\hat{\phi}_k(N)\right|^2}\right| \left(\left|w_{e,k-1}\right| |e_0(N+1)| + \sum_{i=0}^{N-1}\left|\hat{\phi}_k(i)\right| |\Delta u_k(i)| \right. \\ &\quad \left. + \left|w_{e,k-1}\right| \left(\sum_{i=0}^{N-1}\left|\phi_{k-1,0}(i)\right| \left|u_{k-1}(i)\right| + \sum_{i=0}^{N}\left|\phi_{k-1,0}(i)\right| |u_0(i)|\right)\right) \\ &< 2C_2\left(b_e(N) + b_\phi \sum_{i=0}^{N}|u_0(i)| + \left(b_\phi + b_{\hat{\phi}}\right) \sum_{i=0}^{N-1} b_u(i)\right)\end{aligned} \tag{7.41}$$

that is, $p_{2,k}(N)$ is bounded, where $b_e(N)$, $b_\phi$ and $b_{\hat{\phi}}$ are some proper positive constants, respectively, and $|u_k(t)| \leq b_u(t)$ is bounded, $\forall t \in \{0, \ldots, N-1\}$.

For the same reason, $\hat{\phi}_k(N)\phi_{k-1,0}(N)$ is positive. Therefore, $0 \leq |a_{2,k}(N)| < 1$ because the parameters are selected to satisfy $\lambda > \rho^2 b_{\hat{\phi}}^2/4$. Then, according to Lemma 7.1, we can get from (7.41) that $|u_k(N)| \leq b_u(N)$ is bounded where $b_u(N) > 0$ is a proper constant. Thus, the hypothesis made for $t \in \{0, 1, \ldots, N-1\}$ in Step (2) also applies to $t = N$.

In other words, we can prove that $\lim\limits_{k\to\infty} e_k(t+1) = 0$ and $|u_k(t)| \leq b_u(t)$ is bounded, by combining Steps (1) and (2), and $|e_k(t+1)| \leq b_e$ is also bounded, where $b_u = \max\limits_{t\in\{0,1,\ldots,N\}} \{b_u(t)\}$ and $b_e = \max\limits_{t\in\{0,1,\ldots,N\}} \{b_e(t)\}$ are two positive constants.

*Part 3. The boundedness of system output and control input*

The boundedness of the control input has been shown in Part 2. Now, considering the definition of tracking error and the boundedness of the desired output $y_d(t+1)$, one can obtain $|y_k(t+1)| \leq b_e + b_d$ is bounded, where $b_d = \max\limits_{t\in\{0,1,\ldots,N\}} |y_d(t+1)|$.

## 7.5 Simulation Study

**Example 7.1** Consider a nonlinear discrete-time system,

$$y_k(t+1) = \frac{y_k(t)}{1 + y_k^2(t)} + u_k^3(t)$$

where $t \in \{0, 1, \ldots, 40\}$.

The desired trajectory is

$$y_d(t+1) = 0.5 \sin(\pi t/10) + 0.3 \cos(\pi t/10) \tag{7.42}$$

In the simulation, we set the initial values as $y_k(0) = 0$ and $u_0(t) = 0$, and the initial estimation as $\hat{\varphi}_0(t) = 0.01$; the controller parameters are chosen as $\eta$=0.8, $\mu$= 2, $\lambda = 0.1$, $\rho$= 0.9, $\varepsilon$=$10^{-5}$; set $r_e = r_\Delta = 0.9$ and $h_0 = 7$.

Applying the proposed QDDILC method (7.9)–(7.11), the results are illustrated in Figs. 7.1 and 7.2. Figure 7.1 shows the output tracking performance at the 100th iteration, where the vertical axis denotes the system output and the horizontal axis denotes the time instant. Figure 7.2 shows the convergence performance of tracking error, where the Y-axis represents mean square error, defined as $e_{k,\text{mean}}^* = \sum_{t=1}^{N} |y_d(t) - y_k(t)|^2/N$, and the X-axis denotes the iteration numbers. It is seen from Figs. 7.1 and 7.2 that the proposed QDDILC achieves a good tracking performance even though only the quantized measurements are available. Further, the tracking error converges iteratively.

For the comparison, the modified quantized DDILC (MQDDILC) proposed in Bu et al. (2018) is also applied,

$$u_k(t) = u_{k-1}(t) + \frac{\rho_2 \hat{\phi}_k(t)}{\lambda_2 + \left|\hat{\phi}_k(t)\right|^2} q(e_{k-1}(t+1)) \tag{7.43}$$

$$\begin{aligned} \hat{\phi}_k(t) = & \frac{\eta_2 \Delta u_{k-1}(t)}{\mu_2 + |\Delta u_{k-1}(t)|^2} \Big(q(\Delta y_{k-1}(t+1)) \\ & - \hat{\phi}_{k-1}(t) \Delta u_{k-1}(t)\Big) + \hat{\phi}_{k-1}(t) \end{aligned} \tag{7.44}$$

$$\begin{aligned} \hat{\phi}_k(t) = \hat{\phi}_0(t), \text{ if } \operatorname{sgn}(\hat{\phi}_k(i)) \neq \operatorname{sgn}(\hat{\phi}_0(i)), \\ \text{or } \left|\hat{\phi}_k(t)\right| \leq \varepsilon,\ i \in \{0, 1, \ldots, t\} \end{aligned} \tag{7.45}$$

where $\rho_2$, $\lambda_2$, $\eta_2$ and $\mu_2$ are the controller parameters; $\hat{\phi}_0(t)$ is the initial value of $\hat{\phi}_k(t)$; $q(e_k(t+1))$ and $q(\Delta y_{k-1}(t+1))$ are quantized data.

Under the same settings, applying the MQDDILC (Bu et al. 2018), the results are shown as the solid lines in Figs. 7.1 and 7.2. It is clear that the MQDDILC method (7.43)–(7.45) can also achieve a satisfied tracking performance using the quantized data. However, the tracking performance of MQDDILC is inferior to that of the proposed QDDILC method. The reason is that the proposed QDDILC utilizes additional control information in the learning control law to improve tracking performance.

To verify the effectiveness of the proposed data-driven method against both nonlinearities and uncertainties, a disturbance, $d_k(t)$, that varies randomly within $[0, 0.1]$ is considered in the simulation. Under the same simulation background, applying the proposed QDDILC method, the results are shown in Fig. 7.3. One can see that the proposed method is also effective in the presence of random exogenous disturbances.

**Example 7.2** A realistic manipulator system (Tan et al. 2001) is considered as below,

$$\begin{cases} \dot{x}_1 = x_2 \\ \dot{x}_2 = \frac{u - f x_2 - (z/2 + M) g l \sin(x_1)}{J} \\ y = x_2 \end{cases}$$

where $x_1 = \theta$, $x_2 = \dot{\theta}$, and $\ddot{\theta}$ denote the angular displacement, velocity, and acceleration of the manipulator, respectively; $y(t)$ denotes the system output, i.e., the angular velocity (m/s); $u(t)$ is the driving torque of the manipulator ($\mathrm{kg} \cdot \mathrm{m}^3/\mathrm{s}^2$); the sampling time is 0.1 s and the running time is 4 s; $z = 1\,\mathrm{kg}$, $M = 2\,\mathrm{kg}$, $l = 0.5\,\mathrm{m}$, $f = 3\,\mathrm{kg} \cdot \mathrm{m}^3/\mathrm{s}^2$, and $g = 9.8\,\mathrm{m/s}^2$ are the mass, tip load, length, damping coefficient, and the gravity acceleration of the manipulator, respectively.

Note that the mathematical model of the manipulator plant is not available in the process of the controller design but it only serves for producing I/O data to simulate a realistic plant.

The control task is to track a desired reference given as $y_d(t+1) = t^3 - 8t^2 + 16t$.

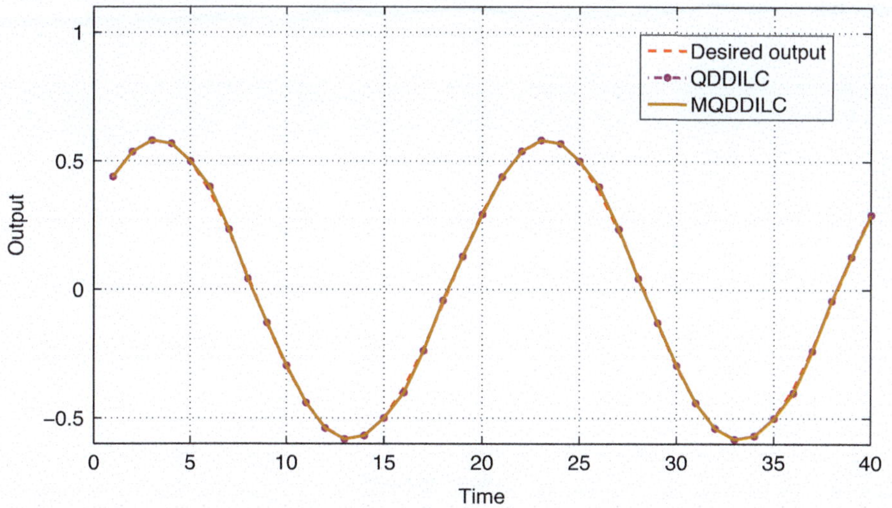


**Fig. 7.1** Profiles of system output at the 100th iteration

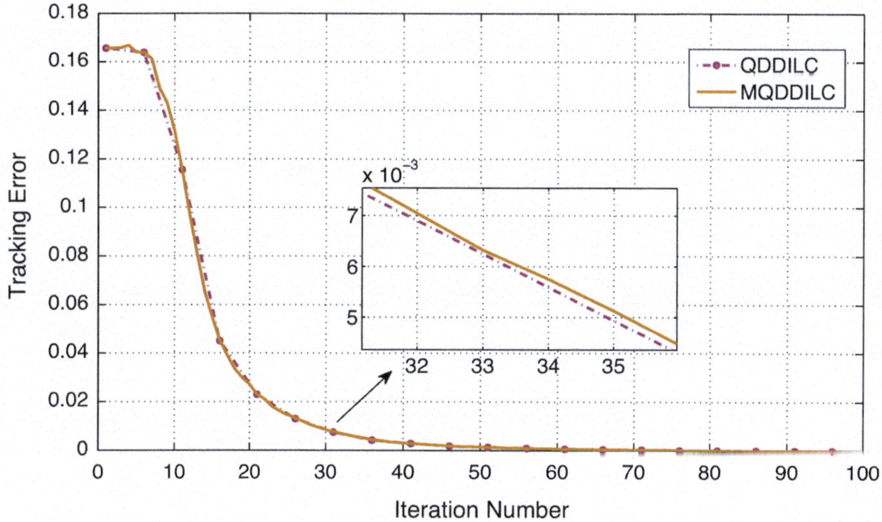


**Fig. 7.2** Profiles of tracking error along with iterations

Set controller parameters as $\eta = 0.01$, $\rho = 0.2$, $\mu = 0.01$, $\lambda = 0.003$, $\varepsilon = 10^{-}5$; the quantizer parameters are chosen as $r_e = r_\Delta = 0.9$ and $h_0 = 7$; the initial values are set as $\hat{\varphi}_0(t) = 0.006$, $\theta_k(0) = 0.6$, $\dot{\theta}_k(0) = 0$ and $u_0(t) = 0$. Applying the QDDILC scheme (7.9)–(7.11), the results are given in Figs. 7.4 and 7.5. It is observed that the QDDILC under data quantization realizes a good control performance.

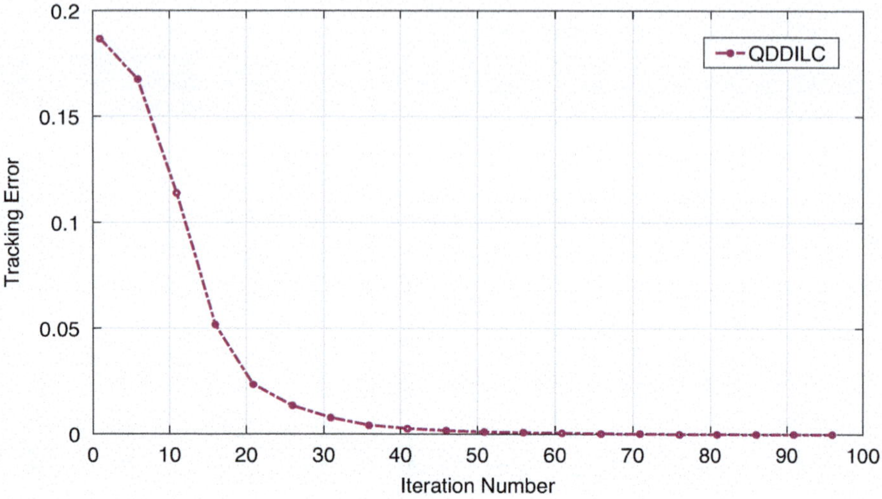


**Fig. 7.3** Profiles of tracking error with random disturbances

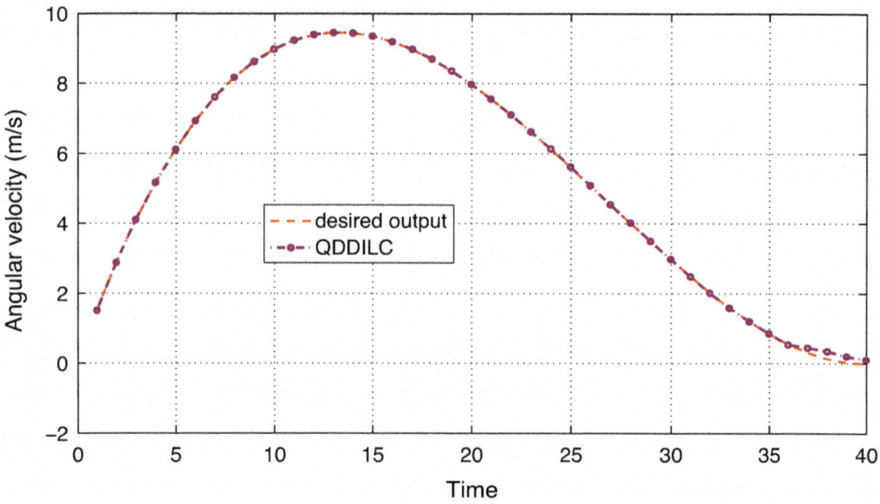


**Fig. 7.4** Profiles of angular velocity at the 100th iteration

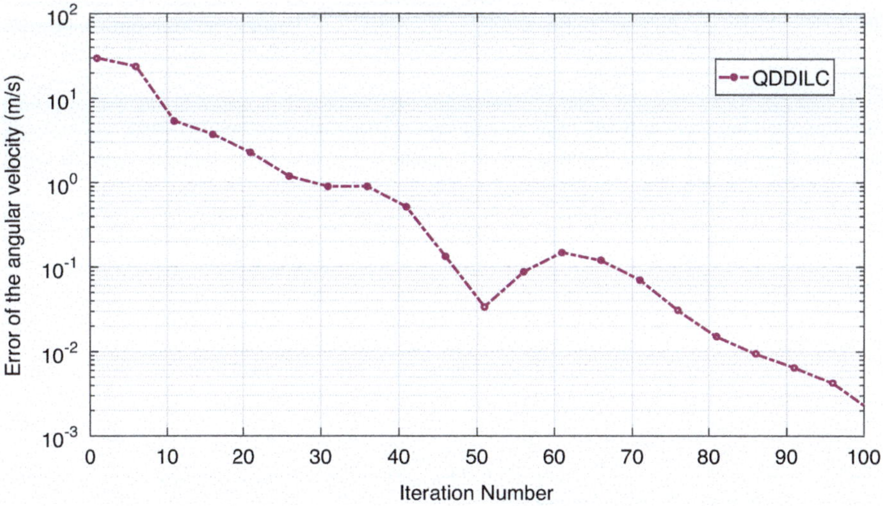


**Fig. 7.5** Tracking error profiles of angular velocity with iterations

## 7.6 Summary

Aiming for addressing the quantization problem in ILC, this chapter proposes a QDDILC method for nonlinear nonaffine systems. The nonlinear nonaffine system is transferred to a new LDM with much more parameters for the subsequent algorithm design and analysis under a data-driven framework. The proposed QDDAILC can utilize additional input information to enhance control performance. The parameter estimation method derived from the adaptive updating law makes the learning gain of the proposed method adjustable using the real-time measurements, which improves the robustness against uncertainties. Both theoretical analysis and simulation tests verify the effectiveness of the proposed method.

## References

Bu X, Wang T, Hou Z, Chi R (2015) Iterative learning control for discrete-time systems with quantized measurements. IET Control Theory A 9(9):1455–1460

Bu X, Hou Z, Cui L, Yang J (2017) Stability analysis of quantized iterative learning control systems using lifting representation. Int J Adapt Control Signal Process 31(9):1327–1336

Bu X, Hou Z, Yu Q, Yang Y (2018) Quantized data driven iterative learning control for a class of nonlinear systems with sensor saturation. IEEE Trans Syst Man Cybern B Syst 1(11):1473–1480

Chi R, Hou Z, Huang B, Jin S (2015) A unified data-driven design framework of optimality-based generalized iterative learning control. Comput Chem Eng 77:10–23

Chi R, Hou Z, Jin S, Huang B (2018) Computationally-light non-lifted data-driven norm-optimal iterative learning control. Asian J Control 20(1):115–124

Chi R, Hui Y, Huang B, Hou Z, Bu X (2022) Spatial linear dynamic relationship of strongly connected multiagent systems and adaptive learning control for different formations. IEEE Trans Cybern 52(1):531–543

Chi R, Hou Z, Jin S, Huang B (2018) Computationally efficient data-driven higher order optimal iterative learning control. IEEE Trans Neural Netw Learn Syst 29(12):5971–5980

Fu M, Xie L (2005) The sector bound approach to quantized feedback control. IEEE Trans Automat Contr 50(11):1698–1711

Gao M, Sheng L, Zhou D, Gao F (2017) Iterative learning fault-tolerant control for networked batch processes with multirate sampling and quantization effects. Ind Eng Chem Res 56(9):2515–2525

Hui Y, Chi R, Huang B, Hou Z (2020) 3-D learning-enhanced adaptive ILC for iteration-varying formation tasks. IEEE Trans Neural Netw Learn Syst 31(1):89–99

Huo N, Shen D (2019) Improving boundary level calculation in quantized iterative learning control with encoding and decoding mechanism. IEEE ACCESS 7:66623–66632

Huo N, Shen D (2020) Encoding-decoding mechanism-based finite-level quantized iterative learning control with random data dropouts. IEEE Trans Autom Sci Eng 17(3):1343–1360

Janssens P, Pipeleers G, Swevers J (2013) A data-driven constrained norm-optimal iterative learning control framework for LTI systems. IEEE Trans Control Syst Technol 21(2):546–551

Jin X (2016) Adaptive iterative learning control for high-order nonlinear multi-agent systems consensus tracking. Syst Control Lett 89:16–23

Jin X (2018) Iterative learning control for output-constrained nonlinear systems with input quantization and actuator faults. Int J Robust Nonlinear Control 28(2):729–741

Li J, Li J (2014) Adaptive iterative learning control for coordination of second-order multi-agent systems. Int J Robust Nonlinear Control 24(18):3282–3299

Shen D, Zhang C, Xu J (2019) Distributed learning consensus control based on neural networks for heterogeneous nonlinear multiagent systems. Int J Robust Nonlinear Control 29(13):4328–4347

Shen D, Xu J (2017) Zero-error tracking of iterative learning control using probabilistically quantized measurements. In: The 11th Asian control conference. IEEE. https://ieeexplore.ieee.org/document/8287312

Tan KK, Lee TH, Huang SN, Leu FM (2001) Adaptive-predictive control of a class of SISO nonlinear systems. Dyn Control 11(2):151–174

Xiong W, Yu X, Patel R, Yu W (2016) Iterative learning control for discrete-time systems with event-triggered transmission strategy and quantization. Automatica 72:84–91

Xu Y, Shen D, Bu X (2017) Zero-error convergence of iterative learning control using quantized error information. IMA J Mathema Control Inform 34(3):1061–1077

Yin Y, Bu X, Liang J (2018) Quantized iterative learning control design for linear systems based on a 2-D roesser model. Asian J Control 20(3):1058–1066

Zhang T, Li J (2017) Iterative learning control for multi-agent systems with finite-leveled sigma-delta quantization and random packet losses. IEEE Trans Circuits Syst I-Regul Pap 64(8):2171–2181

# Chapter 8
# Data-Driven ILC for Specified Point Tracking

## 8.1 Introduction

In real industries, the tracking tasks are normally classified into three control scenarios, i.e., tracking an entire reference trajectory, tracking specified multiple reference points, and tracking a single terminal point at the endpoint. The first scenario is most common in practical applications. The examples of the second scenario include robotic "pick and place" tasks, crane positioning, production line automation, and so on (Freeman and Tan 2012; Son et al. 2013; Dinh et al. 2012). The third scenario mainly focuses on the practical applications in process industry, such as rapid thermal processing (RTP) systems for chemical vapor deposition (Xu et al. 1999) and batch to batch processes (Gauthier and Boulet 2009; Chi et al. 2012; Flores-Cerrillo and MacGregor 2005). It is often that for the later two scenarios, the measurements at the other time instants than the specified ones are not available. So, the traditional ILC methods (Arimoto et al. 1984; Sun and Wang 2001; Chi et al. 2008; Chien 2008; Barton and Alleyne 2010) are not applicable for the specified points tracking tasks since they require the exact errors at all of the points of a desired reference trajectory. Therefore, some point-to-point ILC (PTPILC) (Freeman and Tan 2012; Son et al. 2013; Owens et al. 2012; Freeman 2012) and terminal ILC (TILC) (Xu et al. 1999; Gauthier and Boulet 2009; Chi et al. 2012; Flores-Cerrillo and MacGregor 2005, Mondal et al. 2010; Xu and Huang 2008; Hou et al. 2011; Chi et al. 2016) methods are proposed to track the intermediate pass points rather than all the points of a trajectory by using the errors at the given points only. Note that the TILC (Xu et al. 1999; Gauthier and Boulet 2009; Chi et al. 2012; Flores-Cerrillo and MacGregor 2005; Mondal et al. 2010; Xu and Huang 2008; Hou et al. 2011; Chi et al. 2016) only uses the information at the terminal point to update the learning law.

On the other hand, if all of the system states and outputs are measurable, the standard ILCs could be applied to the second and the third control scenarios by employing a designed reference passing through all the desired points. However, it

 
R. Chi et al., *Data-Driven Iterative Learning Control for Discrete-Time Systems*, Intelligent Control and Learning Systems 2,
https://doi.org/10.1007/978-981-19-5950-9_8

is difficult to select an optimal reference trajectory passing the specified points for the second control scenario and the optimal reference may no longer be the optimal one if a change occurs in the controlled plant. Another drawback of using standard ILCs in the second and the third scenarios is that it fails to utilize the extra freedom available to secure additional performance demands. If removing the unnecessary constraints that the plant has to follow, some additional control performance such as the reduced control effort and faster convergent speed could be expected. In addition, the size of memory and computation time can be reduced greatly by choosing the necessary control index only at certain time instants.

Therefore, the PTPILC and TILC approaches that only track the specified intermediate points are not a simple extension of the traditional ILC, but a new attractive problem with many challenges in theory. However, it is generally unable to add feedback information at the previous time instants in the current iteration in the PTPILC and TILC methods to enhance control performance because the system output/state is unmeasurable at the intermediate points other than the designated points of interest and thus only the errors at the designated points are incorporated into the control mechanism.

This chapter explores the ILC design and analysis for specified points tracking tasks. Since the TILC can be regarded as a special case of PTPILC tracking only one terminal point, this chapter only discusses the design and analysis problem of PTPILC under a data-driven framework. Consequently, a data-driven PTPILC (DDPTPILC) method is proposed in this chapter consisting of a control input iterative learning law and a gradient parameter updating law. The presented method is data-driven without requiring any model information of the systems. Further, an improved DDPTPILC method is also proposed for the multiple point-to-point tracking task by using additional input knowledge to enhance the control performance. It should be noted that the input signals play an important role in improving control performance since they are obtained by calculating the tracking errors in previous time instants and thus reflect the system real-time dynamics. Meanwhile, the proposed DDPTPILC methods for specified points tracking do not need to track the unnecessary output points and thus the control performance can be improved by removing redundant index constraints. The simulation study shows the effectiveness of the proposed approaches.

This chapter is organized as follows. Section 8.2 is the problem formulation with a point-to-point tracking task. The DDPTPILC design and analysis are presented in Sects. 8.3 and 8.4, respectively. Section 8.5 provides a simulation study to show the effectiveness of the proposed DDPTPILC. Further, Sect. 8.6 proposes an improved DDPTPILC method by using continuous input information, where the design procedure, convergence analysis, and simulation results are all provided in detail. Section 8.7 provides some conclusions.

## 8.2 Problem Formulation

Consider a nonlinear nonaffine SISO system

$$\begin{aligned} y_k(t+1) = f(&y_k(t), y_k(t-1), \ldots, y_k(t-n_y), \\ &u_k(t), u_k(t-1), \ldots, u_k(t-n_u)) \end{aligned} \tag{8.1}$$

where $y_k(t)$ and $u_k(t)$ are the system output and input, respectively; $y_k(t) = 0$ and $u_k(t) = 0$, $\forall t < 0$; $n_y$ and $n_u$ are two unknown orders; $t \in \{0, 1, \ldots, N\}$ is time instant and $N \in Z^+$ represents the endpoint; $k \in \{0, 1, \ldots\}$ indicates the iteration number; $f(\cdots) : R^{n_u+n_y+2} \mapsto R$ is an unknown nonlinear scalar function; $f(0, 0) = 0$.

Assume there are certain desired output points $y_d(t_1), y_d(t_2), \ldots, y_d(t_M)$ at time instants $\{t_1, t_2, \ldots, t_M\}$ in the finite time interval $\{0, 1, \ldots, N\}$ during the system operation, where $M$ is the number of intermediate desired point, $N$ is the endpoint of the finite time interval, and $t_M \leq N$. The control task is to construct a learning law that drives the outputs through, or at least close to, these points.

Consider nonlinear system (8.1) satisfying Assumptions 4.1, 4.2, and 4.3.

Define $\bar{\boldsymbol{Y}}_k = [y_k(t_1), y_k(t_2), \ldots, y_k(t_M)]^{\mathrm{T}} \in R^M$. Following the similar steps in Theorem 4.1, the nonlinear system (8.1) can be transformed into the following STIDL-based multi-point LDM

$$\begin{aligned} \Delta\bar{\boldsymbol{Y}}_k &= \begin{bmatrix} \frac{\partial g_{t_1-1}}{\partial u_k(0)} & \frac{\partial g_{t_1-1}}{\partial u_k(1)} & \cdots & \frac{\partial g_{t_1-1}}{\partial u_k(t_1-1)} & 0 & 0 & \cdots & 0 & 0 \cdots 0 \\ \frac{\partial g_{t_2-1}}{\partial u_k(0)} & \frac{\partial g_{t_2-1}}{\partial u_k(1)} & \cdots & \frac{\partial g_{t_2-1}}{\partial u_k(t_1-1)} & \cdots & \frac{\partial g_{t_2-1}}{\partial u_k(t_2-1)} & \cdots & 0 & 0 \cdots 0 \\ \vdots & \vdots & \ddots & \vdots & & \vdots & \ddots & \vdots & \vdots \ddots \vdots \\ \frac{\partial g_{t_M-1}}{\partial u_k(0)} & \frac{\partial g_{t_M-1}}{\partial u_k(1)} & \cdots & \frac{\partial g_{t_M-1}}{\partial u_k(t_1-1)} & \cdots & \frac{\partial g_{t_M-1}}{\partial u_k(t_2-1)} & \cdots & \frac{\partial g_{t_M-1}}{\partial u_k(t_M-1)} & 0 \cdots 0 \end{bmatrix} \Delta\boldsymbol{U}_k \\ &= \bar{\boldsymbol{\Phi}}_k \Delta\boldsymbol{U}_k \end{aligned} \tag{8.2}$$

where

$$\bar{\boldsymbol{\Phi}}_k = \begin{bmatrix} \frac{\partial g_{t_1-1}}{\partial u_k(0)} & \frac{\partial g_{t_1-1}}{\partial u_k(1)} & \cdots & \frac{\partial g_{t_1-1}}{\partial u_k(t_1-1)} & 0 & 0 & \cdots & 0 & 0 \cdots 0 \\ \frac{\partial g_{t_2-1}}{\partial u_k(0)} & \frac{\partial g_{t_2-1}}{\partial u_k(1)} & \cdots & \frac{\partial g_{t_2-1}}{\partial u_k(t_1-1)} & \cdots & \frac{\partial g_{t_2-1}}{\partial u_k(t_2-1)} & \cdots & 0 & 0 \cdots 0 \\ \vdots & \vdots & \ddots & \vdots & & \vdots & \ddots & \vdots & \vdots \ddots \vdots \\ \frac{\partial g_{t_M-1}}{\partial u_k(0)} & \frac{\partial g_{t_M-1}}{\partial u_k(1)} & \cdots & \frac{\partial g_{t_M-1}}{\partial u_k(t_1-1)} & \cdots & \frac{\partial g_{t_M-1}}{\partial u_k(t_2-1)} & \cdots & \frac{\partial g_{t_M-1}}{\partial u_k(t_M-1)} & 0 \cdots 0 \end{bmatrix} \in R^{M \times N},$$

$\boldsymbol{U}_k = [u_k(0), u_k(1), \ldots, u_k(N-1)]^{\mathrm{T}} \in R^N$, $\Delta\bar{\boldsymbol{Y}}_k = \bar{\boldsymbol{Y}}_k - \bar{\boldsymbol{Y}}_{k-1}$, and $\Delta\boldsymbol{U}_k = \boldsymbol{U}_k - \boldsymbol{U}_{k-1}$.

## 8.3 Controller Design

Let $\bar{\boldsymbol{Y}}_d = [y_d(t_1), y_d(t_2), \ldots, y_d(t_M)]^{\mathrm{T}} \in R^M$ and define a tracking error of multiple pass points as $\bar{\boldsymbol{E}}_k = \bar{\boldsymbol{Y}}_d - \bar{\boldsymbol{Y}}_k$. Consider an objective function as follows:

$$J(\bar{\boldsymbol{U}}_k) = \|\bar{\boldsymbol{E}}_k\|^2 + \lambda\|\boldsymbol{U}_k - \boldsymbol{U}_{k-1}\|^2 \tag{8.3}$$

where $\lambda > 0$ is a weighting factor.

Following the similar steps that lead to (4.15), one has

$$\boldsymbol{U}_k = \boldsymbol{U}_{k-1} + \frac{\rho \bar{\boldsymbol{\Phi}}_k^{\mathrm{T}} \bar{\boldsymbol{E}}_{k-1}}{\lambda + \|\bar{\boldsymbol{\Phi}}_k\|^2} \tag{8.4}$$

where $\rho > 0$ is a step-size factor.

Since $\bar{\boldsymbol{\Phi}}_k$ in (8.4) is unknown, a parameter estimation algorithm is required. According to the prior knowledge about $\bar{\boldsymbol{\Phi}}_k$, one only needs to estimate its nonzero elements.

Let $\bar{\boldsymbol{\varphi}}_k(t_m - 1) = \left[\frac{\partial g_{t_m-1}}{\partial u_k(0)}, \frac{\partial g_{t_m-1}}{\partial u_k(1)}, \ldots, \frac{\partial g_{t_m-1}}{\partial u_k(t_m-1)}\right]$ be the $m$th row nonzero entry of matrix $\bar{\boldsymbol{\Phi}}_k$, and $\bar{\boldsymbol{u}}_k(t_m - 1) = [u_k(0), u_k(1), \ldots, u_k(t_m - 1)]^{\mathrm{T}} \in R^{t_m}, m = 1, \ldots, M$. Then

$$\begin{aligned}
\Delta y_k(t_m) &= \frac{\partial g_{t_m-1}}{\partial u_k(0)}(u_k(0) - u_{k-1}(0)) + \frac{\partial g_{t_m-1}}{\partial u_k(1)}(u_k(1) - u_{k-1}(1)) \\
&\quad + \cdots + \frac{\partial g_{t_m-1}}{\partial u_k(t_m - 1)}(u_k(t_m - 1) - u_{k-1}(t_m - 1)) \\
&= \bar{\boldsymbol{\varphi}}_k(t_m - 1)\Delta\bar{\boldsymbol{u}}_k(t_m - 1)
\end{aligned} \tag{8.5}$$

where $\Delta\bar{\boldsymbol{u}}_k(t_m - 1) = \bar{\boldsymbol{u}}_k(t_m - 1) - \bar{\boldsymbol{u}}_{k-1}(t_m - 1)$.

Define an objective function

$$\begin{aligned}
J\left(\hat{\bar{\boldsymbol{\varphi}}}_k(t_m - 1)\right) &= \left\|\Delta y_{k-1}(t_m) - \hat{\bar{\boldsymbol{\varphi}}}_k(t_m - 1)\Delta\bar{\boldsymbol{u}}_{k-1}(t_m - 1)\right\|^2 \\
&\quad + \mu\left\|\hat{\bar{\boldsymbol{\varphi}}}_k(t_m - 1) - \hat{\bar{\boldsymbol{\varphi}}}_{k-1}(t_m - 1)\right\|^2
\end{aligned} \tag{8.6}$$

where $\mu > 0$ is a weighting factor, and $\hat{\bar{\boldsymbol{\varphi}}}_k(t_m - 1)$ is the estimation of $\bar{\boldsymbol{\varphi}}_k(t_m - 1)$.

Minimizing (8.6) with respect to $\hat{\bar{\boldsymbol{\varphi}}}_k(t_m - 1)$ gives the following estimation algorithm,

$$\begin{aligned}\hat{\bar{\boldsymbol{\varphi}}}_k(t_m-1) &= \hat{\bar{\boldsymbol{\varphi}}}_{k-1}(t_m-1)\\ &+ \frac{\eta\left(\Delta y_{k-1}(t_m) - \hat{\bar{\boldsymbol{\varphi}}}_{k-1}(t_m-1)\Delta\bar{\boldsymbol{u}}_{k-1}(t_m-1)\right)\Delta\bar{\boldsymbol{u}}_{k-1}^{\mathrm{T}}(t_m-1)}{\mu + \|\Delta\bar{\boldsymbol{u}}_{k-1}(t_m-1)\|^2},\\ m &= 1,\ldots,M \end{aligned} \tag{8.7}$$

where $0 < \eta < 2$ is a step-size factor.

By combining the control algorithm (8.4) and the parameter estimation algorithm (8.7), the proposed data-driven PTPILC (DDPTPILC) scheme is constructed as

$$\boldsymbol{U}_k = \boldsymbol{U}_{k-1} + \frac{\rho\hat{\bar{\boldsymbol{\Phi}}}_k^{\mathrm{T}}\bar{\boldsymbol{E}}_{k-1}}{\lambda + \left\|\hat{\bar{\boldsymbol{\Phi}}}_k\right\|^2} \tag{8.8}$$

$$\begin{aligned}\hat{\bar{\boldsymbol{\varphi}}}_k(t_m-1) &= \hat{\bar{\boldsymbol{\varphi}}}_{k-1}(t_m-1)\\ &+ \frac{\eta\left(\Delta y_{k-1}(t_m) - \hat{\bar{\boldsymbol{\varphi}}}_{k-1}(t_m-1)\Delta\bar{\boldsymbol{u}}_{k-1}(t_m-1)\right)\Delta\bar{\boldsymbol{u}}_{k-1}^{\mathrm{T}}(t_m-1)}{\mu + \|\Delta\bar{\boldsymbol{u}}_{k-1}(t_m-1)\|^2},\\ m &= 1,\ldots,M \end{aligned} \tag{8.9}$$

$$\begin{aligned}&\hat{\bar{\boldsymbol{\varphi}}}_k(t_m-1) = \hat{\bar{\boldsymbol{\varphi}}}_0(t_m-1), \quad \text{if } \operatorname{sgn}\left(\hat{\bar{\phi}}_k(i)\right) \neq \operatorname{sgn}\left(\hat{\bar{\phi}}_0(i)\right) \text{ or } \left\|\hat{\bar{\boldsymbol{\varphi}}}_k(t_m-1)\right\| \leq \varepsilon\\ &i \in \{0,1,\ldots,t_m\}, \quad m \in \{1,\ldots,M\}\end{aligned} \tag{8.10}$$

where $\varepsilon$ is a small positive constant; $\hat{\bar{\boldsymbol{\Phi}}}_k$ is the estimation of $\bar{\boldsymbol{\Phi}}_k$; $\hat{\bar{\boldsymbol{\varphi}}}_0(t_m-1)$ is the initial vector value of $\hat{\bar{\boldsymbol{\varphi}}}_k(t_m-1)$; $\hat{\bar{\phi}}_k(i)$ is the $(i+1)$th element of vector $\hat{\bar{\boldsymbol{\varphi}}}_k(t_m-1)$.

**Remark 8.1** Compared with the proposed DDILC (4.18)–(4.20), the proposed DDPTPILC (8.8)–(8.10) only utilizes the error information at the specified points, thus it can be applied to the scenario where only the system outputs at the given time instants are measured, and the control performance could be improved with a faster convergence speed, lower control energy, and less memory size by removing the unnecessary constraint on the tracking.

## 8.4 Convergence Analysis

The properties of the presented DDPTPILC (8.8)–(8.10) are given by the following theorem.

**Theorem 8.1** *For a class of nonlinear nonaffine discrete-time systems (8.1), if Assumptions 4.1, 4.2, and 4.3 hold, then the proposed DDPTPILC (8.8)–(8.10) guarantees that*

*(i)* *the estimation of gradient matrix* $\hat{\bar{\boldsymbol{\Phi}}}_k$ *is bounded,* $\forall k$;
*(ii)* *the tracking error* $\bar{\boldsymbol{E}}_k$ *converges to zero iteratively, i.e.,* $\lim_{k\to\infty} \bar{\boldsymbol{E}}_k = 0$;
*(iii)* *the system output* $y_k(t)$ *and the control input signal* $u_k(t)$ *are bounded,* $\forall t \in \{0, 1, \ldots, N\}$, $\forall k$.

***Proof*** *Part 1. The boundedness of* $\hat{\bar{\boldsymbol{\Phi}}}_k$

Define $\tilde{\bar{\boldsymbol{\varphi}}}_k(t_m-1) = \bar{\boldsymbol{\varphi}}_k(t_m-1) - \hat{\bar{\boldsymbol{\varphi}}}_k(t_m-1)$ as the parameter estimation error. Subtracting $\bar{\boldsymbol{\varphi}}_k(t_m-1)$ from both sides of (8.9) and using (8.5), we have

$$\begin{aligned}\tilde{\bar{\boldsymbol{\varphi}}}_k(t_m-1) = {}& \tilde{\bar{\boldsymbol{\varphi}}}_{k-1}(t_m-1)\left(I - \frac{\eta \Delta\bar{\boldsymbol{u}}_{k-1}(t_m-1)\,\Delta\bar{\boldsymbol{u}}_{k-1}^{\mathrm{T}}(t_m-1)}{\mu + \|\Delta\bar{\boldsymbol{u}}_{k-1}(t_m-1)\|^2}\right)\\ &+ \bar{\boldsymbol{\varphi}}_k(t_m-1) - \bar{\boldsymbol{\varphi}}_{k-1}(t_m-1).\end{aligned} \tag{8.11}$$

According to Assumption 4.2, $\|\bar{\boldsymbol{\varphi}}_k(t_m-1)\| \le b_{\bar{\varphi}}$, which leads to $\|\bar{\boldsymbol{\varphi}}_k(t_m-1) - \bar{\boldsymbol{\varphi}}_{k-1}(t_m-1)\| \le 2b_{\bar{\varphi}}$. Following the similar steps that lead to (4.25), one can obtain that

$$\begin{aligned}\left\|\tilde{\bar{\boldsymbol{\varphi}}}_k(t_m-1)\right\| &\le d_1\left\|\tilde{\bar{\boldsymbol{\varphi}}}_{k-1}(t_m-1)\right\| + 2b_{\bar{\varphi}}\\ &\le \cdots \le d_1^k\left\|\tilde{\bar{\boldsymbol{\varphi}}}_0(t_m-1)\right\| + \frac{2b_{\bar{\varphi}}}{1-d_1}\end{aligned} \tag{8.12}$$

where $0 < d_1 < 1$ is a constant.

Hence, $\tilde{\bar{\boldsymbol{\varphi}}}_k(t_m-1)$ is bounded because $\hat{\bar{\boldsymbol{\varphi}}}_0(t_m-1)$ is chosen bounded and $\bar{\boldsymbol{\varphi}}_k(t_m-1)$ is also bounded. Therefore, it is apparent that $\hat{\bar{\boldsymbol{\varphi}}}_k(t_m-1)$ is bounded too. Consequently, conclusion (i) of Theorem 8.1 is obtained directly, i.e., $\hat{\bar{\boldsymbol{\Phi}}}_k$ is bounded.

*Part 2. The convergence of tracking error* $\bar{\boldsymbol{E}}_k$

From Eq. (8.2) and control law (8.8), one has

$$\begin{aligned}\bar{\boldsymbol{E}}_k &= \bar{\boldsymbol{Y}}_d - \bar{\boldsymbol{Y}}_{k-1} - \bar{\boldsymbol{\Phi}}_k \Delta\boldsymbol{U}_k\\ &= \bar{\boldsymbol{E}}_{k-1} - \frac{\rho\bar{\boldsymbol{\Phi}}_k\hat{\bar{\boldsymbol{\Phi}}}_k^{\mathrm{T}}\bar{\boldsymbol{E}}_{k-1}}{\lambda + \|\hat{\bar{\boldsymbol{\Phi}}}_k\|^2}\\ &= \left(I - \frac{\rho\bar{\boldsymbol{\Phi}}_k\hat{\bar{\boldsymbol{\Phi}}}_k^{\mathrm{T}}}{\lambda + \|\hat{\bar{\boldsymbol{\Phi}}}_k\|^2}\right)\bar{\boldsymbol{E}}_{k-1}.\end{aligned} \tag{8.13}$$

We have shown that both of $\bar{\boldsymbol{\Phi}}_k$ and $\hat{\bar{\boldsymbol{\Phi}}}_k$ are bounded, and $\bar{\boldsymbol{\Phi}}_k\hat{\bar{\boldsymbol{\Phi}}}_k^{\mathrm{T}}$ is positive definite according to the reset algorithm (8.10) and Assumption 4.3. So by properly setting controller parameters $\rho$ and $\lambda$, one can guarantee that there exists a positive constant $0 < d_2 < 1$ such that

$$0 < \left\| \boldsymbol{I} - \frac{\rho \bar{\boldsymbol{\Phi}}_k \hat{\bar{\boldsymbol{\Phi}}}_k^{\mathrm{T}}}{\lambda + \left\| \hat{\bar{\boldsymbol{\Phi}}}_k \right\|^2} \right\| \leq d_2 < 1. \tag{8.14}$$

And thus, in terms of (8.13), one has

$$\left\| \bar{\boldsymbol{E}}_k \right\| \leq d_2 \left\| \bar{\boldsymbol{E}}_{k-1} \right\| \leq \cdots \leq d_2^k \left\| \bar{\boldsymbol{E}}_0 \right\| \tag{8.15}$$

which implies $\lim_{k\to\infty} \bar{\boldsymbol{E}}_k = 0$. As a result, conclusion (ii) of Theorem 8.1 is obtained.

*Part 3. The boundedness of system output and control input*

By using the similar steps leading to (4.31), one obtains that

$$\|\boldsymbol{U}_k\| \leq \|\boldsymbol{U}_0\| + \frac{\rho}{(1-d_2)2\sqrt{\lambda}} \|\bar{\boldsymbol{E}}_0\|. \tag{8.16}$$

Since the initial tracking error $\bar{\boldsymbol{E}}_0$ and initial control input $\boldsymbol{U}_0$ are given bounded, the boundedness of $\boldsymbol{U}_k$ is evident.

In the following, the boundedness of system output at every operation point is analyzed.

For nonlinear system (8.1) satisfying Assumptions 4.1, 4.2, and 4.3, we also have (4.7). Since $\boldsymbol{Y}_k = \boldsymbol{Y}_k - \boldsymbol{Y}_{k-1} + \boldsymbol{Y}_{k-1} - \cdots + \boldsymbol{Y}_1 - \boldsymbol{Y}_0 + \boldsymbol{Y}_0 = \boldsymbol{Y}_0 + \sum_{j=1}^{k} \Delta \boldsymbol{Y}_j$, then according to (4.7) and (8.8), we have

$$\|\boldsymbol{Y}_k\| \leq \|\boldsymbol{Y}_0\| + \sum_{j=1}^{k} \left\| \frac{\rho \boldsymbol{\Phi}_j \hat{\bar{\boldsymbol{\Phi}}}_j^{\mathrm{T}} \bar{\boldsymbol{E}}_{j-1}}{\lambda + \left\| \hat{\bar{\boldsymbol{\Phi}}}_j \right\|^2} \right\|. \tag{8.17}$$

According to Assumption 4.2, we know that $\sup_k \|\boldsymbol{\Phi}_k\| \leq b_{\Phi}$ is bounded, so we can obtain that

$$\left\| \frac{\rho \boldsymbol{\Phi}_j \hat{\bar{\boldsymbol{\Phi}}}_j^{\mathrm{T}}}{\lambda + \left\| \hat{\bar{\boldsymbol{\Phi}}}_j \right\|^2} \right\| \leq \frac{\rho \left\| \boldsymbol{\Phi}_j \right\| \left\| \hat{\bar{\boldsymbol{\Phi}}}_j \right\|}{2\sqrt{\lambda} \left\| \hat{\bar{\boldsymbol{\Phi}}}_j \right\|} \leq \frac{\rho b_{\Phi}}{2\sqrt{\lambda}}. \tag{8.18}$$

In view of (8.14), (8.17) and (8.18), one can obtain that

$$\|\boldsymbol{Y}_k\| \leq \|\boldsymbol{Y}_0\| + \frac{1}{(1-d_2)} \frac{\rho b_{\Phi}}{2\sqrt{\lambda}} \left\| \bar{\boldsymbol{E}}_0 \right\|. \tag{8.19}$$

Thus the system output signal $\boldsymbol{Y}_k$ is bounded with the given bounded $\boldsymbol{Y}_0$ and $\bar{\boldsymbol{E}}_0$. The conclusion (iii) of Theorem 8.1 is obtained. □

## 8.5 Simulation Study

Consider a nonlinear discrete-time system,

$$y_k(t+1) = \frac{y_k(t)}{1 + y_k(t)^3} + u_k(t)^3 \tag{8.20}$$

where $t \in \{0, 1, \ldots, 20\}$ is the time instant; $k \in \{0, 1, \ldots\}$ is the iteration number.

The desired trajectory is given as follows:

$$y_d(t) = 0.5\sin(t\pi/10) + 0.3\cos(t\pi/10), \;\; t \in \{0, 1, \ldots, 20\}. \tag{8.21}$$

The control task is to drive the system output through the given desired points $y_d(4)$, $y_d(9)$, $y_d(14)$, $y_d(20)$ of the reference trajectory (8.21) at time instants $\{4, 9, 14, 20\}$, instead of the entire desired trajectory.

Set $\rho = 2$, $\eta = 1$, $\lambda = 1$, $\mu = 1$, $y_k(0) = 0$, $u_0(t) = 0$, $\hat{\boldsymbol{\varphi}}_0 = 0.5$. Applying the proposed DDPTPILC (8.8)–(8.10), the results are shown in Figs. 8.1 and 8.2. Figure 8.1 shows the output tracking performance at the 100th iteration, where the vertical axis denotes the system output and the horizontal axis denotes the time instant. Figure 8.2 shows the convergence performance of tracking error, where the Y-axis denotes the maximum absolute value of the output tracking error, defined as $e_{\max}(k) = \max_{t \in \{4,9,14,20\}}\{|e_k(t)|\}$, and the X-axis denotes the iteration numbers. One can see from Fig. 8.1 that system output can track the desired points accurately at the 100th iteration. It is seen from Fig. 8.2 that the maximum output tracking error asymptotically converges along iterative direction.

## 8.6 DDPTPILC Using Continuous Input Information

### 8.6.1 Problem Formulation

Consider nonlinear system (8.1) satisfying Assumptions 4.1, 4.2 and 4.3.

The control objective is to drive the system output through the specified multiple intermediate desired points, $y_d(t_1), y_d(t_2), \cdots, y_d(t_M)$ at time instants $\{t_1, t_2, \ldots, t_M\}$, $t_M \leq N$.

Similar to (4.4), the system output at the specified points $\{t_1, t_2, \ldots, t_M\}$ can be expressed as

$$y_k(t_i) = g_{t_i-1}\left(y_k(0), u_k(0), \ldots, u_k(t_i - 1)\right), i = 1, \ldots, M \tag{8.22}$$

where $g_{t_i-1}(\cdot)$ is a proper continuously differentiable nonlinear function.

Let $\boldsymbol{u}_k(t_i - 1) = [u_k(0)\ u_k(1)\ \cdots\ u_k(t_i - 1)]^{\mathrm{T}}$. We can rewrite (8.22) as

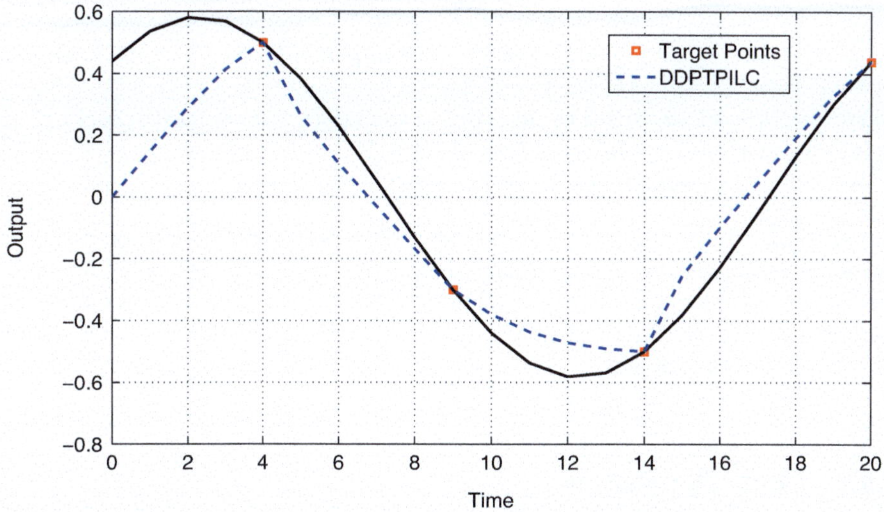


Fig. 8.1 System output

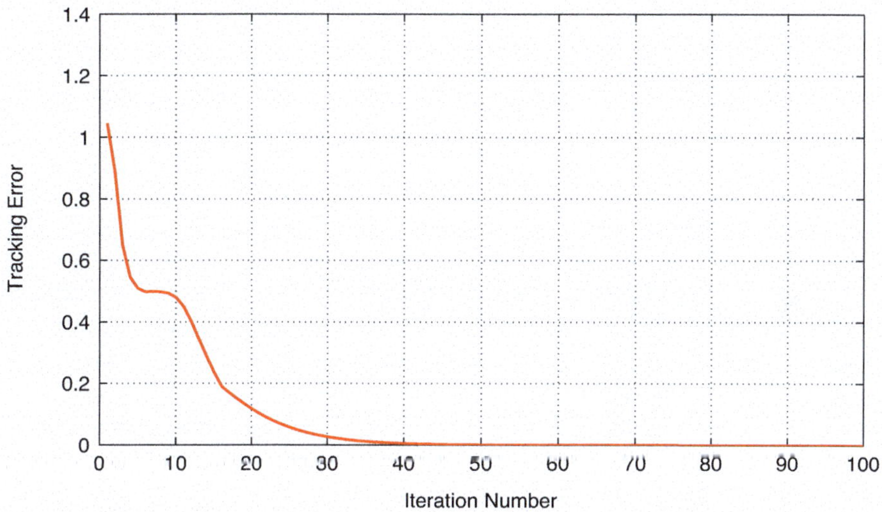


Fig. 8.2 Convergence of tracking error

$$y_k(t_i) = g_{t_i-1}\left(y_k(0), \boldsymbol{u}_k^{\mathrm{T}}(t_i - 1)\right), i = 1, \ldots, M. \tag{8.23}$$

Then, similar to (8.5), a nonlifted STIDL-based multi-point LDM can be obtained as follows:

$$\begin{aligned}\Delta y_k(t_i) &= y_k(t_i) - y_{k-1}(t_i) \\ &= \boldsymbol{\varphi}_k(t_i - 1)\,\Delta \boldsymbol{u}_k(t_i - 1)\end{aligned} \tag{8.24}$$

where $\Delta \boldsymbol{u}_k(t_i - 1) = \boldsymbol{u}_k(t_i - 1) - \boldsymbol{u}_{k-1}(t_i - 1)$ and $\boldsymbol{\varphi}_k(t_i - 1) = [\phi_k(0), \phi_k(1), \ldots, \phi_k(t_i - 1)] = [(\partial g_{t_i-1})/(\partial \boldsymbol{u}_k(t_i - 1))]^{\mathrm{T}} \in R^{1\times t_i}$.

According to Assumption 4.2, one has $\|\boldsymbol{\varphi}_k(t_i - 1)\| \le b_\varphi$.

**Remark 8.2** It is seen from (8.24) that the system output at the specified points is linear with respect to the control input, where $\boldsymbol{\varphi}_k(t_i - 1)$ can be estimated by using the I/O data of the nonlinear process. Further, (8.24) is an accurate formulation of the nonlinear system (8.23), which is completely equivalent to the nonlinear system (8.1) at the specified points.

## 8.6.2 Controller Design

Since $\boldsymbol{\varphi}_k(t_i - 1)$ in (8.24) is unknown and it is slowly time-iteration-varying, a modified project algorithm is utilized for its iterative estimation

$$\begin{aligned}\hat{\boldsymbol{\varphi}}_k(t_i - 1) &= \hat{\boldsymbol{\varphi}}_{k-1}(t_i - 1) \\ &+ \frac{\eta\left(\Delta y_{k-1}(t_i) - \hat{\boldsymbol{\varphi}}_{k-1}(t_i - 1)\,\Delta \boldsymbol{u}_{k-1}(t_i - 1)\right)\Delta \boldsymbol{u}_{k-1}^{\mathrm{T}}(t_i - 1)}{\mu + \|\Delta \boldsymbol{u}_{k-1}(t_i - 1)\|^2}\end{aligned} \tag{8.25}$$

$$\begin{aligned}&\hat{\boldsymbol{\varphi}}_k(t_i - 1) = \hat{\boldsymbol{\varphi}}_0(t_i - 1), \\ &\text{if } \operatorname{sgn}\left(\hat{\phi}_k(\tau)\right) \ne \operatorname{sgn}(\phi_0(\tau)) \text{ or } \left\|\hat{\boldsymbol{\varphi}}_k(t_i - 1)\right\| \le \varepsilon\end{aligned} \tag{8.26}$$

where $\tau \in \{0, 1, \ldots, t_i - 1\}$, $\hat{\boldsymbol{\varphi}}_k(t_i - 1)$ is the estimation of $\boldsymbol{\varphi}_k(t_i - 1)$, $\mu > 0$ denotes a weighting factor, $0 < \eta < 2$ is a step-size factor, and $\varepsilon$ is a small positive scalar. The initial value of $\hat{\boldsymbol{\varphi}}_0(t_i - 1)$ is selected such that all its elements have the same signs as that of $\boldsymbol{\varphi}_k(t_i - 1)$ whose elements' signs can be identified by using the I/O data produced from the deterministic control plant, where the exogenous noises are neglected.

To utilize more control information in the following controller design, the control input series $\{u(0), u(1), \ldots, u(N - 1)\}$ over the finite time interval is divided into multiple subintervals according to the designated time instants, i.e., $\{u(0), u(1), \ldots, u(t_1 - 1)\}$, $\{u(t_1), u(t_1 + 1), \ldots, u(t_2 - 1)\}$, $\cdots$, $\{u(t_{M-1}), u(t_{M-1} + 1), \ldots, u(t_M - 1)\}$. Then, a time-subinterval formulation of the system output is presented

$$\boldsymbol{u}_k(t_i - t_{i-1}) = \begin{bmatrix} u_k(t_{i-1}) & u_k(t_{i-1} + 1) & \cdots & u_k(t_i - 1)\end{bmatrix}^{\mathrm{T}} \in R^{t_i - t_{i-1}}$$

and

$$\boldsymbol{\varphi}_k(t_i - t_{i-1}) = \left[\phi_k(t_{i-1}) \quad \phi_k(t_{i-1}+1) \quad \ldots \quad \phi_k(t_i - 1)\ \right] \in R^{1\times(t_i - t_{i-1})}$$

at the time subinterval $\{t_{i-1}, t_{i-1}+1, \ldots, t_i - 1\}$, whose horizon is varying with different $i \in \{1, \ldots, M\}$.

Assume $t_0 = 1$. Then output difference at the specified time instant $t_i$ between two consecutive iterations can be expressed with varying receding-horizon as

$$\begin{aligned}\Delta y_k(t_i) &= \boldsymbol{\varphi}_k(t_i - 1)\,\Delta\boldsymbol{u}_k(t_i - 1)\\ &= \boldsymbol{\varphi}_k(t_1 - 1)\,\Delta\boldsymbol{u}_k(t_1 - 1) + \boldsymbol{\varphi}_k(t_2 - t_1)\,\Delta\boldsymbol{u}_k(t_2 - t_1)\\ &+ \cdots + \boldsymbol{\varphi}_k(t_i - t_{i-1})\,\Delta\boldsymbol{u}_k(t_i - t_{i-1})\\ &= \sum_{j=1}^{i} \boldsymbol{\varphi}_k\left(t_j - t_{j-1}\right)\boldsymbol{u}_k\left(t_j - t_{j-1}\right).\end{aligned} \tag{8.27}$$

Consider an objective function as follows:

$$J\left(\boldsymbol{u}_k(t_i - t_{i-1})\right) = |e_k(t_i)|^2 + \lambda\,\|\boldsymbol{u}_k(t_i - t_{i-1}) - \boldsymbol{u}_{k-1}(t_i - t_{i-1})\|^2 \tag{8.28}$$

where $e_k(t_i) = y_d(t_i) - y_k(t_i)$ is the output tracking error at the $t_i$th time point, and $\lambda > 0$ is a penalty factor to restrict the iterative changing rate of the control input such that $\boldsymbol{\varphi}_k(t_i - 1)$ varies slowly along the iteration direction.

Substituting (8.27) into (8.28), yields

$$\begin{aligned}J\left(\boldsymbol{u}_k(t_i - t_{i-1})\right) =& \mid e_{k-1}(t_i) - \boldsymbol{\varphi}_k(t_1 - 1)\,\Delta\boldsymbol{u}_k(t_1 - 1)\\ &- \boldsymbol{\varphi}_k(t_2 - t_1)\,\Delta\boldsymbol{u}_k(t_2 - t_1) - \cdots\\ &- \boldsymbol{\varphi}_k(t_i - t_{i-1})\,\Delta\boldsymbol{u}_k(t_i - t_{i-1})\big|^2\\ &+ \lambda\,\|\boldsymbol{u}_k(t_i - t_{i-1}) - \boldsymbol{u}_{k-1}(t_i - t_{i-1})\|^2.\end{aligned} \tag{8.29}$$

Let $t_0 = 1$. Differentiating (8.29) with respect to $\boldsymbol{u}_k(t_i - t_{i-1})$, and letting it be zero, gives

$$\begin{aligned}\boldsymbol{u}_k(t_i - t_{i-1}) &= \boldsymbol{u}_{k-1}(t_i - t_{i-1}) + \frac{\rho_i e_{k-1}(t_i)\,\boldsymbol{\varphi}_k^{\mathrm{T}}(t_i - t_{i-1})}{\lambda + \left\|\boldsymbol{\varphi}_k(t_i - t_{i-1})\right\|^2}\\ &- \frac{\boldsymbol{\varphi}_k^{\mathrm{T}}(t_i - t_{i-1})\sum_{j=1}^{i-1}\rho_j\boldsymbol{\varphi}_k\left(t_j - t_{j-1}\right)\Delta\boldsymbol{u}_k\left(t_j - t_{j-1}\right)}{\lambda + \left\|\boldsymbol{\varphi}_k(t_i - t_{i-1})\right\|^2}\end{aligned} \tag{8.30}$$

where $0 < \rho_j \leq 1$, $j \in \{0, 1, \ldots, i\}$ is a step-size constant, which is added to make the control algorithm more flexible.

Since $\boldsymbol{\varphi}_k(t_i - 1)$ is unknown, replacing it with $\hat{\boldsymbol{\varphi}}_k(t_i - 1)$, then the learning control law becomes

$$\boldsymbol{u}_k(t_i - t_{i-1}) = \boldsymbol{u}_{k-1}(t_i - t_{i-1}) + \frac{\rho_i e_{k-1}(t_i)\hat{\boldsymbol{\varphi}}_k^{\mathrm{T}}(t_i - t_{i-1})}{\lambda + \left\|\hat{\boldsymbol{\varphi}}_k(t_i - t_{i-1})\right\|^2} - \frac{\hat{\boldsymbol{\varphi}}_k^{\mathrm{T}}(t_i - t_{i-1})\sum_{j=1}^{i-1}\rho_j\hat{\boldsymbol{\varphi}}_k\left(t_j - t_{j-1}\right)\Delta\boldsymbol{u}_k\left(t_j - t_{j-1}\right)}{\lambda + \left\|\hat{\boldsymbol{\varphi}}_k(t_i - t_{i-1})\right\|^2}. \quad (8.31)$$

So, the improved DDPTPILC (iDDPTPILC) is constructed by (8.32), (8.33), and (8.34), shown as follows:

$$\boldsymbol{u}_k(t_i - t_{i-1}) = \boldsymbol{u}_{k-1}(t_i - t_{i-1}) + \frac{\rho_i e_{k-1}(t_i)\hat{\boldsymbol{\varphi}}_k^{\mathrm{T}}(t_i - t_{i-1})}{\lambda + \left\|\hat{\boldsymbol{\varphi}}_k(t_i - t_{i-1})\right\|^2} - \frac{\hat{\boldsymbol{\varphi}}_k^{\mathrm{T}}(t_i - t_{i-1})\sum_{j=1}^{i-1}\rho_j\hat{\boldsymbol{\varphi}}_k\left(t_j - t_{j-1}\right)\Delta\boldsymbol{u}_k\left(t_j - t_{j-1}\right)}{\lambda + \left\|\hat{\boldsymbol{\varphi}}_k(t_i - t_{i-1})\right\|^2} \quad (8.32)$$

$$\hat{\boldsymbol{\varphi}}_k(t_i - 1) = \hat{\boldsymbol{\varphi}}_{k-1}(t_i - 1) + \frac{\eta\left(\Delta y_{k-1}(t_i) - \hat{\boldsymbol{\varphi}}_{k-1}(t_i - 1)\Delta\boldsymbol{u}_{k-1}(t_i - 1)\right)\Delta\boldsymbol{u}_{k-1}^{\mathrm{T}}(t_i - 1)}{\mu + \|\Delta\boldsymbol{u}_{k-1}(t_i - 1)\|^2} \quad (8.33)$$

$$\hat{\boldsymbol{\varphi}}_k(t_i - 1) = \hat{\boldsymbol{\varphi}}_0(t_i - 1), \quad \text{if } \operatorname{sgn}\left(\hat{\phi}_k(\tau)\right) \neq \operatorname{sgn}\left(\phi_0(\tau)\right) \text{ or } \left\|\hat{\boldsymbol{\varphi}}_k(t_i - 1)\right\| \leq \varepsilon. \quad (8.34)$$

**Remark 8.3** The proposed iDDPTPILC (8.32)–(8.34) is a data-driven control approach because no explicit process model information is included in the control mechanism other than the I/O measurements only. In addition, only the system output information at the specified points is used in both the learning law and the parameter estimation algorithm, and thus it is still operational when the system outputs at the other time instants are not measured.

**Remark 8.4** Additional control input information from previous time subintervals of the same iteration is included in the learning control law (8.32) to enhance control performance and the control input is updated pointwisely. As a comparison, however, only the control input in the immediately previous single iteration is utilized in the learning control law proposed in Sect. 8.3, and the control inputs over the finite time interval are denoted as a vector calculated in a batchwise.

**Remark 8.5** Note that the tracking errors at the previous time instants of the current iteration cannot be added directly to the learning control law (8.32) as a feedback term because the system outputs/states are not measurable at intermediate points other than the designated points. And thus there is no performance constraint on these intermediate points for the point-to-point tracking tasks.

### 8.6.3 Convergence Analysis

**Theorem 8.2** *For a class of nonlinear nonaffine discrete-time systems (8.1), if Assumptions 4.1–4.3 hold, then the proposed iDDPTPILC scheme (8.32)–(8.34) guarantees that*

*(i) the estimation $\hat{\varphi}_k(t_i-1)$ is bounded along the iteration direction, $\forall i = 1,\ldots,M$, $\forall k$;*
*(ii) the tracking error $e_k(t_i)$ at the specified point approaches to zero monotonically along with iterations, i.e., $\lim_{k\to\infty} e_k(t_i)=0$, $\forall i=1,\ldots,M$;*
*(iii) the system output $y_k(t)$ and $u_k(t)$ are bounded, $\forall t\in\{0,1,\ldots,N\}$, $\forall k$.*

***Proof*** *Part 1. The boundedness of* $\hat{\varphi}_k(t_i-1)$

The boundedness of $\hat{\varphi}_k(t_i-1)$ can be easily obtained following the similar steps in Theorem 8.1. Therefore, we assume that $\|\hat{\varphi}_k(t_i-1)\|\le b_{\hat{\varphi}}$ for all $i=1,\ldots,M$, where $b_{\hat{\varphi}}$ is a positive constant.

*Part 2. The convergence of output tracking error* $e_k(t_i)$

Define $\boldsymbol{A}_k(t_i-1)$ as follows.

$$\boldsymbol{A}_k(t_i-1)=\begin{bmatrix} 0 & I & 0 & \cdots & 0\\ 0 & 0 & I & \cdots & 0\\ \vdots & \vdots & \vdots & \ddots & \vdots\\ 0 & 0 & 0 & \cdots & I\\ 0 & -\frac{\rho_1\hat{\varphi}_k^{\mathrm{T}}(t_i-t_{i-1})\hat{\varphi}_k(t_1-1)}{\lambda+\|\hat{\varphi}_k(t_i-t_{i-1})\|^2} & -\frac{\rho_2\hat{\varphi}_k^{\mathrm{T}}(t_i-t_{i-1})\hat{\varphi}_k(t_2-t_1)}{\lambda+\|\hat{\varphi}_k(t_i-t_{i-1})\|^2} & \cdots & -\frac{\rho_{i-1}\hat{\varphi}_k^{\mathrm{T}}(t_i-t_{i-1})\hat{\varphi}_k(t_{i-1}-t_{i-2})}{\lambda+\|\hat{\varphi}_k(t_i-t_{i-1})\|^2} \end{bmatrix}_{t_i\times t_i}.$$

Define

$$\boldsymbol{B}_k(t_i-1)=[\underbrace{0\ \cdots\ 0}_{t_{i-1}}\ \frac{\rho_i\hat{\varphi}_k(t_i-t_{i-1})}{\lambda+\|\hat{\varphi}_k(t_i-t_{i-1})\|^2}]^{\mathrm{T}}\in R^{t_i}$$

$$\boldsymbol{U}_k(t_i-1)=\left[\boldsymbol{u}_k(t_1-1)^{\mathrm{T}},\boldsymbol{u}_k(t_2-t_1)^{\mathrm{T}},\ldots,\boldsymbol{u}_k(t_i-t_{i-1})^{\mathrm{T}}\right]^{\mathrm{T}}\in R^{t_i}$$
$$\boldsymbol{U}_k(t_{i-1}-1)=[\underbrace{0,\ldots,0}_{t_i-t_{i-1}},\boldsymbol{u}_k(t_1-1)^{\mathrm{T}},\ldots,\boldsymbol{u}_k(t_{i-1}-t_{i-2})^{\mathrm{T}}]^{\mathrm{T}}\in R^{t_i}.$$

The only difference between $\boldsymbol{U}_k(t_i-1)$ and $\boldsymbol{u}_k(t_i-1)$ is that the dimension of $\boldsymbol{U}_k(t_i-1)$ is uniformly fixed as $t_i$ for all $\boldsymbol{U}_k(t_j-1)$, $j\le i$, while the dimension of $\boldsymbol{u}_k(t_j-1)$ varies with different $t_j$, $j\le i$. Then according to (8.32)

$$\Delta \boldsymbol{U}_k(t_i-1)=\boldsymbol{A}_k(t_i-1)\,\Delta \boldsymbol{U}_k(t_{i-1}-1)+\boldsymbol{B}_k(t_i-1)\,e_{k-1}(t_i). \quad (8.35)$$

It is evident that

$$\sum_{j=1}^{i-1}\left\|\frac{\rho_j\hat{\boldsymbol{\varphi}}_k^{\mathrm{T}}(t_i-t_{i-1})\,\hat{\boldsymbol{\varphi}}_k(t_j-t_{j-1})}{\lambda+\left\|\hat{\boldsymbol{\varphi}}_k(t_i-t_{i-1})\right\|^2}\right\|\le\sum_{j=1}^{i-1}\frac{\rho_{\max}\left\|\hat{\boldsymbol{\varphi}}_k(t_j-t_{j-1})\right\|}{2\sqrt{\lambda}}. \quad (8.36)$$

It is also apparent that $\|\hat{\boldsymbol{\varphi}}_k(t_j-t_{j-1})\|\le\|\hat{\boldsymbol{\varphi}}_k(t_i-1)\|\le b_{\hat{\varphi}}$ for all $j=1,\ldots,i$. Therefore, for all $i\in\{0,1,\ldots,M\}$, by properly selecting $\lambda$ and $\rho_{\max}$, $\rho_{\max}=\max_{j=1,\ldots,i,i=1,\ldots,M}\{\rho_j\}$, one can find a positive constant $0<d_1<0.5$ such that

$$0<\sum_{j=1}^{i-1}\left\|\frac{\rho_j\hat{\boldsymbol{\varphi}}_k^{\mathrm{T}}(t_i-t_{i-1})\,\hat{\boldsymbol{\varphi}}_k(t_j-t_{j-1})}{\lambda+\left\|\hat{\boldsymbol{\varphi}}_k(t_i-t_{i-1})\right\|^2}\right\|\le\frac{\rho_{\max}b_{\hat{\varphi}}M}{2\sqrt{\lambda}}=d_1<0.5. \quad (8.37)$$

In view of Lemma 4.2, (8.37) implies that the spectral radius $s(\boldsymbol{A}_k(t_i-1))<0.5$. Then according to Lemma 3.2, there exists an arbitrary small positive constant $\delta$ such that

$$0<\|\boldsymbol{A}_k(t_i-1)\|_v\le s(\boldsymbol{A}_k(t_i-1))+\delta\le d_1+\delta=d_2<0.5 \quad (8.38)$$

where $\|\boldsymbol{A}_k(t_i-1)\|_v$ denotes a proper matrix norm of $\boldsymbol{A}_k(t_i-1)$ on the normed vector space $v$.

Taking norms on both sides of (8.35), yields

$$\begin{aligned}\|\Delta \boldsymbol{U}_k(t_i-1)\|_v\le\|\boldsymbol{A}_k(t_i-1)\|_v\|\Delta \boldsymbol{U}_k(t_{i-1}-1)\|_v\\+\|\boldsymbol{B}_k(t_i-1)\|_v\,|e_{k-1}(t_i)|.\end{aligned} \quad (8.39)$$

By properly selecting $\lambda$ and $\rho_{\max}$, one can guarantee that

$$\|\boldsymbol{B}_k(t_i-1)\|_v=\frac{\rho_i\left\|\hat{\boldsymbol{\varphi}}_k(t_i-t_{i-1})\right\|_v}{\lambda+\left\|\hat{\boldsymbol{\varphi}}_k(t_i-t_{i-1})\right\|_v^2}\le\frac{\rho_{\max}}{2\sqrt{\lambda}}=d_3<1 \quad (8.40)$$

where $d_3$ is a positive constant.

Hence, (8.38)–(8.40) together imply that

$$\begin{aligned}\|\Delta \boldsymbol{U}_k(t_i-1)\|_v&\le d_2\|\Delta \boldsymbol{U}_k(t_{i-1}-1)\|_v+d_3\,|e_{k-1}(t_i)|\\&\le d_2^i\|\Delta \boldsymbol{U}_k(t_0-1)\|_v+d_2^{i-1}d_3\,|e_{k-1}(t_1)|\\&\quad+d_2^{i-2}d_3\,|e_{k-1}(t_2)|+\cdots+d_3\,|e_{k-1}(t_i)|.\end{aligned} \quad (8.41)$$

Because the initial input signal is set to be identical for all iterations, so $\|\Delta \boldsymbol{U}_k(t_0-1)\|_v=|\Delta u_k(0)|=0$. Then, (8.41) implies

$$\|\Delta \boldsymbol{U}_k(t_i-1)\|_v \le d_3 \sum_{j=1}^{i} d_2^{i-j} \left|e_{k-1}(t_j)\right|. \tag{8.42}$$

By virtue of (8.24) and (8.35), the dynamics of the tracking error at the prespecified points is

$$\begin{aligned} e_k(t_i) = & \; y_d(t_i) - y_k(t_i) = y_d(t_i) - y_{k-1}(t_i) - \boldsymbol{\varphi}_k(t_i-1)\Delta \boldsymbol{U}_k(t_i-1) \\ = & \; e_{k-1}(t_i) - \boldsymbol{\varphi}_k(t_i-1)\Delta \boldsymbol{U}_k(t_i-1) \\ = & \; \left(1-\boldsymbol{\varphi}_k(t_i-1)\boldsymbol{B}_k(t_i-1)\right) e_{k-1}(t_i) \\ & -\boldsymbol{\varphi}_k(t_i-1)\boldsymbol{A}_k(t_i-1)\Delta \boldsymbol{U}_k(t_{i-1}-1) \\ = & \left(1-\frac{\rho_i \boldsymbol{\varphi}_k(t_i-t_{i-1})\hat{\boldsymbol{\varphi}}_k^{\mathrm{T}}(t_i-t_{i-1})}{\lambda+\left\|\hat{\boldsymbol{\varphi}}_k(t_i-t_{i-1})\right\|^2}\right) e_{k-1}(t_i) \\ & -\boldsymbol{\varphi}_k(t_i-1)\boldsymbol{A}_k(t_i-1)\Delta \boldsymbol{U}_k(t_{i-1}-1). \end{aligned} \tag{8.43}$$

Taking norms on both sides of (8.43), yields

$$\begin{aligned} |e_k(t_i)| \le & \left|1-\frac{\rho_i \boldsymbol{\varphi}_k(t_i-t_{i-1})\hat{\boldsymbol{\varphi}}_k^{\mathrm{T}}(t_i-t_{i-1})}{\lambda+\left\|\hat{\boldsymbol{\varphi}}_k(t_i-t_{i-1})\right\|^2}\right| |e_{k-1}(t_i)| \\ & + \left\|\boldsymbol{\varphi}_k(t_i-1)\right\|_v \|\boldsymbol{A}_k(t_i-1)\|_v \|\Delta \boldsymbol{U}_k(t_{i-1}-1)\|_v. \end{aligned} \tag{8.44}$$

According to Assumption 4.3 and the reset algorithm (8.34), $\boldsymbol{\varphi}_k(t_i-1)\hat{\boldsymbol{\varphi}}_k^{\mathrm{T}}(t_i-1) > 0$. Hence, by selecting $\rho_{\max}$ and $\lambda$ properly, one can guarantee that

$$\begin{aligned} 0 < & \frac{\rho_i \boldsymbol{\varphi}_k(t_i-t_{i-1})\hat{\boldsymbol{\varphi}}_k^{\mathrm{T}}(t_i-t_{i-1})}{\lambda+\left\|\hat{\boldsymbol{\varphi}}_k(t_i-t_{i-1})\right\|^2} \\ \le & \frac{\rho_{\max}\left\|\boldsymbol{\varphi}_k(t_i-t_{i-1})\right\|}{2\sqrt{\lambda}} \\ \le & \frac{\rho_{\max} b_\varphi}{2\sqrt{\lambda}} = d_4 < 1 \end{aligned} \tag{8.45}$$

where $d_4 = b_\varphi d_3$ is a positive scalar

$$\begin{aligned} |e_k(t_i)| & \le d_5 |e_{k-1}(t_i)| + d_2 L \|\Delta \boldsymbol{U}_k(t_{i-1}-1)\|_v \\ & \le d_5 |e_{k-1}(t_i)| + b_\varphi d_3 \sum_{j=1}^{i} d_2^{i-j+1} \left|e_{k-1}(t_j)\right| \end{aligned} \tag{8.46}$$

where $0 < d_5 = 1 - d_4 < 1$.

Define $e_{\max}(k) = |e_k(t_\tau)| = \max_{i\in\{1,\ldots,M\}}\{|e_k(t_i)|\}$, where $t_\tau$ is some time instant over $\{t_1, \ldots, t_M\}$, $\tau \in \{1, \ldots, M\}$. Then according to (8.46), the following inequality holds

$$\begin{aligned} |e_{\max}(k)| &= |e_k(t_\tau)| \le d_5 |e_{k-1}(t_\tau)| + b_\varphi d_3 \sum_{j=1}^{\tau} d_2^{\tau-j+1} \left|e_{k-1}\left(t_j\right)\right| \\ &\le \left(d_5 + d_3 \frac{d_2\left(1-d_2^{\tau-1}\right)}{1-d_2} b_\varphi\right) |e_{\max}(k-1)| . \end{aligned} \tag{8.47}$$

Since $d_2 \in (0, 0.5)$, one can get

$$0 < \frac{d_2}{1-d_2} = \hbar < 1 \tag{8.48}$$

and

$$\begin{aligned} 0 < d_5 + d_3 \tfrac{d_2(1-d_2^{\tau-1})}{1-d_2} b_\varphi &\le \left(d_5 + \tfrac{d_2 d_3 b_\varphi}{1-d_2}\right) \\ &= \left(1 - d_4 + \tfrac{d_2 d_4}{1-d_2}\right) = d_6 \end{aligned} \tag{8.49}$$

where $\hbar$ and $d_6$ is a positive constant.

Consequently

$$0 < d_6 = 1 - d_4 + \frac{d_2 d_4}{1-d_2} = 1 - (1-\hbar) d_4 < 1. \tag{8.50}$$

Then inequality (8.47) and the condition (8.50) together imply

$$|e_{\max}(k)| \le d_6 |e_{\max}(k-1)| \le \cdots \le d_6^k |e_{\max}(0)| . \tag{8.51}$$

Apparently, $\lim_{k\to\infty} e_{\max}(k) = 0$. Then, conclusion (ii) of Theorem 8.2 is derived directly from (8.51), i.e., $\lim_{k\to\infty} e_k(t_i) = 0$, $i \in \{1, \ldots, M\}$.

*Part 3. The boundedness of system output and control input*

Following the same procedure of Sect. 8.3, one can obtain that

$$\|\boldsymbol{U}_k(t_i-1)\|_v \le \sum_{l=1}^{k} \|\Delta\boldsymbol{U}_l(t_i-1)\|_v + \|\boldsymbol{U}_0(t_i-1)\|_v . \tag{8.52}$$

According to (8.42), it is easy to get

$$\begin{aligned} \|\boldsymbol{U}_k(t_i-1)\|_v \le \tfrac{d_3\left(1-d_2^{i-1}\right)}{1-d_2} & \textstyle\sum_{l=1}^{k} |e_{\max}(l-1)| \\ &+ \|\boldsymbol{U}_0(t_i-1)\|_v . \end{aligned} \tag{8.53}$$

In terms of (8.51), inequality (8.53) implies

$$\|\boldsymbol{U}_k(t_i-1)\|_v \leq \frac{d_3}{(1-d_2)(1-d_6)}|e_{\max}(0)| + \|\boldsymbol{U}_0(t_i-1)\|_v. \tag{8.54}$$

Because both the initial maximum tracking error $e_{\max}(0)$ and initial control input $\boldsymbol{U}_0(t_i-1)$ are bounded, the inequality (8.54) leads to a bounded control input sequence directly, that is, $\|\boldsymbol{U}_k(t_i-1)\|_v \leq b_U$ where $b_U$ is a positive constant.

In general, $t_M = N$. Then

$$\|\boldsymbol{U}_k(N-1)\|_v = \|\boldsymbol{U}_k(t_M-1)\|_v \leq b_U.$$

Otherwise, one can let the system run freely over the rest interval of $\{t_M, t_{M+1}, \ldots, N\}$ without requiring additional control action. Therefore, $\|\boldsymbol{U}_k(N-1)\|_v$ is still bounded.

The system outputs for all time instants over the entire time interval can be expressed as

$$\Delta\boldsymbol{Y}_k = \boldsymbol{\Phi}_k(N-1)\Delta\boldsymbol{U}_k(N-1) \tag{8.55}$$

where $\boldsymbol{Y}_k = [y_k(1), y_k(2), \ldots, y_k(N)]^{\mathrm{T}}$, $\boldsymbol{U}_k = [u_k(0), u_k(1), \ldots, u_k(N-1)]^{\mathrm{T}}$, and

$$\boldsymbol{\Phi}_k(N-1) = \begin{bmatrix} \frac{\partial g_0}{\partial u_k(0)} & 0 & \cdots & 0 \\ \frac{\partial g_1}{\partial u_k(0)} & \frac{\partial g_1}{\partial u_k(1)} & \cdots & 0 \\ \vdots & \vdots & \ddots & \vdots \\ \frac{\partial g_{N-1}}{\partial u_k(0)} & \frac{\partial g_{N-1}}{\partial u_k(1)} & \cdots & \frac{\partial g_{N-1}}{\partial u_k(N-1)} \end{bmatrix}.$$

According to Assumption 4.2, it clear that $\|\boldsymbol{\Phi}_k(N-1)\|$ is bounded. Assume $\|\boldsymbol{\Phi}_k(N-1)\| \leq b_\Phi$, where $b_\Phi$ is a positive finite constant. Then, according to (8.42), (8.51), and (8.55), we have

$$\begin{aligned} \|\boldsymbol{Y}_k\|_v &\leq \|\boldsymbol{Y}_0\|_v + \sum_{l=1}^{k}\|\Delta\boldsymbol{Y}_l\|_v \\ &\leq \|\boldsymbol{Y}_0\|_v + b_\Phi\sum_{l=1}^{k}\|\Delta\boldsymbol{U}_l(t_M-1)\|_v \\ &\leq \|\boldsymbol{Y}_0\|_v + b_\Phi\sum_{l=1}^{k}d_3\sum_{j=1}^{M}d_2^{M-j}\left|e_{l-1}(t_j)\right| \\ &\leq \|\boldsymbol{Y}_0\|_v + b_\Phi\frac{d_3}{1-d_2}\frac{1}{1-d_6}|e_{\max}(0)|. \end{aligned} \tag{8.56}$$

Thus the system outputs $\boldsymbol{Y}_k$ at all points are bounded because the initial values of $\boldsymbol{Y}_0$ and $e_{\max}(0)$ are bounded. The conclusion (iii) of Theorem 8.2 is obtained consequently. □

### 8.6.4 Simulation Study

For the purpose of validating the effectiveness of the proposed iDDPTPILC and comparing it with the DDPTPILC proposed in Sect. 8.3, a nonlinear batch reactor with the well-tested dynamical equation (Logsdon and Biegler 1989) is given as

$$\begin{aligned} dx_1/dt &= -k_1 \exp\left(-E_1/\left(uT_{\text{ref}}\right)\right) x_1^2 \\ dx_2/dt &= k_1 \exp\left(-E_1/\left(uT_{\text{ref}}\right)\right) x_1^2 - k_2 \exp\left(-E_2/\left(uT_{\text{ref}}\right)\right) x_2 \\ y &= x_2 \end{aligned} \tag{8.57}$$

where $x_1$ and $x_2$ are the dimensionless concentrations of the raw material and the product, respectively; $u = T/T_{\text{ref}}$ is the dimensionless temperature of the reactor. The system parameters are given as $T_{\text{ref}} = 348\text{K}$, $E_1 = 2500\text{K}$, $E_2 = 5000\text{K}$, $k_1 = 4000$, $k_2 = 620000$, the running time $t_f = 1h$, and the sampling time $h = 0.1h$ listed. Hence, the discrete-time instant $t \in \{0, 1, \ldots, 10\}$.

Note that the above mathematical model of the batch reactor in this simulation is assumed to be unavailable, and the model just serves to produce I/O data for the controlled plants. No information of models will be utilized in the proposed iDDPTPILC controller.

The desired product reference points are

$$y_d\left(t_i\right) = 0.61 \sin\left(\frac{\pi t_i}{2N}\right), t_1 = 2,\ t_2 = 4,\ t_3 = 6,\ t_4 = 10$$

and the initial values are set as $x_1(0) = 1$, $x_2(0) = 0$; $u_0(t) = 0.92$. The controller parameters are set as: $\hat{\varphi}_0(j-1) = 0.01$, $\rho_j = 0.1$, for $j = 1, \ldots, t_i$, $i = 1, 2, 3, 4$; $\eta = 0.05$, $\lambda = 0.001$, and $\mu = 0.02$.

Applying the proposed iDDPTPILC scheme (8.32)-(8.34), the results are shown in Figs. 8.3 and 8.4. Figure 8.3 shows the output tracking performance, where the vertical axis denotes the system output and the horizontal axis denotes the time instant. Figure 8.4 shows the convergence performance of tracking error, where the Y-axis denotes the maximum absolute value of the output tracking error, defined as $e_{\max}(k) = \max_{i\in\{1,\ldots,4\}}\{|e_k(t_i)|\}$, and the X-axis denotes the iteration numbers.

From Figs. 8.3 and 8.4, the applicability and effectiveness of the proposed iDDPTPILC approach is verified with a good tracking performance for the multiple intermediate target points.

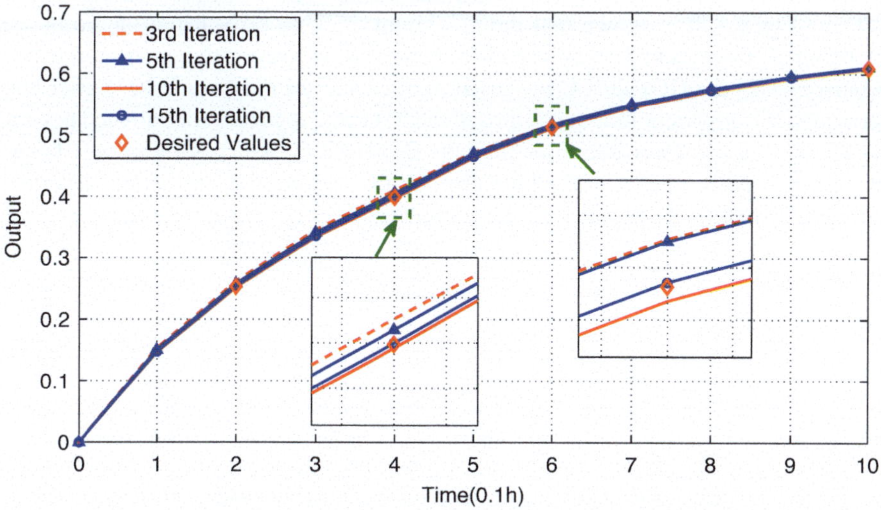


**Fig. 8.3** Tracking performance by applying the proposed iDDPTPILC into the batch reactor

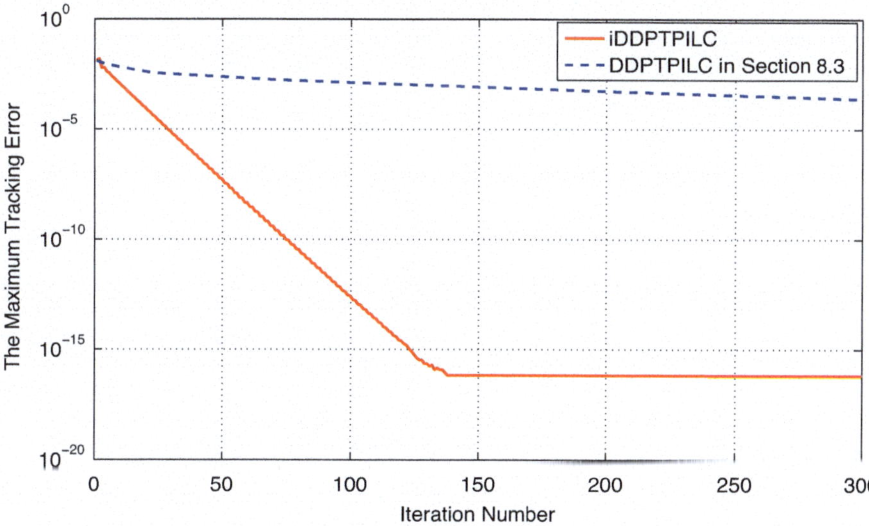


**Fig. 8.4** Comparative results of error convergence

For the comparison, the DDPTPILC scheme (8.8)–(8.10) proposed in Sect. 8.3 is also applied to such a multiple intermediate points tracking task.

Applying the DDPTPILC under the same simulation condition for a comparison, i.e., $x_1(0) = 1$, $x_2(0) = 0$; $u_0(t) = 0.92$, $\hat{\varphi}_0(j-1) = 0.01$, setting parameters as $\rho_j = 0.1$, $\eta = 0.02$, $\lambda = 0.01$, $\mu = 0.0001$, the maximum tracking error at the specified points is shown as the dashed line in Fig. 8.4.

From Fig. 8.4, it is shown that the maximum tracking error using the proposed iDDPTPILC has reached $10^{-16}$ after 100 iterations. However, the maximum tracking error using the DDPTPILC proposed in Sect. 8.3 can only reach $10^{-4}$ after 300 iterations. It is apparent that the proposed iDDPTPILC achieves the satisfied performance after much fewer iterations than the DDPTPILC, which shows that a better convergence performance of tracking error can be achieved by utilizing additional control information from previous time subintervals of the same iteration.

## 8.7 Summary

This chapter shows how the DDILC is tackled for the tasks of tracking specified intermediate points. The presented DDPTPILC method and its improvement with additional input signals only employ the output tracking errors at the specified points, instead of the whole output trajectory tracking error. All of the methods proposed in this chapter are data-driven and only depend on the measured I/O data of the plant without requiring any other model information. The learning gains of the learning control laws are updated iteratively that makes the DDPTPILC and its improvement robust against uncertainties. Rigorous mathematical analysis is presented to show the monotonic convergence. Extensive simulation results illustrate the effectiveness and the applicability of the proposed methods.

## References

Arimoto S, Kawamura S, Miyazaki F (1984) Bettering operation of robots by learning. J Robot Syst 1(2):123–140

Barton KL, Alleyne AG (2010) A norm optimal approach to time-varying ILC with application to a multi-axis robotic testbed. IEEE Trans Control Syst Technol 19(1):166–180

Chi R, Hou Z, Xu J (2008) Adaptive ILC for a class of discrete-time systems with iteration-varying trajectory and random initial condition. Automatica 44(8):2207–2213

Chi R, Wang D, Hou Z, Jin S (2012) Data-driven optimal terminal iterative learning control. J Process Control 22(10):2026–2037

Chi R, Huang B, Wang D, Zhang R, Feng Y (2016) Data-driven optimal terminal iterative learning control with initial value dynamic compensation. IET Control Theory A 10(12):1357–1364

Chien CJ (2008) A combined adaptive law for fuzzy iterative learning control of nonlinear systems with varying control tasks. IEEE Trans Fuzzy Syst 16(1):40–51

Dinh TV, Freeman CT, Lewin PL, Tan Y (2012) Convergence and robustness of a point-to-point iterative learning control algorithm. In: 2012 IEEE 51st IEEE conference on decision and control. IEEE. https://ieeexplore.ieee.org/document/6426908

Flores-Cerrillo J, MacGregor JF (2005) Iterative learning control for final batch product quality using partial least squares models. Ind Eng Chem Res 44(24):9146–9155

Freeman CT (2012) Constrained point-to-point iterative learning control with experimental verification. Control Eng Pract 20(5):489–498

Freeman CT, Tan Y (2012) Iterative learning control with mixed constraints for point-to-point tracking. IEEE Trans Control Syst Technol 21(3):604–616

Gauthier G, Boulet B (2009) Terminal iterative learning control design with singular value decomposition decoupling for thermoforming ovens. In: 2009 American control conference. IEEE. https://ieeexplore.ieee.org/document/5160445

Hou Z, Wang Y, Yin C, Tang T (2011) Terminal iterative learning control based station stop control of a train. Int J Control 84(7):1263–1274

Logsdon JS, Biegler LT (1989) Accurate solution of differential-algebraic optimization problems. Ind Eng Chem Res 28(11):1628–1639

Mondal S, Yun Y, Chung WK (2010) Terminal iterative learning control for calibrating systematic odometry errors in mobile robots. In: 2010 IEEE/ASME international conference on advanced intelligent mechatronics. IEEE. https://ieeexplore.ieee.org/document/5695734

Owens DH, Freeman CT, Van Dinh T (2012) Norm-optimal iterative learning control with intermediate point weighting: theory, algorithms, and experimental evaluation. IEEE Trans Control Syst Technol 21(3):999–1007

Son TD, Ahn HS, Moore KL (2013) Iterative learning control in optimal tracking problems with specified data points. Automatica 49(5):1465–1472

Sun M, Wang D (2001) Sampled-data iterative learning control for nonlinear systems with arbitrary relative degree. Automatica 37(2):283–289

Xu J, Huang D (2008) Initial state iterative learning for final state control in motion systems. Automatica 44(12):3162–3169

Xu J, Chen Y, Lee TH, Yamamoto S (1999) Terminal iterative learning control with an application to RTPCVD thickness control. Automatica 35(9):1535–1542

# Chapter 9
# Higher Order Data-Driven Iterative Learning Control

## 9.1 Introduction

High-order ILC methods (Bien and Huh 1989; Chen et al. 1997, 1998; Hätönen et al. 2006; Gunnarsson and Norrl 2006) have attracted continuous attention for achieving a better control performance by using additional error information from previous cycles, although more memory storage is required for the high-order algorithms as a trade-off. The basic idea in Bien and Huh (1989), Chen et al. (1997), Chen et al. (1998), Hätönen et al. (2006), Gunnarsson and Norrl (2006) is to utilize more historical data collected from previous iterations to enhance the online control performance. However, it is noteworthy that online data in the current iteration, compared with the historical data in the previous iterations, can reflect varying characteristics of a process and the process disturbances in real time. Therefore, whenever possible, one should incorporate the historical data as well as the online data to achieve a better control performance. Furthermore, the existing higher order methods mainly focused on PID-type and optimization-based schemes, and most of their design and analysis are also limited to linear systems.

To evaluate the performance of high-order ILC, some works (Bien and Huh 1989; Chen et al. 1997, 1998; Hätönen et al. 2006; Gunnarsson and Norrl 2006; Chi et al. 2015a, b) have been done by simulation study to illustrate that the high-order ILC can achieve a better performance with enhanced robustness than the low-order one. However, in theory, the question whether the control performance of high-order ILC is superior or inferior to that of the low-order ILC is still challenging. In Xu and Tan (2002a, 2003), it was shown that the low-order ILC was faster than the high-order ILC, but the discussion only focused on the linear learning controller for linear systems in the sense of time-weighted norm. Xu and Tan (2002a, 2002b, 2003) have also pointed out that based on the heuristic knowledge, it is possible to choose a high-order ILC scheme to achieve faster convergence for a certain norm. The reference Schmid (2007) has pointed out that up to now there is no theoretical proof about whether the high-order ILC works are better with faster convergence than the low-

 
R. Chi et al., *Data-Driven Iterative Learning Control for Discrete-Time Systems*, Intelligent Control and Learning Systems 2,
https://doi.org/10.1007/978-981-19-5950-9_9

order one. Therefore, how to utilize more control information of previous iterations to guarantee that the high-order algorithm outperforms the low-order one is significant and challenging in the existing ILC theory.

This chapter explores a novel higher order DDILC (HODDILC) method using both historical and online data for a nonlinear nonaffine discrete-time system. A non-lifted IDL is used to equivalently reformulate the original nonlinear plant without any approximation. Owing to the use of additional historical and online data, that is, more tracking errors from the previous iterations and more control inputs at previous time instants within the current iteration, a better control performance can be achieved by applying the proposed HODDILC method. It is the first time to show in theory that under certain conditions, the convergence speed of the higher order learning law is faster than that of the lower order one by selecting the higher order factors and controller parameters properly. Meanwhile, due to the avoidance of the matrix inverse calculation in the control law, an efficient computation can be achieved even though the number of samples in the trial and the trajectory length are larger. The controller design and analysis in this chapter does not use any explicit model information of the controlled plant except for the measured I/O data. Simulations in this chapter verify the correctness of the theoretical results.

This chapter is organized as follows. Section 9.2 shows the problem formulation. The HODDILC design and analysis are presented in Sects. 9.3 and 9.4, respectively. Simulation study is given in Sect. 9.5. Section 9.6 gives some conclusions.

## 9.2 Problem Formulation

Consider a class of nonlinear nonaffine discrete-time systems

$$y_k(t+1) = f\left(y_k(t), \ldots, y_k\left(t-n_y\right), u_k(t), \ldots, u_k\left(t-n_u\right)\right) \tag{9.1}$$

where $y_k(t)$ and $u_k(t)$ are system output and input, respectively; $y_k(t) = 0$ and $u_k(t) = 0$, $\forall t < 0$; $n_y$ and $n_u$ are two unknown orders; $t \in \{0, 1, \ldots, N\}$ is time instant and $N \in Z^+$ represents the endpoint; $k \in \{0, 1, \ldots\}$ indicates the iteration number; $f(\ldots) : R^{n_u+n_y+2} \mapsto R$ is an unknown nonlinear scalar function and $f(0, 0) = 0$.

Consider nonlinear system (9.1) satisfying Assumptions 4.1, 4.2, and 4.3.

From Theorem 4.1, one has

$$\Delta y_k(t+1) = \boldsymbol{\varphi}_k(t)\Delta \boldsymbol{u}_k(t) \tag{9.2}$$

where $\Delta y_k(t+1) = y_k(t+1) - y_{k-1}(t+1)$, $\Delta \boldsymbol{u}_k(t) = \boldsymbol{u}_k(t) - \boldsymbol{u}_{k-1}(t)$; $\boldsymbol{\varphi}_k(t) = [\phi_k(0), \phi_k(1) \ldots, \phi_k(t)]$, $\boldsymbol{u}_k(t) = [u_k(0), u_k(1), \ldots, u_k(t)]^{\mathrm{T}}$; $\phi_k(i)$ and $u_k(i)$ are the $(i+1)$th element of $\boldsymbol{\varphi}_k(t)$ and $\boldsymbol{u}_k(t)$, respectively, $i = 0, \ldots, t$.

In terms of Assumption 4.2, it follows that $\left\|\boldsymbol{\varphi}_k(t)\right\| \le b_\varphi$, where $b_\varphi > 0$ is a constant.

## 9.3 Controller Design

Assume that $y_d(t), t \in \{0, \ldots, N\}$ is a target trajectory and is bounded for all time instants. The control objective is to make the tracking error $e_k(t) = y_d(t) - y_k(t)$ converge to zero asymptotically along the iterations.

Since $\boldsymbol{\varphi}_k(t)$ in (9.2) is unknown but it is slowly time-iteration-varying, a modified projection algorithm (Chi et al. 2015) is utilized for its iterative estimation

$$\hat{\boldsymbol{\varphi}}_k(t) = \hat{\boldsymbol{\varphi}}_{k-1}(t) + \frac{\eta\left(\Delta y_{k-1}(t+1) - \hat{\boldsymbol{\varphi}}_{k-1}(t)\Delta \boldsymbol{u}_{k-1}(t)\right)\Delta \boldsymbol{u}^{\mathrm{T}}_{k-1}(t)}{\mu + \|\Delta \boldsymbol{u}_{k-1}(t)\|^2} \tag{9.3}$$

$$\hat{\boldsymbol{\varphi}}_k(t) = \hat{\boldsymbol{\varphi}}_0(t), \text{ if } \operatorname{sgn}\left(\hat{\phi}_k(i)\right) \neq \operatorname{sgn}\left(\hat{\phi}_0(i)\right) \text{ or } \left\|\hat{\boldsymbol{\varphi}}_k(t)\right\| \leq \varepsilon \tag{9.4}$$

where $i = 0, \ldots, t$, $\hat{\boldsymbol{\varphi}}_k(t)$ is the estimation of $\boldsymbol{\varphi}_k(t)$, $\mu > 0$ denotes a weighting factor, $0 < \eta < 2$ is a step-size factor, $\hat{\boldsymbol{\varphi}}_0(t)$ is the initial value of $\hat{\boldsymbol{\varphi}}_k(t)$, and $\varepsilon$ is a small positive scalar.

The initial value of $\hat{\boldsymbol{\varphi}}_0(t)$ is selected such that its elements have the same signs as that of $\boldsymbol{\varphi}_k(t)$, whose elements' signs can be identified by using the I/O data sampled from the controlled plant.

Consider an objective function with higher order factors

$$J\left(u_k(t), \boldsymbol{\alpha}\right) = \left(\sum_{m=1}^{M} \alpha_m e_{k-m+1}(t+1)\right)^2 + \lambda\left(u_k(t) - u_{k-1}(t)\right)^2 \tag{9.5}$$

where $\lambda > 0$ is a weighting factor; and $\boldsymbol{\alpha} = [\alpha_1, \ldots, \alpha_M]$ denotes the higher order factors and $\sum_{m=1}^{M} \alpha_m = 1, 0 < \alpha_m \leq 1, \alpha_1 + \alpha_2 - \sum_{m=3}^{M} \alpha_m = \bar{\alpha} > 0$, in which $M$ is a positive integer.

Rewrite (9.2) as

$$y_k(t+1) = y_{k-1}(t+1) + \phi_k(t)\Delta u_k(t) + \sum_{i=0}^{t-1} \phi_k(i)\Delta u_k(i). \tag{9.6}$$

Substituting (9.6) into (9.5), yields

$$\begin{aligned} &J\left(u_k(t), \boldsymbol{\alpha}\right) \\ &= \left\{\alpha_1\left(e_{k-1}(t+1) - \sum_{i=0}^{t-1} \phi_k(i)\Delta u_k(i) - \phi_k(t)\Delta u_k(t)\right) + \sum_{m=2}^{M} \alpha_m e_{k-m+1}(t+1)\right\}^2 \\ &\quad + \lambda\left|u_k(t) - u_{k-1}(t)\right|^2 \end{aligned} \tag{9.7}$$

where $\Delta u_k(i) = u_k(i) - u_{k-1}(i), i \in \{0, 1, \ldots, t\}$.

Minimizing the objective function (9.5) with respect to $u_k(t)$ by replacing the unknown $\phi_k(t)$ with $\hat{\phi}_k(t)$, a learning control law is derived as

$$\begin{aligned} u_k(t) = u_{k-1}(t) &- \frac{\rho\alpha_1^2\hat{\phi}_k(t)\sum_{i=0}^{t-1}\hat{\phi}_k(i)\Delta u_k(i)}{\lambda + \alpha_1^2\hat{\phi}_k(t)^2} \\ &+ \frac{\rho\alpha_1\hat{\phi}_k(t)\left[\alpha_1 e_{k-1}(t+1) + \sum_{m=2}^{M}\alpha_m e_{k-m+1}(t+1)\right]}{\lambda + \alpha_1^2\hat{\phi}_k(t)^2} \end{aligned} \quad (9.8)$$

where $\rho > 0$ is a positive factor.

By integrating the control algorithm (9.8) and the parameter estimation algorithm (9.3) and (9.4), the proposed higher order DDILC (HODDILC) scheme is constructed as follows:

$$\begin{aligned} u_k(t) = u_{k-1}(t) &- \frac{\rho\alpha_1^2\hat{\phi}_k(t)\sum_{i=0}^{t-1}\hat{\phi}_k(i)\Delta u_k(i)}{\lambda + \alpha_1^2\hat{\phi}_k(t)^2} \\ &+ \frac{\rho\alpha_1\hat{\phi}_k(t)\left[\alpha_1 e_{k-1}(t+1) + \sum_{m=2}^{M}\alpha_m e_{k-m+1}(t+1)\right]}{\lambda + \alpha_1^2\hat{\phi}_k(t)^2} \end{aligned} \quad (9.9)$$

$$\hat{\boldsymbol{\varphi}}_k(t) = \hat{\boldsymbol{\varphi}}_{k-1}(t) + \frac{\eta\left(\Delta y_{k-1}(t+1) - \hat{\boldsymbol{\varphi}}_{k-1}(t)\Delta \boldsymbol{u}_{k-1}(t)\right)\Delta \boldsymbol{u}_{k-1}^{\mathrm{T}}(t)}{\mu + \|\Delta \boldsymbol{u}_{k-1}(t)\|^2} \quad (9.10)$$

$$\hat{\boldsymbol{\varphi}}_k(t) = \hat{\boldsymbol{\varphi}}_0(t), \text{ if } \operatorname{sgn}\left(\hat{\phi}_k(i)\right) \neq \operatorname{sgn}\left(\hat{\phi}_0(i)\right) \text{ or } \left\|\hat{\boldsymbol{\varphi}}_k(t)\right\| \leq \varepsilon. \quad (9.11)$$

**Remark 9.1** From the proposed HODDILC (9.9)–(9.11), it is clearly seen that (i) there is no matrix inverse calculation included, so an improved computational efficiency can be achieved; (ii) the unknown parameters in (9.9) are updated in terms of (9.10) and (9.11) such that the proposed method is more flexible for the modifications/or expansions of the controlled plant; and (iii) there is no model information used except for the input and output measurements, so the proposed method is data-driven and suitable for complex nonlinear processes in practice.

## 9.4 Convergence Analysis

Before proceeding to the analysis, a lemma is given as follows.

**Lemma 9.1** (Huang (1984)) *Let* $\|\boldsymbol{A}\|_v$ *and* $\|\boldsymbol{A}\|_\mu$ *be the proper matrix norm of* $\boldsymbol{A}$ *on the normed vector spaces* $v$ *and* $\mu$, *respectively. There exists* $\gamma \geq \sigma > 0$ *such that*

$$\gamma\|\boldsymbol{A}\|_v \geq \|\boldsymbol{A}\|_\mu \geq \sigma\|\boldsymbol{A}\|_v, \quad \forall \boldsymbol{A} \in C^{m\times n}.$$

Then, the convergence property of the proposed HODDILC (9.9)–(9.11) is summarized in the following theorem.

**Theorem 9.1** *For nonlinear system (9.1), if Assumptions 4.1, 4.2, and 4.3 hold, by properly selecting the controller parameters, the proposed HODDILC (9.9)–(9.11) guarantees that*

*i) the parameter estimation* $\hat{\boldsymbol{\varphi}}_k(t)$ *is bounded over the finite time interval* $t \in \{0, 1, \ldots, N\}$, $\forall k$;
*ii) the tracking error* $e_k(t+1)$ *converges to zero pointwisely over the finite time interval* $t \in \{0, 1, \ldots, N\}$ *as* $k$ *approaches to infinity, i.e.,* $\lim_{k\to\infty} e_k(t+1) = 0$;
*iii) the system output* $y_k(t)$ *and the control input* $u_k(t)$ *are bounded,* $\forall t \in \{0, 1, \ldots, N\}$, $\forall k$.

***Proof*** *Part 1. The boundedness of* $\hat{\boldsymbol{\varphi}}_k(t)$

The boundedness of $\hat{\boldsymbol{\varphi}}_k(t)$ can be easily obtained following the similar steps in Theorem 4.3. Therefore, we assume that $\|\hat{\boldsymbol{\varphi}}_k(t)\| \le b_{\hat{\varphi}}$, where $b_{\hat{\varphi}}$ is a positive constant.

*Part 2. The convergence of tracking error* $e_k(t+1)$

Two matrices $\boldsymbol{A}_k^u(t)$ and $\boldsymbol{B}_k^u(t)$ are defined as follows.

$$\boldsymbol{A}_k^u(t) = \begin{bmatrix} 0 & 1 & 0 & \cdots & 0 \\ 0 & 0 & 1 & \cdots & 0 \\ \vdots & \vdots & \vdots & \ddots & \vdots \\ 0 & 0 & 0 & \cdots & 1 \\ 0 & -\frac{\rho\alpha_1^2\hat{\phi}_k(t)\hat{\phi}_k(0)}{\lambda+\alpha_1^2\left|\hat{\phi}_k(t)\right|^2} & -\frac{\rho\alpha_1^2\hat{\phi}_k(t)\hat{\phi}_k(1)}{\lambda+\alpha_1^2\left|\hat{\phi}_k(t)\right|^2} & \cdots & -\frac{\rho\alpha_1^2\hat{\phi}_k(t)\hat{\phi}_k(t-1)}{\lambda+\alpha_1^2\left|\hat{\phi}_k(t)\right|^2} \end{bmatrix} \in R^{(t+1)\times(t+1)}$$

$$\boldsymbol{B}_k^u(t) = \left[ \frac{\rho\alpha_1\hat{\phi}_k(t)\alpha_M}{\lambda+\alpha_1^2\left|\hat{\phi}_k(t)\right|^2} \quad \frac{\rho\alpha_1\hat{\phi}_k(t)\alpha_{M-1}}{\lambda+\alpha_1^2\left|\hat{\phi}_k(t)\right|^2} \quad \cdots \quad \frac{\rho\alpha_1\hat{\phi}_k(t)\alpha_3}{\lambda+\alpha_1^2\left|\hat{\phi}_k(t)\right|^2} \quad \frac{\rho\alpha_1\hat{\phi}_k(t)\left(\alpha_1+\alpha_2\right)}{\lambda+\alpha_1^2\left|\hat{\phi}_k(t)\right|^2} \right] \in R^{1\times(M-1)}.$$

Define $\boldsymbol{C}^u = [0, \ldots, 0, 1]^{\mathrm{T}} \in R^{t+1}$, and $\bar{\boldsymbol{e}}_{k-1}(t+1) = [e_{k-M+1}(t+1), e_{k-M+2}(t+1), \ldots, e_{k-1}(t+1)]^{\mathrm{T}} \in R^{M-1}$. Furthermore, we define an expended vector with fixed $(t+1)$ dimensions as $\bar{\boldsymbol{u}}_k(t) = \boldsymbol{u}_k(t) = [u_k(0)\ u_k(1)\ \cdots\ u_k(t)]^{\mathrm{T}}$, then $\bar{\boldsymbol{u}}_k(t-1) = [u_k(-1)\ u_k(0)\ \cdots\ u_k(t-1)]^{\mathrm{T}}$, where the input signal $u_k(t)$ is set as zero for $t < 0$.

Then, according to (9.9), one has

$$\Delta\bar{\boldsymbol{u}}_k(t) = \boldsymbol{A}_k^u(t)\Delta\bar{\boldsymbol{u}}_k(t-1) + \boldsymbol{C}^u\boldsymbol{B}_k^u(t)\bar{\boldsymbol{e}}_{k-1}(t+1). \tag{9.12}$$

Note that the following inequality holds

$$\sum_{i=0}^{t-1}\left|\frac{\rho\alpha_1^2\hat{\phi}_k(t)\hat{\phi}_k(i)}{\lambda+\alpha_1^2\left|\hat{\phi}_k(t)\right|^2}\right| \leq \sum_{i=0}^{t-1}\frac{\rho\alpha_1^2\left|\hat{\phi}_k(t)\right|\left|\hat{\phi}_k(i)\right|}{2\sqrt{\lambda}\alpha_1\left|\hat{\phi}_k(t)\right|} \leq \sum_{i=0}^{t-1}\frac{\rho\alpha_1\left|\hat{\phi}_k(i)\right|}{2\sqrt{\lambda}}. \tag{9.13}$$

Since $\hat{\varphi}_k(t)$ is bounded, we have $|\hat{\phi}_k(t)| \leq b_{\hat{\phi}}$, where $b_{\hat{\phi}} > 0$ is a constant. Since $t$ is finite over $\{0, 1, \ldots, N\}$, by properly selecting $\lambda$ and $\rho$, such $\lambda > (\rho\alpha_1 b_{\hat{\phi}}N)^2$, there exists a positive constants $0 < d_1 < 0.5$ such that the following inequality holds, i.e.,

$$\begin{aligned} 0 &\leq \sum_{i=0}^{t-1}\left|\frac{\rho\alpha_1^2\hat{\phi}_k(t)\hat{\phi}_k(i)}{\lambda+\alpha_1^2\left|\hat{\phi}_k(t)\right|^2}\right| \leq \sum_{i=0}^{t-1}\frac{\rho\alpha_1\left|\hat{\phi}_k(i)\right|}{2\sqrt{\lambda}} \\ &\leq \frac{\rho\alpha_1 b_{\hat{\phi}}N}{2\sqrt{\lambda}} = d_1 < 0.5 \end{aligned} \tag{9.14}$$

which implies that the spectral radius $s(\boldsymbol{A}_k^u(t)) < d_1$ holds for all $t$ and $k$ in view of Lemma 4.2. Then, according to Lemma 3.2, there exist an arbitrary small positive constants $\delta_1$ such that

$$\begin{aligned} \left\|\boldsymbol{A}_k^u(t)\right\|_v &\leq s\left(\boldsymbol{A}_k^u(t)\right) + \delta_1 \leq d_1 + \delta_1 \\ &= d_2 < 0.5 \end{aligned} \tag{9.15}$$

where $\|\boldsymbol{A}_k^u(t)\|_v$ denotes a proper matrix norm of $\boldsymbol{A}_k^u(t)$ on the normed vector space $v$ at the $t$th time instant of the $k$th iteration, and $0 < d_2 < 0.5$ is a positive constants.

Taking norms on both sides of (9.12), yields

$$\begin{aligned} \|\Delta\bar{\boldsymbol{u}}_k(t)\|_v \leq{}& \left\|\boldsymbol{A}_k^u(t)\right\|_v \|\Delta\bar{\boldsymbol{u}}_k(t-1)\|_v \\ &+ \left\|\boldsymbol{C}^u\right\|_v \left\|\boldsymbol{B}_k^u(t)\right\|_v \|\bar{\boldsymbol{e}}_{k-1}(t+1)\|_v . \end{aligned} \tag{9.16}$$

According to Lemma 9.1, there exists a constant $\gamma$ such that

$$\left\|\boldsymbol{B}_k^u(t)\right\|_v \leq \gamma\left\|\boldsymbol{B}_k^u(t)\right\|_2 \tag{9.17}$$

Since $\sum_{m=1}^{M}\alpha_m = 1, 0 < \alpha_m \leq 1$, by virtue of (9.17), we have

$$\begin{aligned}\left\|\boldsymbol{B}_k^u(t)\right\|_v^2 &\le \gamma^2\left\|\boldsymbol{B}_k^u(t)\right\|_2^2 \\ &\le \frac{\gamma^2\left(\rho\alpha_1\hat{\phi}_k(t)\right)^2\left[(\alpha_1+\alpha_2)^2+\alpha_3^2+\cdots+\alpha_M^2\right]}{4\lambda\alpha_1^2\hat{\phi}_k(t)^2} \\ &\le \frac{\gamma^2\left(\rho\alpha_1\hat{\phi}_k(t)\right)^2(\alpha_1+\alpha_2+\cdots+\alpha_M)^2}{4\lambda\alpha_1^2\hat{\phi}_k(t)^2} = \frac{\gamma^2\rho^2}{4\lambda}.\end{aligned} \tag{9.18}$$

One can select $\lambda$ properly such that $\lambda > \gamma^2\rho^2/4$, then

$$\begin{aligned}\left\|\boldsymbol{B}_k^u(t)\right\|_v &\le \gamma\left\|\boldsymbol{B}_k^u(t)\right\|_2 \le \gamma\rho/(2\sqrt{\lambda}) \\ &= d_3 < 1\end{aligned} \tag{9.19}$$

where $0 < d_3 < 1$ is a positive constant.

Hence, one can derive from (9.15)–(9.19) that

$$\begin{aligned}\|\Delta\bar{\boldsymbol{u}}_k(t)\|_v &\le d_2\|\Delta\bar{\boldsymbol{u}}_k(t-1)\|_v + d_3\|\bar{\boldsymbol{e}}_{k-1}(t+1)\|_v \\ &\le \cdots \le d_3\sum_{i=0}^{t} d_2^{t-i}\|\bar{\boldsymbol{e}}_{k-1}(i+1)\|_v.\end{aligned} \tag{9.20}$$

By virtue of (9.2) and (9.12), one has

$$\begin{aligned}e_k(t+1) = e_{k-1}(t+1) &- \boldsymbol{\varphi}_k(t)\boldsymbol{C}^u\boldsymbol{B}_k^u(t)\bar{\boldsymbol{e}}_{k-1}(t+1) \\ &-\boldsymbol{\varphi}_k(t)\boldsymbol{A}_k^u(t)\Delta\bar{\boldsymbol{u}}_k(t-1).\end{aligned} \tag{9.21}$$

The second error term in (9.21) can be further expanded as

$$\begin{aligned}\boldsymbol{\varphi}_k(t)\boldsymbol{C}^u\boldsymbol{B}_k^u(t)\bar{\boldsymbol{e}}_{k-1}(t+1) &= \phi_k(t)\boldsymbol{B}_k^u(t)\bar{\boldsymbol{e}}_{k-1}(t+1) \\ &= \frac{\rho\alpha_1\phi_k(t)\hat{\phi}_k(t)(\alpha_1+\alpha_2)}{\lambda+\alpha_1^2\hat{\phi}_k(t)^2}e_{k-1}(t+1) \\ &\quad+\sum_{m=3}^{M}\frac{\rho\alpha_1\phi_k(t)\hat{\phi}_k(t)\alpha_m}{\lambda+\alpha_1^2\hat{\phi}_k(t)^2}e_{k-m+1}(t+1)\end{aligned} \tag{9.22}$$

Let $\vartheta_{k,1}(t) = \rho\alpha_1\phi_k(t)\hat{\phi}_k(t)(\alpha_1+\alpha_2)/\left(\lambda+\alpha_1^2\hat{\phi}_k(t)^2\right)$, and $\vartheta_{k,j}(t) = \rho\alpha_1\phi_k(t)\hat{\phi}_k(t)\alpha_{j+1}/\left(\lambda+\alpha_1^2\hat{\phi}_k(t)^2\right)$, $j = 2, \ldots, M-1$, for notational convenience. According to Assumption 4.3 and reset algorithm (9.11), it is easy to obtain that $\phi_k(t)\hat{\phi}_k(t) > 0$. Since $0 < \alpha_m \le 1, m = 1, \ldots, M$, one can get $\vartheta_{k,j}(t) > 0$, $j = 1, \ldots, M-1$ and

$$0 < \vartheta_{k,1}(t) \leq \frac{\rho\left|\phi_k(t)\right|}{2\sqrt{\lambda}} \text{ and } 0 < \vartheta_{k,j}(t) \leq \frac{\rho\left|\phi_k(t)\right|}{2\sqrt{\lambda}}. \tag{9.23}$$

Since $\|\boldsymbol{\varphi}_k(t)\| \leq b_\varphi$, by properly selecting $\lambda$ and $\rho$ such that $\lambda > \rho^2 b_\varphi^2/4$, one can guarantee that

$$0 < \vartheta_{k,1}(t) \leq \frac{\rho b_\varphi}{2\sqrt{\lambda}} < 1 \text{ and } 0 < \vartheta_{k,j}(t) \leq \frac{\rho b_\varphi}{2\sqrt{\lambda}} < 1. \tag{9.24}$$

Define

$$\boldsymbol{A}_k^e(t) = \begin{bmatrix} 0 & 1 & 0 & \cdots & 0 \\ 0 & 0 & 1 & \cdots & 0 \\ \vdots & \vdots & \vdots & \ddots & \vdots \\ 0 & 0 & 0 & \cdots & 1 \\ -\vartheta_{k,M}(t) & -\vartheta_{k,M-1}(t) & -\vartheta_{k,M-2}(t) & \cdots & 1-\vartheta_{k,1}(t) \end{bmatrix}$$

and $\boldsymbol{C}^e = [0 \ \cdots \ 0 \ 1]^{\mathrm{T}} \in R^{M-1}$. In terms of (9.21), we have

$$\bar{\boldsymbol{e}}_k(t+1) = \boldsymbol{A}_k^e(t)\bar{\boldsymbol{e}}_{k-1}(t+1) - \boldsymbol{C}^e\boldsymbol{\varphi}_k(t)\boldsymbol{A}_k^u(t)\Delta\bar{\boldsymbol{u}}_k(t-1). \tag{9.25}$$

From (9.24)

$$\begin{aligned} &\left|\vartheta_{k,M}(t)\right| + \left|\vartheta_{k,M-1}(t)\right| + \cdots + \left|\vartheta_{k,2}(t)\right| + \left|1-\vartheta_{k,1}(t)\right| \\ &= 1 - \left(\vartheta_{k,1}(t) - \vartheta_{k,2}(t) - \vartheta_{k,3}(t) - \cdots - \vartheta_{k,M}(t)\right) \\ &= 1 - \frac{\rho\alpha_1\left(\alpha_1+\alpha_2-\sum_{m=3}^{M}\alpha_m\right)\phi_k(t)\hat{\phi}_k(t)}{\lambda+\alpha_1^2\hat{\phi}_k(t)^2}. \end{aligned} \tag{9.26}$$

According to the condition $0 < \alpha_1 + \alpha_2 - \sum_{m=3}^{M}\alpha_m = \bar{\alpha} < 1$, as long as $\lambda > \rho^2 b_\varphi^2/4$, the following inequality holds

$$\begin{aligned} 0 < M_1 &\leq \frac{\rho\alpha_1\left(\alpha_1+\alpha_2-\sum_{m=3}^{M}\alpha_m\right)\phi_k(t)\hat{\phi}_k(t)}{\lambda+\alpha_1^2\hat{\phi}_k(t)^2} \\ &\leq \frac{\rho\alpha_1\bar{\alpha}\left|\phi_k(t)\right|\left|\hat{\phi}_k(t)\right|}{2\alpha_1\sqrt{\lambda}\left|\hat{\phi}_k(t)\right|} < \frac{\rho b_\varphi}{2\sqrt{\lambda}} = d_4 < 1 \end{aligned} \tag{9.27}$$

where $0 < M_1 < 1$ and $0 < d_4 < 1$ are two positive constants.

Therefore, it is obtained from (9.26) and (9.27) that

$$
\begin{aligned}
1 - d_4 &< \left|\vartheta_{k,M}(t)\right| + \cdots + \left|\vartheta_{k,2}(t)\right| + \left|1 - \vartheta_{k,1}(t)\right| \\
&\leq 1 - M_1 < 1. 
\end{aligned}
\tag{9.28}
$$

In terms of Lemma 2 and inequality (9.28), one can obtain that $s\left(\boldsymbol{A}_k^e(t)\right) < 1 - M_1 < 1$. Consequently, one can find an arbitrarily small positive constant $\delta_2$ such that

$$
\left\|\boldsymbol{A}_k^e(t)\right\|_v \leq s\left(\boldsymbol{A}_k^e(t)\right) + \delta_2 \leq 1 - M_1 + \delta_2 \leq d_5 < 1 \tag{9.29}
$$

where $0 < d_5 < 1$ is a positive constant.

By virtue of (9.19), (9.20), and (9.29), one can derive that

$$
\begin{aligned}
\|\bar{\boldsymbol{e}}_k(t+1)\|_v \leq& \left\|\boldsymbol{A}_k^e(t)\right\|_v \|\bar{\boldsymbol{e}}_{k-1}(t+1)\|_v \\
&+ \left\|\boldsymbol{C}^e\right\|_v \left\|\boldsymbol{\varphi}_k(t)\right\|_v \left\|\boldsymbol{A}_k^u(t)\right\|_v \|\Delta\bar{\boldsymbol{u}}_k(t-1)\|_v \\
\leq& d_5 \|\bar{\boldsymbol{e}}_{k-1}(t+1)\|_v + b_\varphi d_2 \|\Delta\bar{\boldsymbol{u}}_k(t-1)\|_v \\
\leq& d_5 \|\bar{\boldsymbol{e}}_{k-1}(t+1)\|_v \\
&+ b_\varphi d_3 \sum_{t=0}^{t} d_2^{t-i+1} \|\bar{\boldsymbol{e}}_{k-1}(i+1)\|_v .
\end{aligned}
\tag{9.30}
$$

Assume that $\|\bar{\boldsymbol{e}}_k(\tau+1)\|_v = \max_{t\in\{0,\ldots,N-1\}} \{\|\bar{\boldsymbol{e}}_k(t+1)\|_v\} = \bar{e}_k^{\max}$ is attained at time instant $\tau + 1$ where $\bar{e}_k^{\max}$ denotes the maximum value at the $k$th iteration. From (9.30), we have

$$
\begin{aligned}
\bar{e}_k^{\max} &= \|\bar{\boldsymbol{e}}_k(\tau+1)\|_v \\
&\leq \left(d_5 + \frac{b_\varphi d_3 d_2}{1 - d_2}\right) \|\bar{\boldsymbol{e}}_{k-1}(\tau+1)\|_v \\
&\leq d_6 \bar{e}_{k-1}^{\max}
\end{aligned}
\tag{9.31}
$$

where $d_6 = d_5 + \left(b_\varphi d_3 d_2\right)/(1 - d_2)$. Since $\lambda > \left(\left(\gamma^2\rho^2 b_\varphi^2\right)/4\right)$, one has $0 < b_\varphi d_3 < 1$. Furthermore, $0 < d_5 < 1$ and $0 < d_2/(1 - d_2) < 1$ due to $0 < d_2 < 0.5$. So by selecting $\lambda$ properly, one can get $0 < d_6 < 1$. Then inequality (9.31) implies that

$$
\bar{e}_k^{\max} \leq d_6 \bar{e}_{k-1}^{\max} \leq \cdots \leq d_6^k \bar{e}_0^{\max}. \tag{9.32}
$$

Since the initial value $\bar{e}_0^{\max}$ is bounded, we have

$$
0 \leq \lim_{k\to\infty} \|\bar{\boldsymbol{e}}_k(t+1)\|_v \leq \lim_{k\to\infty} \bar{e}_k^{\max} = 0. \tag{9.33}
$$

So, the asymptotic convergence of tracking error is proved.

*Part 3. The boundedness of system output and control input*

The boundedness of both $y_d(t+1)$ and $e_k(t+1)$ means that $y_k(t+1)$ is bounded. It is also clear that

$$\|\bar{\boldsymbol{u}}_k(t)\|_v \leq \sum_{j=1}^{k} \left\|\Delta\bar{\boldsymbol{u}}_j(t)\right\|_v + \|\bar{\boldsymbol{u}}_0(t)\|_v . \tag{9.34}$$

According to (9.20) and (9.34), we have

$$\|\bar{\boldsymbol{u}}_k(t)\|_v \leq d_3 \sum_{j=1}^{k}\sum_{i=0}^{t} d_2^{t-i} \left\|\bar{\boldsymbol{e}}_{j-1}(i+1)\right\|_v + \|\bar{\boldsymbol{u}}_0(t)\|_v . \tag{9.35}$$

Since $\|\bar{\boldsymbol{e}}_k(t+1)\|_v \leq \bar{e}_k^{\max}$, one can derive that

$$\|\bar{\boldsymbol{u}}_k(t)\|_v \leq \frac{d_3\left(1-d_2^t\right)}{1-d_2} \sum_{j=1}^{k} \bar{e}_k^{\max} + \|\bar{\boldsymbol{u}}_0(t)\|_v . \tag{9.36}$$

Then, according to (9.32)

$$\begin{aligned}\|\bar{\boldsymbol{u}}_k(t)\|_v &\leq M_3\left(d_6^k + d_6^{k-1} + \cdots + d_6^0\right)\bar{e}_0^{\max} + \|\bar{\boldsymbol{u}}_0(t)\|_v \\ &\leq M_3 \frac{1}{1-d_6}\bar{e}_0^{\max} + \|\bar{\boldsymbol{u}}_0(t)\|_v \end{aligned} \tag{9.37}$$

where $M_3 = d_3/(1-d_2)$ is bounded. Because $\bar{e}_0^{\max}$ and $\|\bar{\boldsymbol{u}}_0(t)\|_v$ are bounded, then (9.37) means that the control input is bounded at all-time instants and iterations. □

*Convergence Property Evaluation*: The convergence speed is of interest for practical applications. However, the quantification of the convergence speed is very difficult and little related work can be found in Tong et al. (2015), Ge et al. (2016), Li et al. (2015). In this paper, according to (9.16) and (9.30), it is clear that the convergence performance is affected mainly by the norm values of $\|\boldsymbol{A}_k^u(t)\|_v\|\boldsymbol{B}_k^u(t)\|_v$, and $\|\boldsymbol{A}_k^e(t)\|_v$. Therefore, one can define three indexes as follows to evaluate the convergence property of the proposed higher order learning control law,

$$S_1(\boldsymbol{\alpha}) = \left\|\boldsymbol{A}_k^u(t)\right\|_v,\ S_2(\boldsymbol{\alpha}) = \left\|\boldsymbol{B}_k^u(t)\right\|_v,\ S_3(\boldsymbol{\alpha}) = \left\|\boldsymbol{A}_k^e(t)\right\|_v . \tag{9.38}$$

All the three indices reflect the convergence rate qualitatively in the same direction, that is, the smaller the indexes are, the faster the convergence rate is.

Therefore, in order to guarantee that the higher order algorithm outperforms the lower order one, it is required that $S_n\left(\boldsymbol{\alpha}^H\right) < S_n\left(\boldsymbol{\alpha}^L\right)$, $n = 1, 2, 3$, where $\boldsymbol{\alpha}^H$ denotes the weighting vector for the higher order algorithm, and $\boldsymbol{\alpha}^L$ is for the lower order algorithm.

*Convergence Index 1*: Provided that $\alpha_1^L > \alpha_1^H$, then

$$\sum_{i=0}^{t-1}\left|\frac{\rho\left(\alpha_1^H\right)^2\hat{\phi}_k(t)\hat{\phi}_k(i)}{\lambda+\left(\alpha_1^H\right)^2\hat{\phi}_k(t)^2}\right| \leq \sum_{i=0}^{t-1}\left|\frac{\rho\left(\alpha_1^L\right)^2\hat{\phi}_k(t)\hat{\phi}_k(i)}{\lambda+\left(\alpha_1^L\right)^2\hat{\phi}_k(t)^2}\right|. \tag{9.39}$$

According to (9.14) and (9.15), inequality (9.39) implies that $S_1\left(\boldsymbol{\alpha}^H\right) < S_1\left(\boldsymbol{\alpha}^L\right)$. So, one can select the higher order factors such that $\alpha_1^L > \alpha_1^H$ to make the higher order learning law have a better performance than the lower order one.

*Convergence Index* 2: Because $\alpha_1^H + \alpha_2^H + \cdots + \alpha_H^H = 1$, $\alpha_1^L + \alpha_2^L + \cdots + \alpha_L^L = 1$, $\alpha_1 + \alpha_2 - \sum_{m=3}^{M}\alpha_m = \bar{\alpha} > 0$, and $H > L$, it is obvious that

$$\frac{\left(\rho\alpha_1^H\hat{\phi}_k(t)\right)^2\left(\left(\alpha_1^H+\alpha_2^H\right)^2+\left(\alpha_3^H\right)^2+\cdots+\left(\alpha_H^H\right)^2\right)}{\left(\lambda+\left(\alpha_1^H\right)^2\hat{\phi}_k(t)\right)^2} < \frac{\left(\rho\alpha_1^H\hat{\phi}_k(t)\right)^2\left(\left(\alpha_1^L+\alpha_2^L\right)^2+\left(\alpha_3^L\right)^2+\cdots+\left(\alpha_L^L\right)^2\right)}{\left(\lambda+\left(\alpha_1^H\right)^2\hat{\phi}_k(t)^2\right)^2}. \tag{9.40}$$

According to (9.18), in order to warrant $S_2(\boldsymbol{\alpha}^H) < S_2(\boldsymbol{\alpha}^L)$, it is required that

$$\frac{\alpha_1^H}{\lambda+\left(\alpha_1^H\right)^2\hat{\phi}_k(t)^2} < \frac{\alpha_1^L}{\lambda+\left(\alpha_1^L\right)^2\hat{\phi}_k(t)^2}. \tag{9.41}$$

Solving (9.41) subject to the condition of $\alpha_1^L > \alpha_1^H$, we get $\lambda > \alpha_1^H\alpha_1^L\hat{\phi}_k(t)^2$. That is, the conditions guaranteeing $S_2\left(\boldsymbol{\alpha}^H\right) < S_2\left(\boldsymbol{\alpha}^L\right)$ are $\lambda > \alpha_1^H\alpha_1^L b_{\hat{\phi}}^2$ and $\alpha_1^L > \alpha_1^H$.

*Convergence Index* 3: Define $\bar{\alpha}^H = \left(\alpha_1^H + \alpha_2^H - \sum_{h=3}^{H}\alpha_h^H\right)$, $\bar{\alpha}^L = \left(\alpha_1^L + \alpha_2^L - \sum_{l=3}^{L}\alpha_h^L\right)$.

If $\alpha_1^L\bar{\alpha}^L < \alpha_1^H\bar{\alpha}^H$, then

$$\lambda\left(\alpha_1^L\bar{\alpha}^L - \alpha_1^H\bar{\alpha}^H\right) < \alpha_1^L\alpha_1^H\left(\alpha_1^H\bar{\alpha}^H - \alpha_1^L\bar{\alpha}^L\right)\left|\hat{\phi}_k(t)\right|^2 \tag{9.42}$$

which means

$$\frac{\alpha_1^H\bar{\alpha}^H}{\lambda+\left(\alpha_1^H\right)^2\hat{\phi}_k(t)^2} > \frac{\alpha_1^L\bar{\alpha}^L}{\lambda+\left(\alpha_1^L\right)^2\left|\hat{\phi}_k(t)\right|^2}. \tag{9.43}$$

Hence, according to (9.26)–(9.29), inequality (9.43) implies that $S_3(\boldsymbol{\alpha}^H) < S_3(\boldsymbol{\alpha}^L)$.

Note that the condition of $\alpha_1^L\bar{\alpha}^L < \alpha_1^H\bar{\alpha}^H$ is not applicable to the cases of first-order and second-order algorithms where $\bar{\alpha}^1 = \bar{\alpha}^2 = 1$ are fixed and cannot be manipulated further. So, for the first-order and second-order algorithms, one can only use $S_1(\boldsymbol{\alpha})$ and $S_2(\boldsymbol{\alpha})$ to evaluate the convergence property of the tracking error in theory.

As a summary of the above three cases, for the higher order algorithms with more than third orders, one can select $\boldsymbol{\alpha}$ properly such that $\alpha_1^L > \alpha_1^H$, $\lambda > \alpha_1^H \alpha_1^L b_{\hat{\phi}}^2$, and $\alpha_1^L \bar{\alpha}^L < \alpha_1^H \bar{\alpha}^H$ to guarantee an improved performance of the higher order learning control law.

**Remark 9.2** Note that the controller parameters must be selected to guarantee the convergence of the proposed method at first, and then, the convergence property is evaluated in the sequel. So a faster convergence by selecting the higher order factors properly does not neglect the tracking performance but actually enhances it.

## 9.5 Simulation Study

For comparison with traditional optimal ILC, a linear time-varying (LTV) system is adopted from Hwang et al. (1991). When applying the traditional lifted optimal ILC, the linear model is required to be known exactly. However, it is considered to be unknown and only serves for generating the I/O data for evaluating the proposed HODDILC. The LTV system is shown as follows,

$$\begin{cases} \boldsymbol{x}(t+1) = \boldsymbol{A}(t)\boldsymbol{x}(t) + \boldsymbol{B}(t)u(t) \\ y(t) = \boldsymbol{C}(t)\boldsymbol{x}(t) \end{cases} \tag{9.44}$$

where

$$\boldsymbol{A}(t) = \begin{bmatrix} 0 & 1 \\ -0.5 - 10^{-3}t & -0.5 - 10^{-3}t \end{bmatrix}, \quad \boldsymbol{B}(t) = \begin{bmatrix} 0 \\ 1 \end{bmatrix}$$

and $\boldsymbol{C}(t) = [1\ 0]$; $t \in \{0, \ldots, 200\}$. The control task is to track the following desired trajectory,

$$y_d(t) = 10^{-6}(t-1)^3(4 - 0.03(t-1)), \quad t \in \{0, \ldots, 200\}. \tag{9.45}$$

In the simulation, the initial states $y_k(0)$ in all iterations are set as 0 and the input $u_0(t)$ of the first iteration is 0. The controller parameters are selected as: $\hat{\varphi}_0(t) = 0.7$, $\rho = 0.9$, $\eta = 0.8$, $\lambda = 0.05$, and $\mu = 1$.

For comparison, the first-order, second-order, third-order, and fourth-order forms of the proposed HODDILC are applied respectively with the same simulation conditions except that the higher order factors are selected differently: $\boldsymbol{\alpha}^1 = [1]$ for the first order, $\boldsymbol{\alpha}^2 = [0.9, 0.1]$ for the second-order, and $\boldsymbol{\alpha}^3 = [0.8, 0.14, 0.06]$ for the third-order, and $\boldsymbol{\alpha}^4 = [0.75, 0.23, 0.01, 0.01]$ for the fourth order. Note that $\alpha_1^1 > \alpha_1^2 > \alpha_1^3 > \alpha_1^4$ and $\alpha_1^3 \bar{\alpha}^3 = 0.704 < \alpha_1^4 \bar{\alpha}^4 = 0.72$ satisfy the conditions derived from the convergence indexes.

It should be noted that in order to compare the proposed HODDILC and the traditional OILC under the same simulation conditions, all the learning control laws start from the third iteration because the calculation of the control input in the fourth-order HODDILC uses the error information from the previous three iterations.

The simulation results are shown in Figs. 9.1 and 9.2. Figure 9.1 demonstrates the asymptotic convergence of the proposed non-lifted HODDILC, where the Y-axis is the mean absolute value of tracking error $e_{\text{mean}}(k) = \sum_{t=1}^{200} |y_d(t) - y_k(t)|/200$, and X-axis denotes the iteration numbers. Obviously, the asymptotic convergence of tracking error can be guaranteed by the proposed HODDILC. Meanwhile, it is demonstrated that a faster convergence can be achieved by selecting the higher order factors according to the conditions given in this paper.

Figure 9.2 shows the computation time of the proposed HODDILC by taking the third-order learning control law as an example. The horizontal axis denotes the operating length of the controlled process within each batch. The unit of the horizontal axis is 200 instants per iteration. The vertical axis denotes the calculation time and the unit is second.

The calculation is performed in MATLAB on a Lenovo Thinkbook laptop computer with a 2.40 GHz Intel Core i5 process and 16 GB of RAM. The batch length is changed by selecting different sampling rates. It is verified that the computation time of the proposed HODDILC increases very slowly with the increasing size of data sets because no matrix inverse operation is required for the implementation of the learning control law.

For a comparison, an optimal ILC (OILC) law is selected from Sun and Alleyne (2014)

$$\begin{aligned} \boldsymbol{U}_{k+1} = &\left(\boldsymbol{G}^{\mathrm{T}}\boldsymbol{Q}\boldsymbol{G} + \boldsymbol{R} + \boldsymbol{S}\right)^{-1}\left(\boldsymbol{G}^{\mathrm{T}}\boldsymbol{Q}\boldsymbol{G} + \boldsymbol{R}\right)\boldsymbol{U}_k \\ &+ \left(\boldsymbol{G}^{\mathrm{T}}\boldsymbol{Q}\boldsymbol{G} + \boldsymbol{R} + \boldsymbol{S}\right)^{-1}\boldsymbol{G}^{\mathrm{T}}\boldsymbol{Q}\boldsymbol{E}_k \end{aligned} \tag{9.46}$$

where $\boldsymbol{Q}$, $\boldsymbol{R}$, and $\boldsymbol{S}$ are real-valued symmetric positive definite matrices.

By selecting $\boldsymbol{Q} = \boldsymbol{I}$, $\boldsymbol{R} = 0.2\boldsymbol{I}$, and $\boldsymbol{S} = \boldsymbol{0}$, the simulation results are also shown in Figs. 9.1 and 9.2.

From Fig. 9.1, it is seen that the control performance of the proposed HODDILC is better than that of the traditional OILC because more historical and online control information is employed. In contrast, only the historical control information in the immediate previous one iteration is used in the traditional OILC.

Figure 9.2 shows that the computational time of the traditional OILC approach (9.46) increases dramatically as the trial length increases. The matrix inverse operation in lifted OILC becomes impractical with increasing dimension and data points.

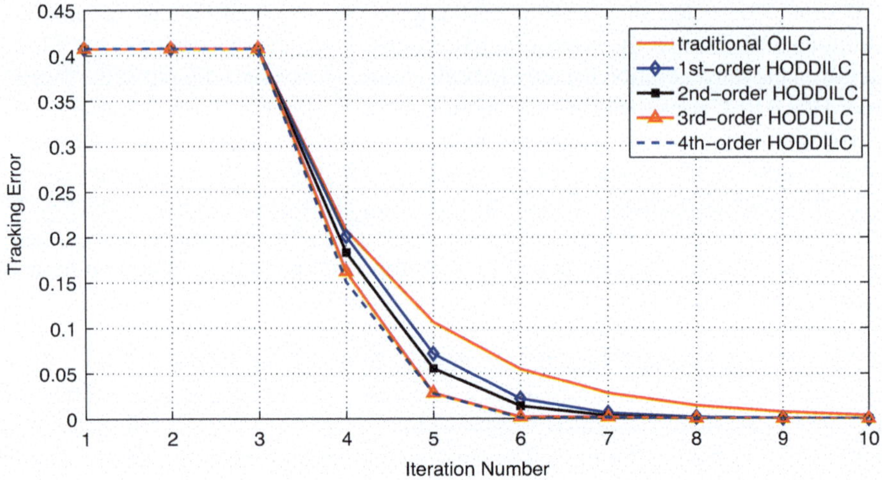


**Fig. 9.1** Tracking errors with respect to iterations

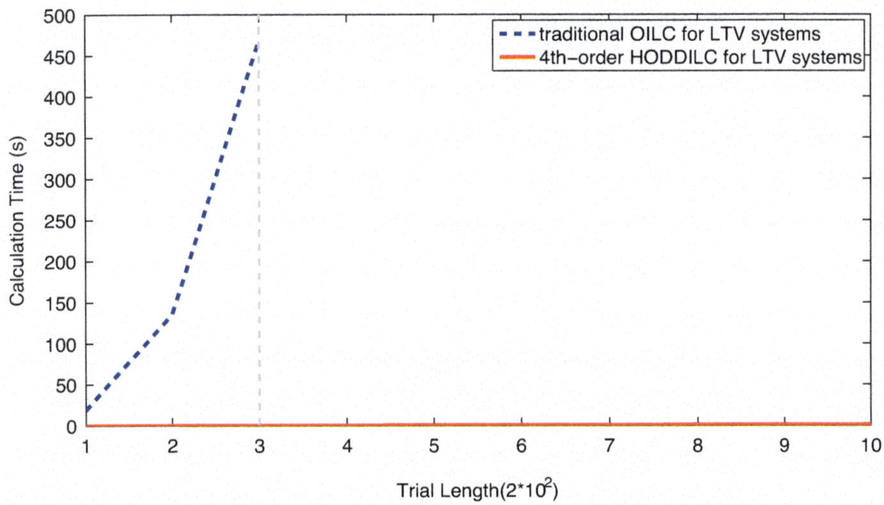


**Fig. 9.2** Computation time

## 9.6 Summary

In this chapter, a new HODDILC is proposed for nonlinear nonaffine discrete-time systems by using historical and online data. A non-lifted IDL is constructed for the nonlinear plant at first, and then the HODDILC is proposed by designing a control objective function with higher order factors. It is shown theoretically that the convergence property of the higher order learning control law can be better than that of the lower order one by properly selecting the higher order factors. Furthermore, the computational complexity of the proposed HODDILC is reduced significantly since no matrix inversion is included. The proposed HODDILC method does not require any explicit model information except for the I/O data and thus can be more applicable to complex nonlinear processes in practice.

## References

Bien Z, Huh KM (1989) Higher-order iterative learning control algorithm. IEE Proc Control Theory Appl 136(3):105–112

Chen Y, Wen C, Sun M (1997) A robust high-order P-type iterative learning controller using current iteration tracking error. Int J Control 68(2):331–342

Chen Y, Gong Z, Wen C (1998) Analysis of a high-order iterative learning control algorithm for uncertain nonlinear systems with state delays. Automatica 34(3):345–353

Chi R, Hou Z, Huang B, Jin S (2015) A unified data-driven design framework of optimality-based generalized iterative learning control. Comput Chem Eng 77:10–23

Chi R, Liu Y, Hou Z, Jin S (2015) Data-driven terminal iterative learning control with high-order learning law for a class of non-linear discrete-time multiple-input-multiple output systems. IET Control Theory A 9(7):1075–1082

Chi R, Liu Y, Hou Z, Jin S (2015) High-order data-driven optimal TILC approach for fed-batch processes. Can J Chem Eng 93(8):1455–1461

Ge X, Stein JL, Ersal T (2016) A frequency domain approach for designing filters for norm-optimal iterative learning control and its fundamental tradeoff between robustness, convergence speed and steady state error. In: Proceedings of American control conference. IEEE. https://ieeexplore.ieee.org/document/7524945

Gunnarsson S, Norrl FM (2006) On the disturbance properties of high order iterative learning control algorithms. Automatica 42(11):2031–2034

Hätönen J, Owens DH, Feng K (2006) Basis functions and parameter optimisation in high-order iterative learning control. Automatica 42:287–294

Huang L (1984) Linear algebra system and control theory. Science Press, Beijing

Hwang DH, Bien Z, Oh SR (1991) Iterative learning control method for discrete-time dynamic systems. IEE Proc Control Theory Appl 138(2):139–144

Li X, Ren Q, Xu J (2015) Precise speed tracking control of a robotic fish via iterative learning control. IEEE Trans Ind Electron 63(4):2221–2228

Schmid R (2007) Comments on "robust optimal design and convergence properties analysis of iterative learning control approaches" and "on the p-type and newton type ILC schemes for dynamic systems with non-affine input factors." Automatica 43(9):1666–1669

Sun H, Alleyne AG (2014) A computationally efficient norm optimal iterative learning control approach for LTV systems. Automatica 50(1):141–148

Tong DS, Pipeleers G, Swevers J (2015) Robust monotonic convergent iterative learning control. IEEE Trans Autom Control 61(4):1063–1068

Xu J, Tan Y (2002) On the p-type and newton-type ILC schemes for dynamic systems with non-affine-in-input factors. Automatica 38(7):1237–1242

Xu J, Tan Y (2002) Robust optimal design and convergence properties analysis of iterative learning control approaches. Automatica 38(11):1867–1880

Xu J, Tan Y (2003) Linear and Nonlinear Iterative Learning Control. Lecture Notes in Control and Information Sciences, Berlin, Heidelberg, New York

# Chapter 10
# Constrained Data-Driven Iterative Learning Control

## 10.1 Introduction

In real process industries, a soft limiter is often added directly into the controller to prevent over large control input signals (Doyle, 1999). However, such a limiter makes the control systems highly nonlinear and may introduce instability. It is also worth pointing out that the input saturation will bring more serious influence on ILC systems because ILC is virtually an integral action along the iteration direction (Xu et al. 2004). Recently, several ILC schemes with input saturation have been proposed under the framework of contraction mapping (Sun and Wang 2001; Driessen and Sadegh 2002; Chu and Owens 2010; Huang et al. 2013). However, the contraction mapping-based ILC may cause some poor transient performance, even divergence at some time instants. Therefore, an adaptive ILC with input saturation has been proposed in Xu et al. (2004) by introducing a composite energy function. In Tan et al. (2011), a 'reference governor'-based ILC is proposed for input constraint systems where the reference signal can be re-designed such that the tracking objectives realizable. The authors in Tan et al. (2012) propose a unified design framework of ILC to deal with nonlinear input uncertainties including input saturations.

Besides the input saturation, the output constraints are also of a great concern from a practical viewpoint. For example, the train operation system is constrained not only by the input force because of its mechanical features, but also by its operation speed because of its over speed limits. The latter seems more important to ensure operation performance and safety. Generally, the position and/or speed constraints are required in most motion systems. In the expository overview of ILC (Moore, 1999), optimal ILC (Amann et al. 1996a, b, c; Lee et al. 2000) is revisited and classified as a connection between ILC and optimal control. By using a predesigned objective function, various practical issues such as constraints on the system output and control input, disturbances, measurement noises, and model errors can be considered in a rigorous and systematic manner. Recently, some model predictive control (MPC)-based ILC schemes are also proposed in the literatures (Mishra et al. 2010; Amann et al. 1998;



R. Chi et al., *Data-Driven Iterative Learning Control for Discrete-Time Systems*, Intelligent Control and Learning Systems 2,
https://doi.org/10.1007/978-981-19-5950-9_10

Lee et al. 1999; Wang et al. 2013; Chin et al. 2004) by incorporating the input and output constraints into the optimization problem. For example, in Chin et al. (2004), a two-stage algorithm has been devised by modifying and combining the existing Q-ILC and BMPC techniques. Further, based on a 2D piecewise linear description of a batch process, Liu and Wang (2012) propose a constrained ILC approach by solving a linear matrix inequality for performance optimization. However, the main limitation is the requirement of an exact linear model of the controlled plant to guarantee a satisfactory control performance. In order to obtain a linear model for the nonlinear system, two open-loop tests have to be carried out where the difference between the two inputs is kept constant throughout the batch.

To handle broader system nonlinearities and uncertainties, adaptive ILC (Chi et al. 2008; Yin et al. 2010; Li et al. 2011; Yu et al. 2013) has been proposed by exploiting the use of energy function approaches. Consequently, an adaptive ILC with state constraints is discussed in Xu and Jin (2012) by using barrier composite energy function. An adaptive ILC is presented in Jin and Xu (2013) for output-constrained systems with both parametric and nonparametric uncertainties by introducing a new barrier composite energy function. However, the model structure of the controlled system must be available as a priori knowledge.

In real world applications, however, the model-based control approaches may encounter many challenges with unsatisfactory performance because it is increasingly difficult to gain an exact model of a complex process with increasing scale and complexity. Recently, a data-driven constrained NOILC is proposed for linear time-invariant systems (Janssens et al. 2012), where the system's impulse response is estimated using I/O measurements from previous iterations. However, the control performance mainly depends on the estimation precision of the impulse response.

This chapter considers input and output constraints and proposes a constrained-DDILC method for nonlinear nonaffine discrete-time systems. Before proceeding to the controller design, the IDL method is revisited with supervector formulation. An objective function of control input is designed with a penalty term on the iterative input change. The constraints on the bounds of control input and system output, as well as the change rate of the control input across iterations, are transferred into a linear matrix inequality (LMI). Then, the constrained-DDILC method is obtained by minimizing the designed objective function under this LMI condition. This chapter also extends the proposed constrained-DDILC to point-to-point motion systems for tracking multiple intermediate points.

The two proposed methods are data-driven, where no explicit mathematical model, either obtained by first-principles or by identification techniques, is needed. The learning gains of the two proposed methods can be iteratively estimated using the real-time I/O data, rather than being fixed as that in the traditional optimal ILC methods, which makes the proposed methods more flexible for the modifications or expansions of the controlled plant. Rigorous analysis shows that both of the proposed approaches can achieve a monotonic convergence along the iteration axis. Simulation results are further provided to verify the effectiveness of the two proposed constrained-DDILC and DDPTPILC methods.

This chapter is organized as follows. Section 10.2 shows the problem formulation. The constraint DDILC design and analysis are presented in Sects. 10.3 and 10.4, respectively. Section 10.5 shows the design and analysis of the constrained DDPTPILC. Simulation study is given in Sect. 10.6. Section 10.7 gives some conclusions.

## 10.2 Problem Formulation

Consider a class of nonlinear nonaffine discrete-time systems

$$y_k(t+1) = f\left(y_k(t), \ldots, y_k\left(t-n_y\right), u_k(t), \ldots, u_k\left(t-n_u\right)\right) \tag{10.1}$$

where $y_k(t)$ and $u_k(t)$ are system output and input, respectively; $y_k(t) = 0$ and $u_k(t) = 0$, $\forall t < 0$; $n_y$ and $n_u$ are two unknown orders; $t \in \{0, 1, \ldots, N\}$ is time instant and $N \in Z^+$ represents the endpoint; $k \in \{0, 1, \ldots\}$ indicates the iteration number; $f(\cdots) : R^{n_u+n_y+2} \mapsto R$ is an unknown nonlinear scalar function, and $f(0, 0) = 0$.

Consider nonlinear system (10.1) satisfying assumptions 4.1-4.2.

Similar to (4.7), one has

$$\Delta \boldsymbol{Y}_k = \boldsymbol{\Phi}_k \Delta \boldsymbol{U}_k \tag{10.2}$$

where $\boldsymbol{U}_k = [u_k(0), u_k(1), \ldots, u_k(N-1)]^{\mathrm{T}} \in R^N$ and $\boldsymbol{Y}_k = [y_k(1), y_k(2), \ldots, y_k(N)]^{\mathrm{T}} \in R^N$; $\Delta \boldsymbol{Y}_k = \boldsymbol{Y}_k - \boldsymbol{Y}_{k-1}$ and $\Delta \boldsymbol{U}_k = \boldsymbol{U}_k - \boldsymbol{U}_{k-1}$; $\boldsymbol{\Phi}_k \in R^{N \times N}$ denotes the partial derivative values of nonlinear function $\boldsymbol{g}(\cdots)$ with respect to $\boldsymbol{U}_k$ at some point in the interval $[\boldsymbol{U}_k, \boldsymbol{U}_{k-1}]$.

In terms of Assumption 4.2, one has $\|\boldsymbol{\Phi}_k\| \leq b_{\Phi}$, where $b_{\Phi} > 0$ is a constant.

In this chapter, three kinds of constraints are taken into consideration.

(1) Constraints on the bounds of input values,

$$\boldsymbol{U}^{\mathrm{low}} \leq \boldsymbol{U}_{k+1} \leq \boldsymbol{U}^{\mathrm{hi}} \tag{10.3}$$

(2) Constraints on the iterative change rate of the control input along with iteration axis,

$$\Delta \boldsymbol{U}^{\mathrm{low}} \leq \Delta \boldsymbol{U}_{k+1} \leq \Delta \boldsymbol{U}^{\mathrm{hi}} \tag{10.4}$$

(3) Constraints on the bounds of system outputs

$$\boldsymbol{Y}^{\mathrm{low}} \leq \boldsymbol{Y}_{k+1} \leq \boldsymbol{Y}^{\mathrm{hi}} \tag{10.5}$$

where $\boldsymbol{U}^{\mathrm{low}}$, $\boldsymbol{U}^{\mathrm{hi}}$, $\Delta \boldsymbol{U}^{\mathrm{low}}$, $\Delta \boldsymbol{U}^{\mathrm{hi}}$, $\boldsymbol{Y}^{\mathrm{low}}$, and $\boldsymbol{Y}^{\mathrm{hi}}$ are some constant vectors, which are known exactly.

Subtracting $\boldsymbol{U}_k$ from both sides of the inequality (10.3), one can obtain

$$\boldsymbol{U}^{\text{low}} - \boldsymbol{U}_k \leq \Delta\boldsymbol{U}_{k+1} \leq \boldsymbol{U}^{\text{hi}} - \boldsymbol{U}_k \tag{10.6}$$

Subtracting $\boldsymbol{Y}_k$ from both sides of (10.5), yields

$$\boldsymbol{Y}^{\text{low}} - \boldsymbol{Y}_k \leq \Delta\boldsymbol{Y}_{k+1} \leq \boldsymbol{Y}^{\text{hi}} - \boldsymbol{Y}_k \tag{10.7}$$

According to (10.2), the following inequalities are satisfied,

$$\boldsymbol{Y}^{\text{low}} - \boldsymbol{Y}_k \leq \boldsymbol{\Phi}_{k+1}\Delta\boldsymbol{U}_{k+1} \leq \boldsymbol{Y}^{\text{hi}} - \boldsymbol{Y}_k \tag{10.8}$$

Combing (10.4), (10.6) and (10.8), the above constraints can be transferred to a linear matrix inequality as follows,

$$\boldsymbol{A}'_k\Delta\boldsymbol{U}_{k+1} \leq \boldsymbol{b}'_k \tag{10.9}$$

where $\boldsymbol{A}'_k = \begin{bmatrix} \boldsymbol{I} \\ -\boldsymbol{I} \\ \boldsymbol{I} \\ -\boldsymbol{I} \\ \boldsymbol{\Phi}_{k+1} \\ -\boldsymbol{\Phi}_{k+1} \end{bmatrix}$ and $\boldsymbol{b}'_k = \begin{bmatrix} \boldsymbol{U}^{\text{hi}} - \boldsymbol{U}_k \\ -\boldsymbol{U}^{\text{low}} + \boldsymbol{U}_k \\ \Delta\boldsymbol{U}^{\text{hi}} \\ -\Delta\boldsymbol{U}^{\text{low}} \\ \boldsymbol{Y}^{\text{hi}} - \boldsymbol{Y}_k \\ -\boldsymbol{Y}^{\text{low}} + \boldsymbol{Y}_k \end{bmatrix}$.

Let $\Delta\boldsymbol{U}_k^{\text{low}*} = \max\left(\boldsymbol{U}^{\text{low}} - \boldsymbol{U}_k, \Delta\boldsymbol{U}^{\text{low}}\right)$, and $\Delta\boldsymbol{U}_k^{\text{hi}*} = \min\left(\boldsymbol{U}^{\text{hi}} - \boldsymbol{U}_k, \Delta\boldsymbol{U}^{\text{hi}}\right)$; then the linear matrix inequality (10.9) becomes

$$\boldsymbol{A}_k\Delta\boldsymbol{U}_{k+1} \leq \boldsymbol{b}_k \tag{10.10}$$

where $\boldsymbol{A}_k = \begin{bmatrix} \boldsymbol{I} \\ -\boldsymbol{I} \\ \boldsymbol{\Phi}_{k+1} \\ -\boldsymbol{\Phi}_{k+1} \end{bmatrix}$ and $\boldsymbol{b}_k = \begin{bmatrix} \Delta\boldsymbol{U}_k^{\text{hi}*} \\ -\Delta\boldsymbol{U}_k^{\text{low}*} \\ \boldsymbol{Y}^{\text{hi}} - \boldsymbol{Y}_k \\ -\boldsymbol{Y}^{\text{low}} + \boldsymbol{Y}_k \end{bmatrix}$.

Given a desired trajectory $y_d(t), t \in \{0, \ldots, N\}$. The objective is to find an optimal control input $u_k(t)$ under the above three kinds of constraints (9) – (11) such that the system output $y_k(t)$ can track $y_d(t)$ exactly for all time instants. That is, the tracking error $\boldsymbol{E}_k = \boldsymbol{Y}_d - \boldsymbol{Y}_k$ converges to zero as the iteration number $k$ approaches to infinity where $\boldsymbol{Y}_d = [y_d(1), y_d(2), \ldots, y_d(N)]^{\text{T}} \in R^N$.

## 10.3 Controller Design

The constraint optimal ILC law is proposed by minimizing the following quadratic performance index function with respect to $\boldsymbol{U}_{k+1}$ under the linear matrix inequality of the constraints (10.10),

$$J\left(\boldsymbol{E}_{k+1}, \boldsymbol{U}_{k+1}\right)=\frac{1}{2}\left\{\boldsymbol{E}_{k+1}^{\mathrm{T}} \boldsymbol{Q} \boldsymbol{E}_{k+1}+\Delta \boldsymbol{U}_{k+1}^{\mathrm{T}} \boldsymbol{R} \Delta \boldsymbol{U}_{k+1}\right\} \tag{10.11}$$

where $\boldsymbol{Q}$ and $\boldsymbol{R}$ are positive definite weighting matrices.

According to (10.2), it is easy to get

$$\boldsymbol{E}_{k+1}=\boldsymbol{Y}_d-\boldsymbol{Y}_{k+1}=\boldsymbol{Y}_d-\boldsymbol{Y}_k-\Delta \boldsymbol{Y}_{k+1}=\boldsymbol{E}_k-\boldsymbol{\Phi}_{k+1} \Delta \boldsymbol{U}_{k+1} \tag{10.12}$$

Substituting (10.12) into (10.11), yields

$$\begin{aligned} J\left(\boldsymbol{E}_{k+1}, \boldsymbol{U}_{k+1}\right)=\frac{1}{2} & \left\{\Delta \boldsymbol{U}_{k+1}^{\mathrm{T}}\left(\boldsymbol{\Phi}_{k+1}^{\mathrm{T}} \boldsymbol{Q} \boldsymbol{\Phi}_{k+1}+\boldsymbol{R}\right) \Delta \boldsymbol{U}_{k+1}\right. \\ & \left.-2 \boldsymbol{E}_k^{\mathrm{T}} \boldsymbol{Q} \boldsymbol{\Phi}_{k+1} \Delta \boldsymbol{U}_{k+1}+\boldsymbol{E}_k^{\mathrm{T}} \boldsymbol{Q} \boldsymbol{E}_k\right\} \end{aligned} \tag{10.13}$$

Then the optimal ILC law can be obtained by minimizing (10.13) with the constraints (10.10), that is,

$$\left\{\begin{aligned} \min_{\boldsymbol{U}_{k+1}} J\left(\boldsymbol{E}_{k+1}, \boldsymbol{U}_{k+1}\right)=\min_{\boldsymbol{U}_{k+1}} \tfrac{1}{2} & \left\{\Delta \boldsymbol{U}_{k+1}^{\mathrm{T}}\left(\boldsymbol{\Phi}_{k+1}^{\mathrm{T}} \boldsymbol{Q} \boldsymbol{\Phi}_{k+1}+\boldsymbol{R}\right) \Delta \boldsymbol{U}_{k+1}\right. \\ & \left.-2 \boldsymbol{E}_k^{\mathrm{T}} \boldsymbol{Q} \boldsymbol{\Phi}_{k+1} \Delta \boldsymbol{U}_{k+1}+\boldsymbol{E}_k^{\mathrm{T}} \boldsymbol{Q} \boldsymbol{E}_k\right\} \\ \text{s.t.} \boldsymbol{A}_k \Delta \boldsymbol{U}_{k+1} \leq \boldsymbol{b}_k & \end{aligned}\right. \tag{10.14}$$

where $\boldsymbol{\Phi}_{k+1}$ is unknown. Therefore, an iterative updating law is presented in this section to estimate $\boldsymbol{\Phi}_{k+1}$ by using the input and output data only. Since $\boldsymbol{\Phi}_{k+1}$ is a lower triangular matrix and all the upper triangular elements are zeros, so only the nonzero elements of $\boldsymbol{\Phi}_{k+1}$ are estimated to reduce the computational complexity. According to Section 4.3, the parameter updating law of $\boldsymbol{\Phi}_{k+1}$ is given as follows,

$$\hat{\boldsymbol{\varphi}}_{k+1}(t)=\hat{\boldsymbol{\varphi}}_k(t)+\frac{\eta\left(\Delta y_k(t+1)-\hat{\boldsymbol{\varphi}}_k(t) \Delta \boldsymbol{u}_k(t)\right) \Delta \boldsymbol{u}_k^{\mathrm{T}}(t)}{\mu+\left\|\Delta \boldsymbol{u}_k(t)\right\|^2} \tag{10.15}$$

where $\hat{\boldsymbol{\varphi}}_{k+1}(t)$ is the estimated value of $\boldsymbol{\varphi}_{k+1}(t)$, and $\boldsymbol{\varphi}_{k+1}(t)=\left[\frac{\partial g_t}{\partial u_{k+1}(0)}, \frac{\partial g_t}{\partial u_{k+1}(1)}, \ldots, \frac{\partial g_t}{\partial u_{k+1}(t)}\right]$ denotes the nonzero vector of the $(t+1)$th row of $\boldsymbol{\Phi}_{k+1}$. Correspondingly, $\boldsymbol{u}_k(t)=\left[u_k(0), u_k(1), \ldots, u_k(t)\right]^{\mathrm{T}} \in R^{t+1}$ is a varying dimension vector; $\mu>0$ is a weighting factor and $0<\eta<2$ is a step-size factor.

Finally, the proposed constrained optimal ILC law can be obtained by solving the following quadratic programming (QP) problem according to the updating law (10.15),

$$\left\{\begin{aligned} \min_{\boldsymbol{U}_{k+1}}\left(\boldsymbol{E}_{k+1}, \boldsymbol{U}_{k+1}\right)=\min_{\boldsymbol{U}_{k+1}} \tfrac{1}{2} & \left\{\Delta \boldsymbol{U}_{k+1}^{\mathrm{T}}\left(\hat{\boldsymbol{\Phi}}_{k+1}^{\mathrm{T}} \boldsymbol{Q} \hat{\boldsymbol{\Phi}}_{k+1}+\boldsymbol{R}\right) \Delta \boldsymbol{U}_{k+1}\right. \\ & \left.-2 \boldsymbol{E}_k^{\mathrm{T}} \boldsymbol{Q} \hat{\boldsymbol{\Phi}}_{k+1} \Delta \boldsymbol{U}_{k+1}+\boldsymbol{E}_k^{\mathrm{T}} \boldsymbol{Q} \boldsymbol{E}_k\right\} \\ \text{s.t.} \hat{\boldsymbol{A}}_k \Delta \boldsymbol{U}_{k+1} \leq \boldsymbol{b}_k & \end{aligned}\right. \tag{10.16}$$

where $\hat{\boldsymbol{A}}_k = \begin{bmatrix} \boldsymbol{I} \\ -\boldsymbol{I} \\ \hat{\boldsymbol{\Phi}}_{k+1} \\ -\hat{\boldsymbol{\Phi}}_{k+1} \end{bmatrix}$.

That is, the proposed constraint DDILC is constructed by (10.15) and (10.16).

**Remark 10.1** From (10.15) and (10.16), it can be seen that no system model information except for the I/O data is used in the algorithms and thus the proposed approach is data-driven. Further, the learning gain of (10.16) can be iteratively updated by using (10.15), so the proposed approach can be more robust to the system uncertainties.

## 10.4 Convergence Analysis

Before proceeding to presenting the convergence theorem, other two assumptions are given for a rigor of analysis.

**Assumption 10.1** For the reference trajectory $\boldsymbol{Y}_d$, there exists a control input $\boldsymbol{U}_\infty$ such that the output of system (10.1) can track $\boldsymbol{Y}_d$ exactly. That is, $\exists \boldsymbol{U}_\infty$ such that $\boldsymbol{Y}_d - \boldsymbol{\Phi}_k \boldsymbol{U}_\infty = \boldsymbol{0}$, $\boldsymbol{U}^{\text{low}} \leq \boldsymbol{U}_\infty \leq \boldsymbol{U}^{\text{hi}}$ and $\boldsymbol{Y}^{\text{low}} \leq \boldsymbol{Y}_d \leq \boldsymbol{Y}^{\text{hi}}$. Moreover, $\Delta \boldsymbol{U}^{\text{low}} < 0$ and $\Delta \boldsymbol{U}^{\text{hi}} > 0$.

**Assumption 10.2** The output constraints are satisfied when $\boldsymbol{E}_k = \boldsymbol{0}$.

**Remark 10.2** The Assumptions 10.1 and 10.2 are similar to that in Lee et al. (2000). They are clearly necessary in order for any algorithm to achieve zero tracking error while satisfying the constraints. Assumption 10.1 is similar to the reachability condition of a control system. If Assumption 10.1 does not hold, it implies that the control task is not achievable and the tracking error can only converge to a bounded range.

**Theorem 10.1** *For a class of nonlinear nonaffine discrete-time systems (10.1) with input and output constraints (10.3)–(10.5), if Assumptions 4.1, 4.2, 10.1, and 10.2 hold, then the proposed constrained-DDILC (10.15)–(10.16) guarantees that:*

*i) the parameter estimation $\hat{\boldsymbol{\Phi}}_k$ is bounded, $\forall k$;*
*ii) The change rate of input along the iteration direction, $\Delta \boldsymbol{U}_k$, converges to zero, i.e., $\lim_{k\to\infty} \Delta \boldsymbol{U}_k = \boldsymbol{0}$;*
*iii) the tracking error $\boldsymbol{E}_k$ also converges to zero iteratively, i.e., $\lim_{k\to\infty} \boldsymbol{E}_k = \boldsymbol{0}$;*
*iv) the system output $\boldsymbol{Y}_k$ and the control input signal $\boldsymbol{U}_k$ are bounded, $\forall k$.*

***Proof*** The proof of the boundedness of $\hat{\boldsymbol{\Phi}}_k$ is similar to that in Theorem 4.3.

According to the performance index function, it is clear that

$$J(\boldsymbol{E}_{k+1}, \boldsymbol{U}_{k+1}) = \frac{1}{2}\left\{\boldsymbol{E}_{k+1}^{\mathrm{T}} \boldsymbol{Q} \boldsymbol{E}_{k+1} + \Delta \boldsymbol{U}_{k+1}^{\mathrm{T}} \boldsymbol{R} \Delta \boldsymbol{U}_{k+1}\right\} \geq 0 \quad (10.17)$$

where $(\boldsymbol{E}_{k+1}, \boldsymbol{U}_{k+1}) \in \boldsymbol{\Omega}_{k+1}$ is the optimal solution of $J(\boldsymbol{E}_{k+1}, \boldsymbol{U}_{k+1})$ and $\boldsymbol{\Omega}_{k+1}$ is the feasible region defined by system constraints (10.10).

According to (10.12), if we fix $\Delta\boldsymbol{U}_{k+1} = \boldsymbol{0}$, one obtains that

$$\boldsymbol{E}_{k+1} = \boldsymbol{E}_k - \boldsymbol{\Phi}_{k+1}\Delta\boldsymbol{U}_{k+1} = \boldsymbol{E}_k \tag{10.18}$$

Therefore, $(\boldsymbol{E}_k, \boldsymbol{0}) \in \boldsymbol{\Omega}_{k+1}$ is also a feasible point. According to the optimality of the index function (10.11) at the point $(\boldsymbol{E}_{k+1}, \boldsymbol{U}_{k+1})$, the following inequality is established,

$$\begin{aligned} J(\boldsymbol{E}_{k+1}, \boldsymbol{U}_{k+1}) &\le J(\boldsymbol{E}_k, \boldsymbol{0}) \\ &= \frac{1}{2}\left\{\boldsymbol{E}_k^{\mathrm{T}}\boldsymbol{Q}\boldsymbol{E}_k\right\} \\ &= J(\boldsymbol{E}_k, \boldsymbol{U}_k) - \frac{1}{2}\Delta\boldsymbol{U}_k^{\mathrm{T}}\boldsymbol{R}\Delta\boldsymbol{U}_k \end{aligned} \tag{10.19}$$

which implies that

$$0 \le J(\boldsymbol{E}_{k+1}, \boldsymbol{U}_{k+1}) \le J(\boldsymbol{E}_0, \boldsymbol{U}_0) - \frac{1}{2}\sum_{i=1}^{k}\Delta\boldsymbol{U}_i^{\mathrm{T}}\boldsymbol{R}\Delta\boldsymbol{U}_i \tag{10.20}$$

Equation (10.20) can be rewritten as,

$$0 \le J(\boldsymbol{E}_{k+1}, \boldsymbol{U}_{k+1}) + \frac{1}{2}\sum_{j=1}^{k}\Delta\boldsymbol{U}_i^{\mathrm{T}}\boldsymbol{R}\Delta\boldsymbol{U}_i \le J(\boldsymbol{E}_0, \boldsymbol{U}_0) \le \infty \tag{10.21}$$

It is clear from (10.21) that $\lim_{k\to\infty}\Delta\boldsymbol{U}_k = 0$. Conclusion (ii) is proved.

Now, we proceed to show that $\left(\boldsymbol{0}, \boldsymbol{U}_\infty^k\right)$ is also a feasible point, where $\boldsymbol{U}_\infty^k = \boldsymbol{U}_\infty - \boldsymbol{U}_k$.

From Assumption 10.1, we know that

$$\boldsymbol{U}^{\mathrm{low}} \le \boldsymbol{U}_\infty \le \boldsymbol{U}^{\mathrm{hi}} \tag{10.22}$$

It is apparent that

$$\boldsymbol{U}^{\mathrm{low}} - \boldsymbol{U}_k \le \boldsymbol{U}_\infty - \boldsymbol{U}_k \le \boldsymbol{U}^{\mathrm{hi}} - \boldsymbol{U}_k \tag{10.23}$$

From conclusion (ii) of Theorem 10.1, $\Delta\boldsymbol{U}_k \to 0$. According to Assumption 10.1, $\Delta\boldsymbol{U}^{\mathrm{low}} < 0$ and $\Delta\boldsymbol{U}^{\mathrm{hi}} > 0$, so we have $\Delta\boldsymbol{U}^{\mathrm{low}} \le \Delta\boldsymbol{U}_{k+1} \le \Delta\boldsymbol{U}^{\mathrm{hi}}$ for a sufficiently large $k$. Therefore, the inequality $\Delta\boldsymbol{U}^{\mathrm{low}} \le \boldsymbol{U}_\infty^k \le \Delta\boldsymbol{U}^{\mathrm{hi}}$ can be established for a large enough $k$.

According to Assumption 10.1, it follows that

$$\boldsymbol{Y}^{\text{low}} - \boldsymbol{Y}_k \leq \boldsymbol{Y}_d - \boldsymbol{Y}_k \leq \boldsymbol{Y}^{\text{hi}} - \boldsymbol{Y}_k \tag{10.24}$$

and

$$\boldsymbol{Y}^{\text{low}} - \boldsymbol{Y}_k \leq \boldsymbol{\Phi}_k \left(\boldsymbol{U}_\infty - \boldsymbol{U}_k\right) \leq \boldsymbol{Y}^{\text{hi}} - \boldsymbol{Y}_k \tag{10.25}$$

Thus, $\left(\mathbf{0}, \boldsymbol{U}_\infty^k\right)$ is also a feasible point.

Moreover, the optimal solution $(\boldsymbol{E}_{k+1}, \boldsymbol{U}_{k+1})$ is a feasible point obviously. By virtue of the convexity of the feasible region $\boldsymbol{\Omega}_{k+1}$, any point between the two feasible points is also feasible.

Define a vector

$$\boldsymbol{l} = \left(\mathbf{0}, \boldsymbol{U}_\infty^k\right)^{\text{T}} - \left(\boldsymbol{E}_{k+1}, \Delta\boldsymbol{U}_{k+1}\right)^{\text{T}} = \left(-\boldsymbol{E}_{k+1}, \boldsymbol{U}_\infty^k - \Delta\boldsymbol{U}_{k+1}\right)^{\text{T}} \tag{10.26}$$

The directional derivative of $J\left(\boldsymbol{E}_{k+1}, \boldsymbol{U}_{k+1}\right)$ along the direction of the partial derivatives $\boldsymbol{l}$ can be obtained as

$$\begin{aligned}\Delta J^{\text{T}}\left(\boldsymbol{E}_{k+1}, \boldsymbol{U}_{k+1}\right)\Big|_{(\boldsymbol{E}_{k+1}, \Delta\boldsymbol{U}_{k+1})} \cdot \boldsymbol{l} &= \left[\boldsymbol{E}_{k+1}^{\text{T}}, \Delta\boldsymbol{U}_{k+1}^{\text{T}}\right] \cdot \begin{bmatrix} -\boldsymbol{E}_{k+1} \\ \boldsymbol{U}_\infty^k - \Delta\boldsymbol{U}_{k+1} \end{bmatrix} \\ &= -\boldsymbol{E}_{k+1}^{\text{T}}\boldsymbol{E}_{k+1} + \Delta\boldsymbol{U}_{k+1}^{\text{T}}\boldsymbol{U}_\infty^k - \Delta\boldsymbol{U}_{k+1}^{\text{T}}\Delta\boldsymbol{U}_{k+1}\end{aligned} \tag{10.27}$$

Since $(\boldsymbol{E}_{k+1}, \Delta\boldsymbol{U}_{k+1})$ is the optimal point to minimize the index function $J(\boldsymbol{E}_{k+1}, \boldsymbol{U}_{k+1})$, clearly the cost of function $J\left(\boldsymbol{E}_{k+1}, \boldsymbol{U}_{k+1}\right)$ at $\left(\mathbf{0}, \boldsymbol{U}_\infty^k\right)$ is no less than that at the point of $(\boldsymbol{E}_{k+1}, \Delta\boldsymbol{U}_{k+1})$. Thus, the directional derivative of $J\left(\boldsymbol{E}_{k+1}, \boldsymbol{U}_{k+1}\right)$ along the direction $\boldsymbol{l}$ is no less than zero. From (10.27), it is derived that

$$\Delta\boldsymbol{U}_{k+1}^{\text{T}}\boldsymbol{U}_\infty^k \geq \boldsymbol{E}_{k+1}^{\text{T}}\boldsymbol{E}_{k+1} + \Delta\boldsymbol{U}_{k+1}^{\text{T}}\Delta\boldsymbol{U}_{k+1} \geq 0 \tag{10.28}$$

It is directly concluded from (10.28) that $\lim_{k\to\infty} \boldsymbol{E}_k = 0$ because $\lim_{k\to\infty} \Delta\boldsymbol{U}_k = 0$ has been shown. So we get conclusion (iii) of Theorem 10.1. □

## 10.5 Constrained Data-Driven PTPILC

### *10.5.1 Controller Design*

In this section, the control objective is to track only certain desired output points $y_d(t_1), y_d(t_2), \ldots, y_d(t_M)$ at time instants of $\{t_1, t_2, \ldots, t_M\}$, $t_M \leq N$. Moreover, the system output at points other than the specified ones $\{t_1, t_2, \ldots, t_M\}$ may be unmeasurable. Therefore, a constrained data-driven point-to-point ILC (constrained-DDPTPILC) law will be designed by only using the error information obtained from the specified points.

Define $\bar{\boldsymbol{Y}}_k = [y_k(t_1), y_k(t_2), \ldots, y_k(t_M)]^{\mathrm{T}} \in R^M$, which is consisting of the $t_i$th row of the row vector $\boldsymbol{Y}_k$, $i = 1, \ldots, M$. Similar to (8.2), one has

$$\Delta\bar{\boldsymbol{Y}}_k = \bar{\boldsymbol{\Phi}}_k \Delta\boldsymbol{U}_k \tag{10.29}$$

where

$$\bar{\boldsymbol{\Phi}}_k = \begin{bmatrix} \frac{\partial g_{t_1-1}}{\partial u_k(0)} & \frac{\partial g_{t_1-1}}{\partial u_k(1)} & \cdots & \frac{\partial g_{t_1-1}}{\partial u_k(t_1-1)} & 0 & 0 & \cdots & 0 & 0\ldots 0 \\ \frac{\partial g_{t_2-1}}{\partial u_k(0)} & \frac{\partial g_{t_2-1}}{\partial u_k(1)} & \cdots & \frac{\partial g_{t_2-1}}{\partial u_k(t_1-1)} & \cdots & \frac{\partial g_{t_2-1}}{\partial u_k(t_2-1)} & \cdots & 0 & 0\ldots 0 \\ \vdots & \vdots & \ddots & \vdots & & \vdots & \ddots & \vdots & \vdots \ddots \vdots \\ \frac{\partial g_{t_M-1}}{\partial u_k(0)} & \frac{\partial g_{t_M-1}}{\partial u_k(1)} & \cdots & \frac{\partial g_{t_M-1}}{\partial u_k(t_1-1)} & \cdots & \frac{\partial g_{t_M-1}}{\partial u_k(t_2-1)} & \cdots & \frac{\partial g_{t_M-1}}{\partial u_k(t_M-1)} & 0\cdots 0 \end{bmatrix} \in R^{M\times N}.$$

Similarly, we also consider three kinds of constraints: bounds on input values, bounds on iterative change rate of the control input, and bounds on system outputs. The first two constraints are the same as that defined in (10.3) and (10.4). The constraints on the system output are required only at the specified points $\{t_1, t_2, \ldots, t_M\}$ for the point-to-point operation systems because the system output at the other points are not necessary. Hence, the constraints on the bounds of system outputs for the point-to-point operation system become

$$\boldsymbol{Y}^{\text{low}} \leq \bar{\boldsymbol{Y}}_{k+1} \leq \boldsymbol{Y}^{\text{hi}} \tag{10.30}$$

By following the same steps from (10.6) to (10.10), the input and output constraints (10.3), (10.4), and (10.30) are transferred into a linear matrix inequality condition

$$\bar{\boldsymbol{A}}_k \Delta\boldsymbol{U}_{k+1} \leq \bar{\boldsymbol{b}}_k \tag{10.31}$$

where $\bar{\boldsymbol{A}}_k = \begin{bmatrix} \boldsymbol{I} \\ -\boldsymbol{I} \\ \bar{\boldsymbol{\Phi}}_{k+1} \\ -\bar{\boldsymbol{\Phi}}_{k+1} \end{bmatrix}$ and $\bar{\boldsymbol{b}}_k = \begin{bmatrix} \Delta\boldsymbol{U}_k^{\text{hi}*} \\ -\Delta\boldsymbol{U}_k^{\text{low}*} \\ \boldsymbol{Y}^{\text{hi}} - \bar{\boldsymbol{Y}}_k \\ -\boldsymbol{Y}^{\text{low}} + \bar{\boldsymbol{Y}}_k \end{bmatrix}$.

Let $\bar{\boldsymbol{Y}}_d = [y_d(t_1), y_d(t_2), \ldots, y_d(t_M)]^{\mathrm{T}} \in R^M$ and $\bar{\boldsymbol{E}}_k = \bar{\boldsymbol{Y}}_d - \bar{\boldsymbol{Y}}_k$ denote the tracking errors at the specified points. Define an index function with respect to $\boldsymbol{U}_{k+1}$,

$$J\left(\bar{\boldsymbol{E}}_{k+1}, \boldsymbol{U}_{k+1}\right) = \frac{1}{2}\left\{\bar{\boldsymbol{E}}_{k+1}^{\mathrm{T}} \boldsymbol{Q} \bar{\boldsymbol{E}}_{k+1} + \Delta\boldsymbol{U}_{k+1}^{\mathrm{T}} \boldsymbol{R} \Delta\boldsymbol{U}_{k+1}\right\} \tag{10.32}$$

where $\boldsymbol{Q}$ and $\boldsymbol{R}$ are positive definite weighting matrices.

Following the same steps from (10.12) to (10.14), the constrained-DDPTPILC law is derived by minimizing (10.32) with the linear matrix inequality constraints (10.31), that is,

$$\begin{cases} \min\limits_{\boldsymbol{U}_{k+1}} J\left(\bar{\boldsymbol{E}}_{k+1}, \boldsymbol{U}_{k+1}\right) = \min\limits_{\boldsymbol{U}_{k+1}} \frac{1}{2}\left\{\Delta\boldsymbol{U}_{k+1}^{\mathrm{T}}\left(\bar{\boldsymbol{\Phi}}_{k+1}^{\mathrm{T}} \boldsymbol{Q} \bar{\boldsymbol{\Phi}}_{k+1} + \boldsymbol{R}\right)\Delta\boldsymbol{U}_{k+1} \right. \\ \qquad\qquad \left. -2\bar{\boldsymbol{E}}_k^{\mathrm{T}} \boldsymbol{Q} \bar{\boldsymbol{\Phi}}_{k+1} \Delta\boldsymbol{U}_{k+1} + \bar{\boldsymbol{E}}_k^{\mathrm{T}} \boldsymbol{Q} \bar{\boldsymbol{E}}_k\right\} \\ \text{s.t.} \bar{\boldsymbol{A}}_k \Delta\boldsymbol{U}_{k+1} \leq \bar{\boldsymbol{b}}_k \end{cases} \tag{10.33}$$

where $\bar{\boldsymbol{\Phi}}_{k+1}$ is unknown and is estimated iteratively by the following updating law,

$$\begin{aligned}\hat{\bar{\varphi}}_k(t_i-1) &= \hat{\bar{\varphi}}_{k-1}(t_i-1) \\ &+ \frac{\eta\left(\Delta y_{k-1}(t_i) - \hat{\bar{\varphi}}_{k-1}(t_i-1)\Delta\bar{\boldsymbol{u}}_{k-1}(t_i-1)\right)}{\mu + \|\Delta\bar{\boldsymbol{u}}_{k-1}(t_i-1)\|^2} \\ &\times \Delta\bar{\boldsymbol{u}}_{k-1}^{\mathrm{T}}(t_i-1),\quad i=1,\ldots,\mathrm{M}\end{aligned} \tag{10.34}$$

where $\mu > 0$ is a weighting factor and $0 < \eta < 2$ is a step-size factor; $\hat{\bar{\varphi}}_k(t_i-1)$ is the estimated value of $\bar{\varphi}_k(t_i-1)$; $\bar{\varphi}_k(t_i-1)$ denotes the nonzero vector of the $i$th row of $\bar{\boldsymbol{\Phi}}_k$.

Then, the proposed constrained-DDPTPILC law can be obtained by solving the following quadratic programming (QP) problem according to the updating law (10.34),

$$\begin{cases}\min\limits_{\boldsymbol{U}_{k+1}} J\left(\bar{\boldsymbol{E}}_{k+1},\boldsymbol{U}_{k+1}\right) = \min\limits_{\boldsymbol{U}_{k+1}} \frac{1}{2}\left\{\Delta\boldsymbol{U}_{k+1}^{\mathrm{T}}\left(\hat{\bar{\boldsymbol{\Phi}}}_{k+1}^{\mathrm{T}}\boldsymbol{Q}\hat{\bar{\boldsymbol{\Phi}}}_{k+1}+\boldsymbol{R}\right)\Delta\boldsymbol{U}_{k+1}\right. \\ \qquad\qquad\qquad\qquad\qquad\left. -2\bar{\boldsymbol{E}}_k^{\mathrm{T}}\boldsymbol{Q}\hat{\bar{\boldsymbol{\Phi}}}_{k+1}\Delta\boldsymbol{U}_{k+1}+\bar{\boldsymbol{E}}_k^{\mathrm{T}}\boldsymbol{Q}\bar{\boldsymbol{E}}_k\right\} \\ \text{s.t.}\,\hat{\bar{\boldsymbol{A}}}_k\Delta\boldsymbol{U}_{k+1}\le\bar{\boldsymbol{b}}_k\end{cases} \tag{10.35}$$

where $\hat{\bar{\boldsymbol{A}}}_k = \begin{bmatrix}\boldsymbol{I}\\ -\boldsymbol{I}\\ \hat{\bar{\boldsymbol{\Phi}}}_{k+1}\\ -\hat{\bar{\boldsymbol{\Phi}}}_{k+1}\end{bmatrix}$.

**Remark 10.3** For the constrained-DDILC proposed in Sect. 10.3, the dimension of $\hat{\boldsymbol{\Phi}}_{k+1}$ in (10.16) is $N \times N$. So we need to store $3N + N \times N$ data in the memory and $N \times N$ dimensional matrix calculation is required in the computation. Comparatively, for the multiple intermediate tracking tasks for $M$ points only, instead of $N$ points ($M << N$), the dimension of $\hat{\bar{\boldsymbol{\Phi}}}_{k+1}$ in (10.35) is $M \times M$. And we need to store $3M + M \times M$ data in the memory, and the dimension of the matrix required for the calculation of the control input is reduced significantly to $M \times M$. It is clear that the constrained-DDPTPILC proposed in this section has less computational burden and memory size than the constrained-DDILC.

### 10.5.2 Convergence Analysis

Two assumptions are given as follows by modifying Assumptions 10.1 and 10.2 slightly.

**Assumption 10.3** For the desired output points $\bar{\boldsymbol{Y}}_d$, there exists a control input $\boldsymbol{U}_\infty$ such that the output of system (10.1) can track $\bar{\boldsymbol{Y}}_d$ exactly. That is, $\exists \boldsymbol{U}_\infty$ such that

$\bar{\boldsymbol{Y}}_d - \boldsymbol{\Phi}_k \boldsymbol{U}_\infty = \mathbf{0}$, $\boldsymbol{U}^{\text{low}} \leq \boldsymbol{U}_\infty \leq \boldsymbol{U}^{\text{hi}}$ and $\boldsymbol{Y}^{\text{low}} \leq \bar{\boldsymbol{Y}}_d \leq \boldsymbol{Y}^{\text{hi}}$. Moreover, $\Delta \boldsymbol{U}^{\text{low}} < 0$ and $\Delta \boldsymbol{U}^{\text{hi}} > 0$.

**Assumption 10.4** The output constraints are satisfied when $\bar{\boldsymbol{E}}_k = \mathbf{0}$.

A similar convergence theorem can be given as follows for the constrained-DDPTPILC (10.34)–(10.35).

**Theorem 10.2** *For a class of nonlinear nonaffine discrete-time systems (10.1) with constraints (10.3), (10.4), and (10.30), if assumptions 4.1, 4.2, 10.3, and 10.4 hold, then the proposed constrained-DDPTPILC (10.34)–(10.35) guarantees that:*

*(i) the parameter estimation $\hat{\bar{\boldsymbol{\Phi}}}_k$ is bounded, $\forall k$;*
*(ii) The change rate of input along the iteration direction, $\Delta \boldsymbol{U}_k$, converges to zero, i.e., $\lim_{k\to\infty} \Delta \boldsymbol{U}_k = \mathbf{0}$;*
*(iii) the tracking error $\bar{\boldsymbol{E}}_k$ also converges to zero iteratively, i.e., $\lim_{k\to\infty} \bar{\boldsymbol{E}}_k = \mathbf{0}$;*
*(iv) the system output $\boldsymbol{Y}_k$ and the control input signal $\boldsymbol{U}_k$ are bounded, $\forall k$.*

***Proof*** The proof of conclusions of Theorem 10.2 is similar to that of Theorem 10.1. □

## 10.6 Simulation Study

***Example 1***. Consider a repetitive nonlinear discrete-time system

$$y_k(t+1) = \frac{y_k(t)}{1 + y_k(t)^3} + u_k(t)^3 \tag{10.36}$$

where $t \in \{0, 1, \ldots, 20\}$ is the time instants; $k \in \{0, 1, \ldots\}$ is the iteration number. The system constraints are set as $-1.0 \leq u_k(t) \leq 1.0$, $-0.25 \leq \Delta u_k(t) \leq 0.25$, and $-0.6 \leq y_k(t) \leq 0.6$.

The desired trajectory is

$$y_d(t) = 0.5 \sin(t\pi/10) + 0.3 \cos(t\pi/10), t \in \{0, 1, \ldots, 20\} \tag{10.37}$$

*Case 1.1: Tracking the desired trajectory (*10.37*) over the entire time interval* $\{0, 1, \ldots, 20\}$. This task means that a good control performance is required for all finite points of the desired trajectory. Hence, we will use the proposed constrained-DDILC (10.15)–(10.16) for this task.

Select the initial values as $y_k(0) = 0$, $u_0(t) = 0.5$, and $\hat{\varphi}_0(t) = 0.5$. The controller parameters are set as $\boldsymbol{Q} = \boldsymbol{R} = \boldsymbol{I}$, $\eta = 1$, and $\mu = 1.5$. The simulation results of the constrained-DDILC are shown in Figs. 10.1, 10.2 and 10.3.

Figure 10.1 is the output tracking performance of the constrained-DDILC at the 100th iteration, where vertical axis denotes the system output and the horizontal axis denotes the time instant. Figure 10.2 shows the convergence of the maximum absolute tracking errors, where the Y-axis denotes the maximum absolute value of the output tracking error, defined as $e_{\max}(k) = \max_{t\in\{1,\ldots,20\}} \{|e_k(t)|\}$, and the X-axis denotes the iteration numbers. Figure 10.3 is the corresponding control input profile, where the vertical axis denotes the system control input and the horizontal axis denotes the time instant.

From the simulation results, one can conclude that: (a) The system output trajectory tracks the desired trajectory exactly with the increasing iteration number (Fig. 10.1); (b) The convergence of the tracking error is achieved along the iteration direction (Fig. 10.2); (c) Both the system output (Fig. 10.1) and the control input (Fig. 10.3) are actively regulated to comply with the constraints.

For the purpose of comparison, the DDILC proposed in Chi et al. (2015) is also applied into the trajectory tracking problem, shown as follows

$$\hat{\boldsymbol{\varphi}}_k(t) = \hat{\boldsymbol{\varphi}}_{k-1}(t) + \frac{\eta\left(\Delta y_{k-1}(t+1) - \hat{\boldsymbol{\varphi}}_{k-1}(t)\Delta \boldsymbol{u}_{k-1}(t)\right)\Delta \boldsymbol{u}_{k-1}^{\mathrm{T}}(t)}{\mu + \|\Delta \boldsymbol{u}_{k-1}(t)\|^2} \tag{10.38}$$

$$\boldsymbol{U}_k = \boldsymbol{U}_{k-1} + \frac{\rho \hat{\boldsymbol{\Phi}}_k^{\mathrm{T}} \boldsymbol{E}_{k-1}}{\lambda + \left\|\hat{\boldsymbol{\Phi}}_k\right\|^2} \tag{10.39}$$

where $\lambda > 0$ and $\rho > 0$ are a weighting factor and a step-size factor respectively.

As it is well-known, in real process industries, a soft limiter is often added directly to prevent large control input signals. For a further comparison, therefore, we also add a soft limiter directly to the above DDILC with consideration of all the system constraints. That is, the conditions of $|u_k(t)| \le 1.0$, $|\Delta u_k(t)| \le 0.25$, and $|y_k(t)| \le 0.6$ are added into the calculation procedure of the DDILC (10.38)–(10.39) as a soft limiter. We define such a control approach as soft limiter based-DDILC in this chapter.

Under the same simulation background, that is, $y_k(0) = 0$, $u_0(t) = 0.5$, $\hat{\varphi}_0(t) = 0.5$, $\eta = 1$, $\mu = 1.5$ and selecting the controller parameters as $\rho = 1$, $\lambda = 0.01$, the simulation results by using the traditional DDILC and soft limiter based-DDILC are shown in Figs. 10.1, 10.2 and 10.3, respectively.

The dashed lines in Figs. 10.1, 10.2 and 10.3 denote the profiles of system output, tracking error, and the control input obtained by using the traditional DDILC (10.38)–(10.39). Although a satisfactory convergence of tracking error can be achieved, the system output exceeds the prescribed bounds such that it cannot be realized through a practical actuator.

The solid lines in Figs. 10.1, 10.2 and 10.3 denote the profiles of system output, tracking error, and the control input by applying the soft limiter based-DDILC defined above, respectively. It is seen that neither the system output nor the control input exceeds the prescribed bounds due to the soft limiter that is added in the calculation

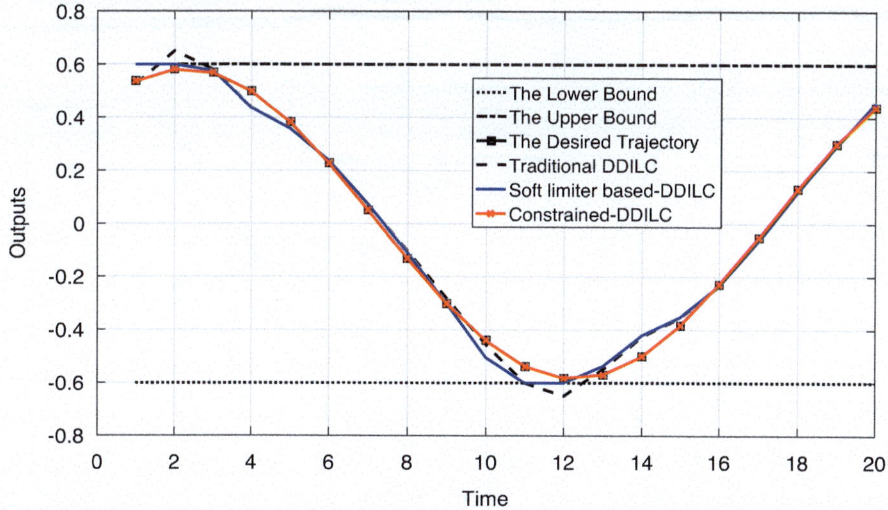


**Fig. 10.1** Output profile at the 100th iteration in Case 1.1

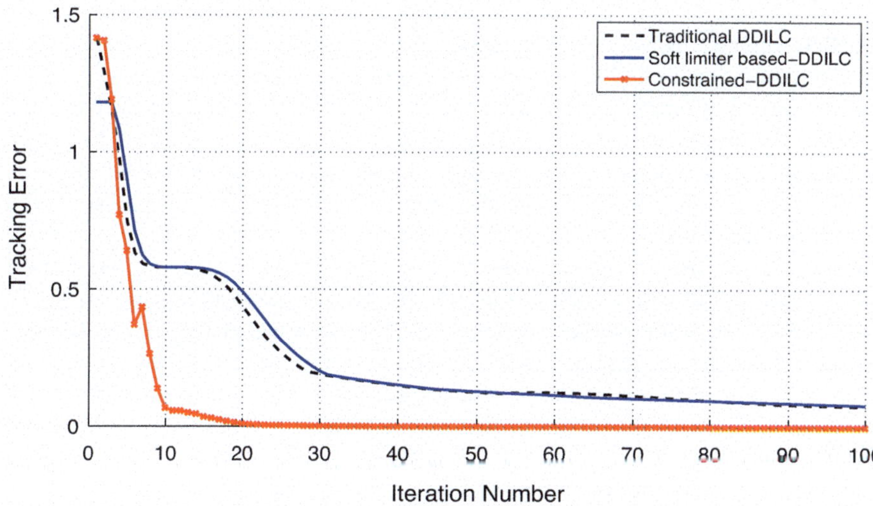


**Fig. 10.2** Maximum absolute tracking error $e_{\max}(k)$ along with iterations in Case 1.1

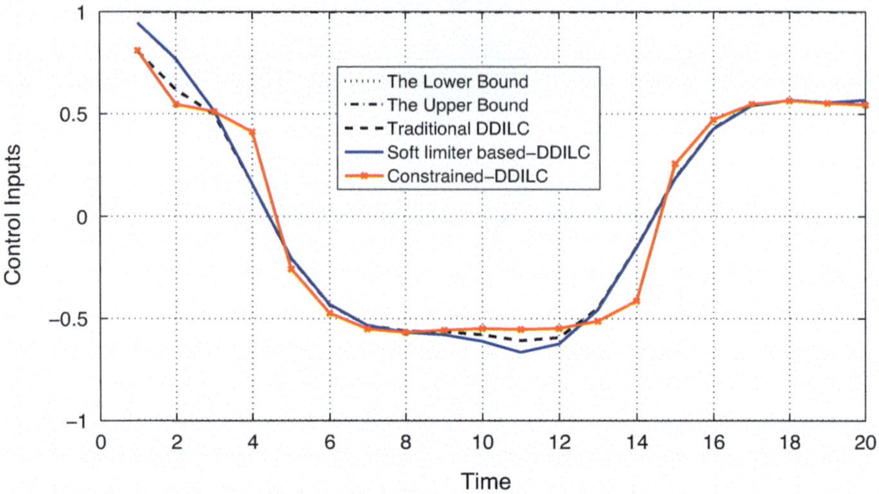


**Fig. 10.3** Input profile at the 100th iteration in Case 1.1

procedure directly. However, the tracking error convergence becomes worse than that of the traditional DDILC because of the addition of the soft limiter.

Compared with both the traditional DDILC and the soft limiter based-DDILC methods, the proposed constrained-DDILC approach attains a better performance while neither system output nor control input violates the prescribed bounds because the I/O constraints are included into the proposed learning controller by solving a QP problem iteratively.

*Case 1.2: Tracking the specified four points of the desired trajectory.* The control task is to drive the system output through the given desired points $y_d(4)$, $y_d(9)$, $y_d(14)$, and $y_d(20)$ of the desired trajectory (10.37) at time instants $\{4, 9, 14, 20\}$ only, instead of the entire desired trajectory. The I/O constraints are also taken into account except that the constraints on the system outputs at the other unspecified points are not required. Applying the proposed constrained-DDPTPILC (10.34)–(10.35) under the same simulation condition, i.e., $y_k(0) = 0$, $u_0(t) = 0.5$, $\hat{\varphi}_0(t) = 0.5$, $\boldsymbol{Q} = \boldsymbol{R} = I$, $\eta = 1$, $\mu = 1.5$, the simulation results are also shown in Figs. 10.4–10.6.

By virtue of the simulation results, it is seen that the system output trajectory, obtained by applying the constrained-DDPTPILC, only passes the desired points $y_d(4)$, $y_d(9)$, $y_d(14)$, $y_d(20)$ precisely and the performance specification at the other points is not required (Fig. 10.4). And both the system output and the control input have been regulated to satisfy the given constraints.

Also, for the purpose of comparison, the traditional DDPTPILC proposed in Section 8.3 is applied.

A soft limiter based-DDPTPILC is also defined by adding the constraint conditions of $|u_k(t)| \leq 1.0$, $|\Delta u_k(t)| \leq 0.25$, and $|y_k(t)| \leq 0.6$ into the calculation procedure of the traditional DDPTPILC.

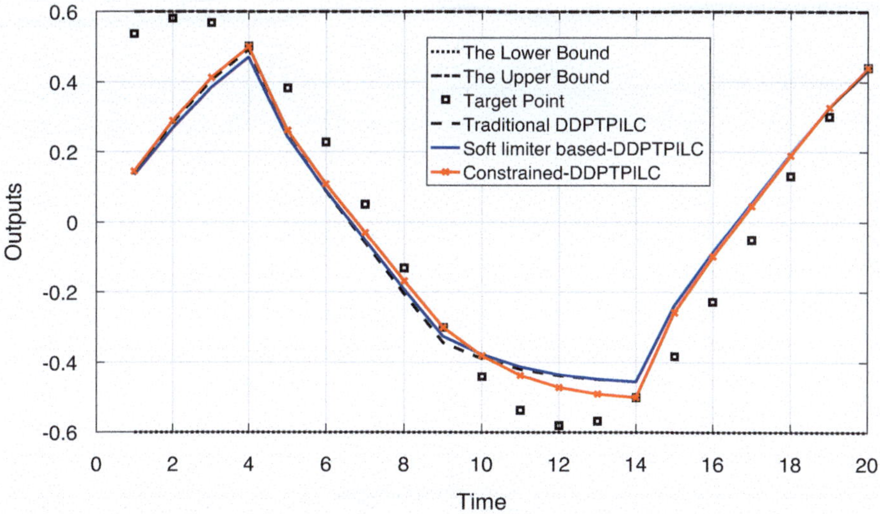


**Fig. 10.4** Output profile at the 100th iteration in Case 1.2

Set the same initial values and estimator parameters as $y_k(0) = 0$, $u_0(t) = 0.5$, $\hat{\varphi}_0(t) = 0.5$, $\eta = 1$, $\mu = 1.5$, and set controller parameters as $\rho = 0.6$, $\lambda = 0.1$. Figures 10.4, 10.5 and 10.6 show the simulation results by applying the traditional DDPTPILC and soft limiter based-DDPTPILC, respectively.

In terms of simulation results, the similar conclusions can be made: (1) the proposed constrained-DDPTPILC achieves the best convergence performance among the three methods, and guarantees the system output and control input being in the predesigned bounds; (2) although a satisfactory convergence of tracking error can be achieved by the traditional DDPTPILC, its control input violates the prescribed bounds (Fig. 10.6); (3) the soft limiter added to the calculation procedure will degrade the control performance though it can make both the control input and the system output be located within the prescribed bounds.

*Case 1.3: Comparison of the proposed constrained-DDILC and constrained-DDPTPILC methods.* As discussed in the Introduction part, the PTPILC approach may outperform ILC because many unnecessary constraints on the free points other than the specified ones are removed. To verify this standpoint, we also provide the following comparison results, shown in Figs. 10.7, 10.8 and 10.9, under the same control parameters and initial values for the both methods, that is, $y_k(0) = 0$, $u_0(t) = 0.5$, $\hat{\varphi}_0(t) = 0.5$, $\boldsymbol{Q} = \boldsymbol{R} = I$, $\eta = 1$, $\mu = 1.5$. Moreover, an index of control effort, defined as $C_E(k) = \|\boldsymbol{U}_k\|^2$, is introduced in this case for comparison.

Although the constrained-DDILC can be applied to the multiple intermediate points tracking problem if all the system outputs are measurable and a desired trajectory can be designed to pass through the specified points, its control performance is inferior to that of the constrained-DDPTPILC because more constraints on the unnecessary points are considered to make the system output exactly track all the desired

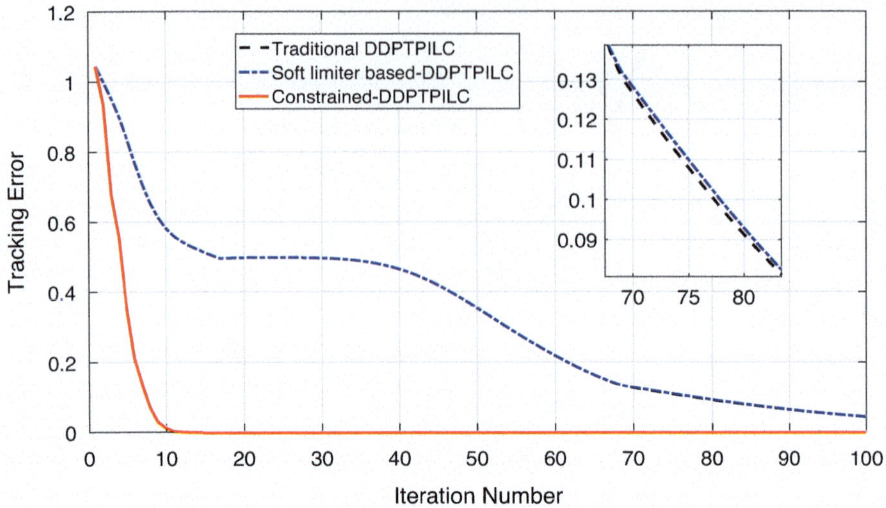


**Fig. 10.5** Maximum absolute tracking error $e_{\max}(k) = \max_{t\in\{4,9,14,20\}}\{|e_k(t)|\}$ along with iterations in Case 1.2

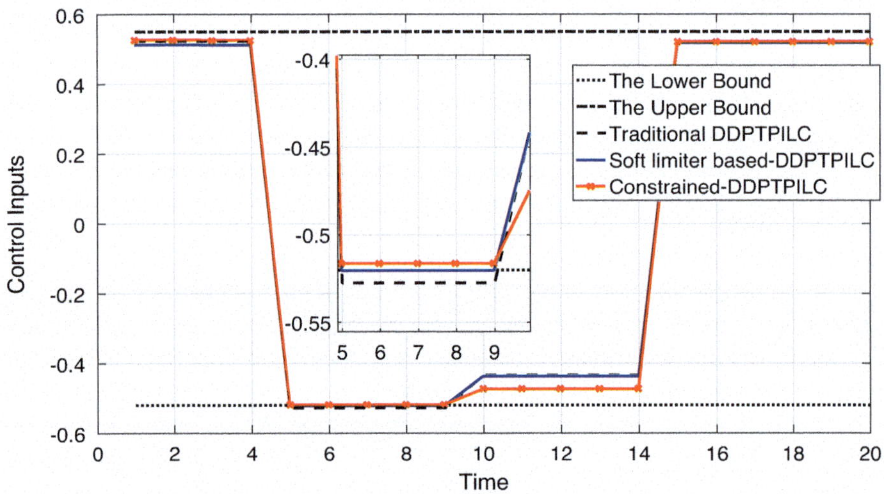


**Fig. 10.6** Input profile at the 100th iteration in Case 1.2

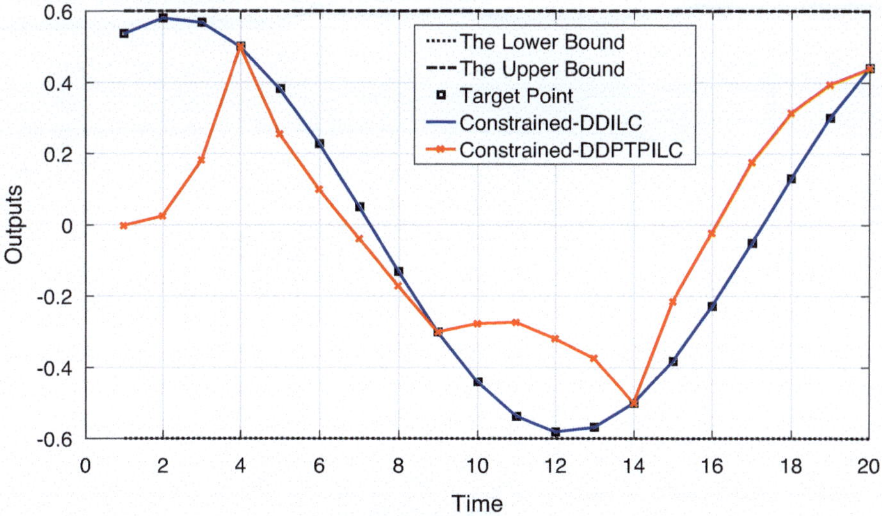


**Fig. 10.7** Tracking performance of constrained-DDILC and constrained-DDPTPILC in case 1.3

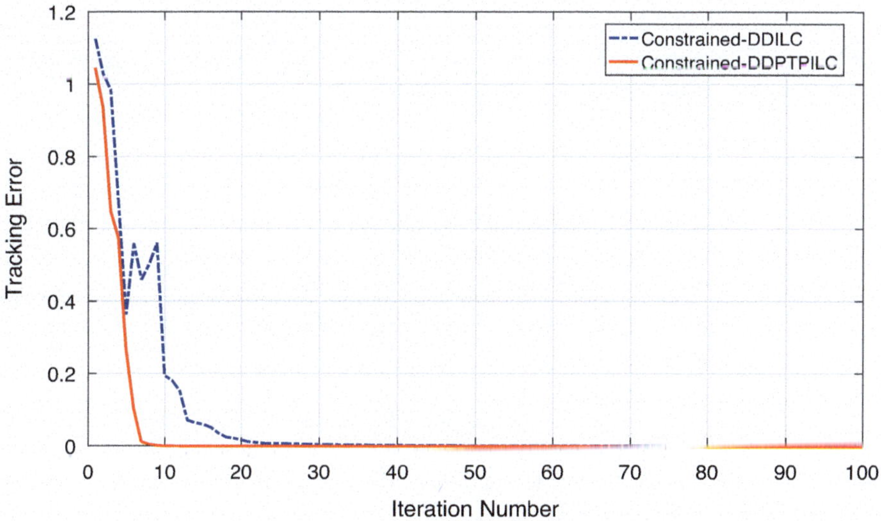


**Fig. 10.8** Error convergence of constrained-DDILC and constrained-DDPTPILC in case 1.3

points of the reference trajectory (Fig. 10.7). As a result, the constrained-DDPTPILC achieves a better control performance with faster convergence (Fig. 10.8) and less control effort (Fig. 10.9).

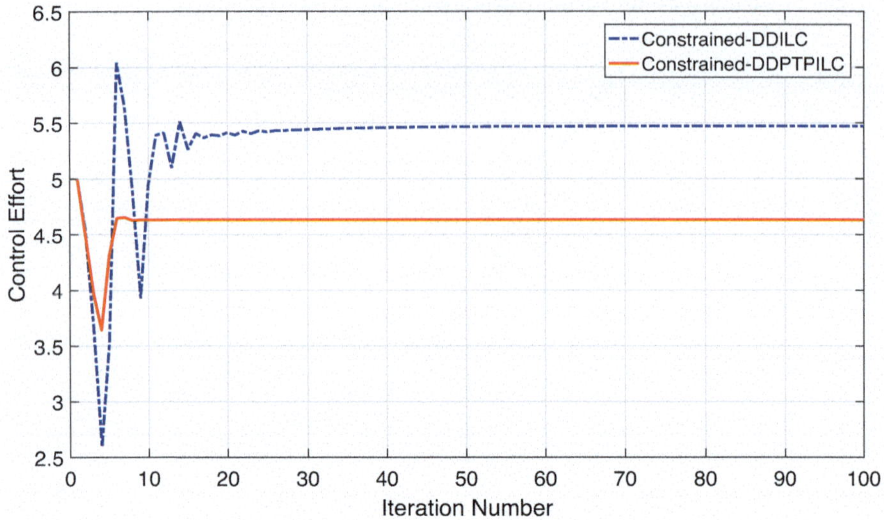


**Fig. 10.9** Control effort of constrained-DDILC and constrained-DDPTPILC in case 1.3

***Example 2***. To show the applicability and effectiveness of the proposed approaches to realistic processes, a nonlinear batch reactor (Raó 1981; Logsdon and Biegler 1992) is considered as follows,

$$\frac{dx_1}{dt} = -k_1 \exp\left(\frac{-E_1}{uT_{\text{ref}}}\right) x_1^2 + d_k(t) \tag{10.40}$$

$$\frac{dx_2}{dt} = k_1 \exp\left(\frac{-E_1}{uT_{\text{ref}}}\right) x_1^2 - k_2 \exp\left(\frac{-E_2}{uT_{\text{ref}}}\right) x_2 + d_k(t) \tag{10.41}$$

$$y = x_2 + \delta_k(t) \tag{10.42}$$

where $x_1$ and $x_2$ are two states denoting the concentrations of the raw material and the product respectively; $u = T/T_{\text{ref}}$ is the control input, where $T_{\text{ref}}$ is the reference temperature and $T$ is the reactor temperature. The operation interval is $t \in \left[0,\ 1.0\right]$ hour. The sampling interval is 0.1 h.

In the following simulation, to mimic the realistic control condition, the system disturbances $d_k(t)$ and output noises $\delta_k(t)$ are added to the batch reactor process, where both of them are randomly varying in the interval of $[-0.01, 0.01]$. The initial states are also assumed varying randomly over different iterations, that is, $\boldsymbol{x}_k(0) = \left[x_{1,k}(0), x_{2,k}(0)\right]^{\mathrm{T}} = [1 + \sigma_k, 0]^T$, where $\sigma_k$ varies with iterations randomly in the interval of $[-0.1, 0.1]$.

The system parameters are listed as $k_1 = 4.0 \times 10^3$, $k_2 = 6.2 \times 10^5$, $E_1 = 2.5 \times 10^3$ K, $E_2 = 5.0 \times 10^3$ K, and $T_{\text{ref}} = 348$ K. Since the reactor temperature $T$ is limited to 298 K $\leq T \leq$ 398 K and $T_{\text{ref}} = 348$ K, the control input is constrained by $298/T_{\mathrm{ref}} \leq u_k(t) \leq 336/T_{\mathrm{ref}}$. The other two constraints are set as $-0.01 \leq \Delta u_k(t) \leq 0.01$ and $-0.1 \leq y_k(t) \leq 0.62$.

The desired product reference trajectory is $y_d(t) = 0.61 \sin(\pi t/20), t \in \{0, 1, \ldots, 10\}$.

It is worth pointing out that the mathematical model (10.40)–(10.42) has been well-tested and widely used in the simulation and control as a typical industrial batch process and it can be regarded as the case of simulated practical industrial application test to show the applicability and efficiency of the proposed method Raó (1981), Logsdon and Biegler (1992). It is also worth noting that the mathematical model of the batch reactor is nonlinear and is assumed to be unavailable in the simulation, and the model simply serves as producing I/O data. Further, more random disturbances in different value-ranges are added in the following simulations to demonstrate the robustness of the proposed approach.

*Case 2.1. Complete trajectory tracking.* The control task is to track all points of the desired trajectory. The initial values and controller parameters are set as $u_0(t) = 0.88$, $\hat{\varphi}_0(t) = 0.1$, $\eta = 0.5$, $\mu = 0.2$, $\boldsymbol{Q} = \boldsymbol{I}$, and $\boldsymbol{R} = 0.5\boldsymbol{I}$. Applying the proposed constrained-DDILC (10.15)–(10.16), the simulation results are shown in Figs. 10.10, 10.12 and 10.12. Obviously, the proposed constrained-DDILC has been verified for its applicability and effectiveness to more practical processes and achieves a satisfactory performance under random disturbances and noises (Figs. 10.10 and 10.11). In addition, both the system output and control input have been confined within the given constraints (Figs. 10.10 and 10.12).

For the comparison purpose, the traditional DDILC (10.38)–(10.39) and the soft limiter based-DDILC defined above are also applied to this task. Under the same simulation conditions by setting $\rho = 1$, $\lambda = 0.1$, the simulation results are shown as the dashed lines and the solid lines in Figs. 10.10, 10.11 and 10.12, respectively.

In terms of the simulation results, it is clear that the traditional DDILC and soft limiter based-DDILC can also achieve a satisfactory performance. However, the control input by using traditional DDILC goes beyond the given bounds and the soft limiter based-DDILC reduces the convergence speed.

*Case 2.2. Multiple points tracking.* The control task is to track four specified points of the desired trajectory at the time instants of $\{2, 4, 6, 10\}$. Under the same simulation conditions as that in Case 2.1, i.e., $u_0(t) = 0.88$, $\hat{\varphi}_0(t) = 0.1$, $\eta = 0.5$, $\mu = 0.2$, $\boldsymbol{Q} = \boldsymbol{I}$, and $\boldsymbol{R} = 0.5\boldsymbol{I}$, by applying the proposed constrained-DDPTPILC (10.34)–(10.35), the simulation results are shown in Figs. 10.13, 10.14 and 10.15.

It is shown that the proposed constrained-DDPTPILC is effective in dealing with constrained processes with a good performance under random disturbances and noises. Both the system output and control input are constrained within the given bounds.

Applying the traditional DDPTPILC and the soft limiter based-DDPTPILC under the same simulation condition for the comparison, the simulation results are also shown in Figs. 10.13, 10.14 and 10.15, which illustrate that the control input violates the prescribed bounds by using traditional DDPTPILC. Figure 10.14 together

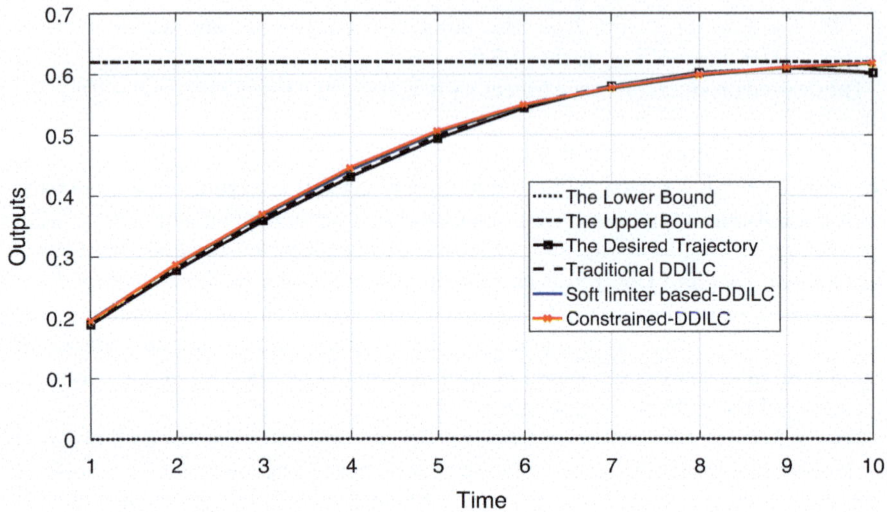


**Fig. 10.10** Output profile at the 100th iteration in Case 2.1

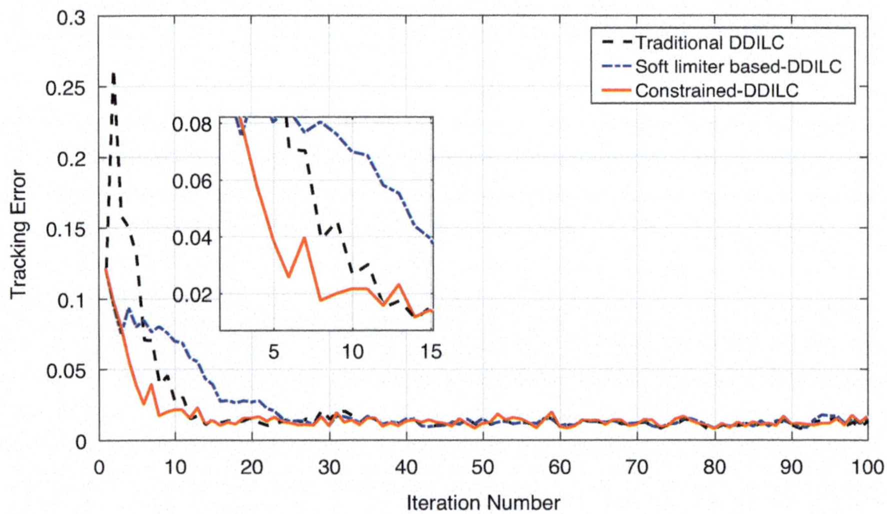


**Fig. 10.11** Maximum absolute tracking error $e_{\max}(k)$ along with iterations in Case 2.1

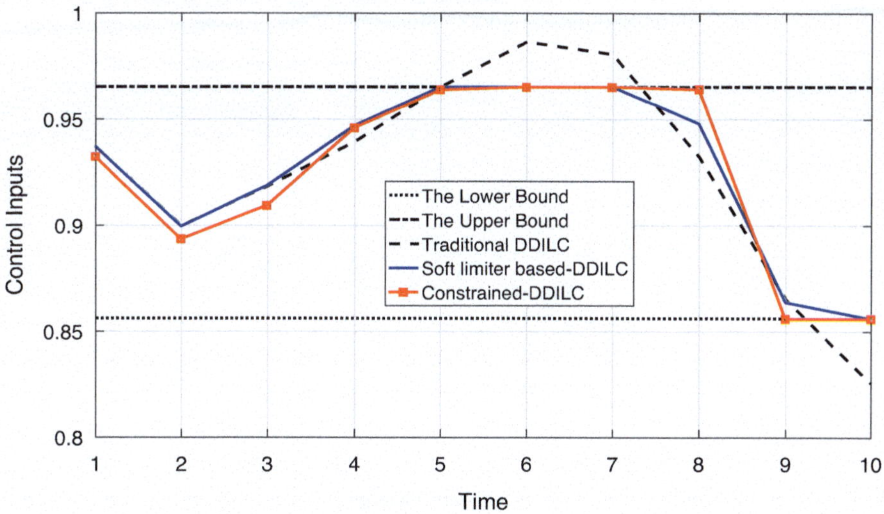


**Fig. 10.12** Input profile at the 100th iteration in Case 2.1

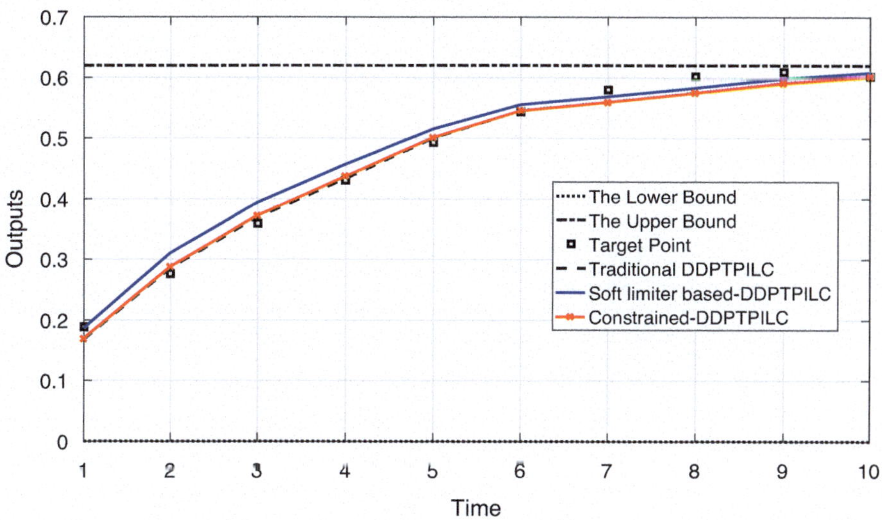


**Fig. 10.13** Output profile at the 100th iteration in Case 2.2

with Fig. 10.15 indicates that the soft limiter based-DDPTPILC can deal with the constrained processes, but the control performance is degraded as a trade-off.

*Case 2.3. Performance comparison of the constrained-DDILC with constrained-DDPTPILC.* In this case, a performance comparison is also provided between the constrained-DDILC and constrained-DDPTPILC when both of them are applied to the realistic nonlinear batch reactor for the multiple points tracking, where the

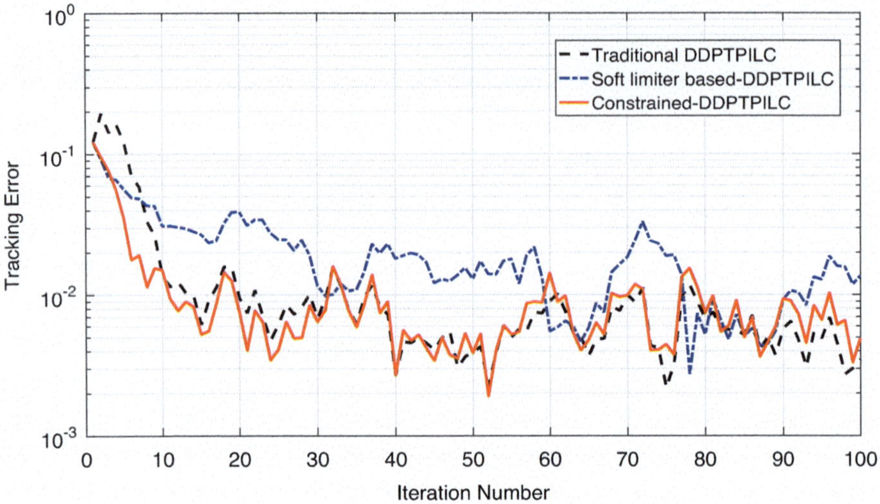


**Fig. 10.14** Maximum absolute tracking error $e_{\max}(k) = \max_{t\in\{4,9,14,20\}}\{|e_k(t)|\}$ along with iterations in Case 2.2

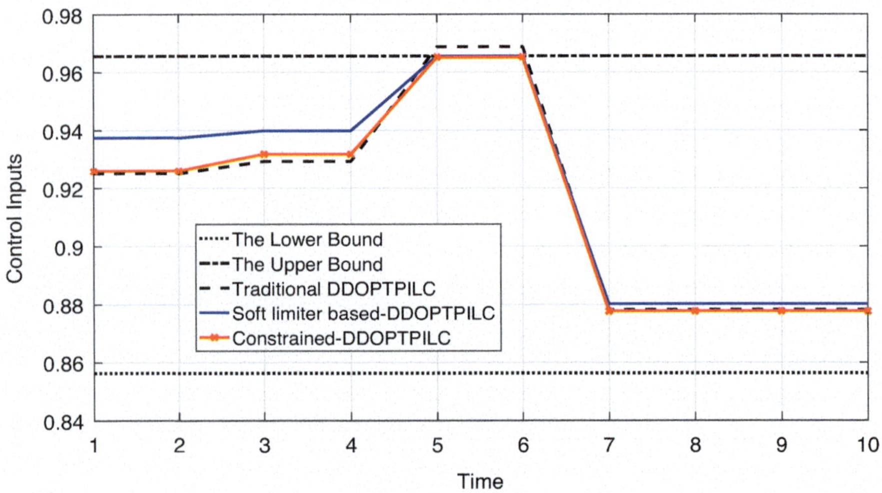


**Fig. 10.15** Input profile at the 100th iteration in Case 2.2

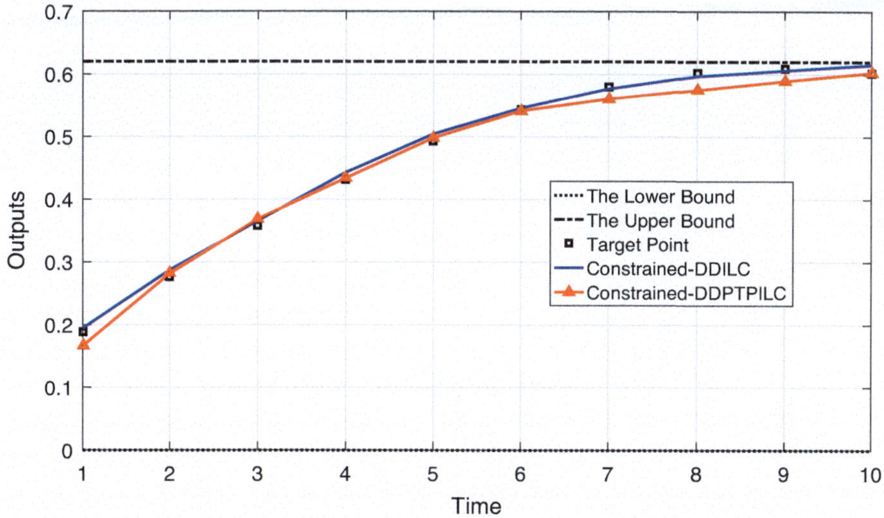


**Fig. 10.16** Tracking performance of constrained-DDILC and constrained-DDPTPILC in case 2.3

constrained-DDILC aims to track all the points of the given desired discrete-time trajectory $y_d(t) = 0.61\sin(\pi t/20)$ which exactly passes through the four points at the time instants of $\{2, 4, 6, 10\}$. The simulations results are shown in Figs. 10.16, 10.17 and 10.18 under the same simulation condition, i.e., $u_0(t) = 0.88$, $\hat{\varphi}_0(t) = 0.1$, $\eta = 0.5$, $\mu = 0.2$, $\boldsymbol{Q} = \boldsymbol{I}$, and $\boldsymbol{R} = 0.5\boldsymbol{I}$. From Fig. 10.16, the system output can exactly track all the desired points by using the constrained-DDILC, while the constrained-DDPTPILC only makes the system output track the prescribed points at $\{2, 4, 6, 10\}$ precisely. Figures 10.17 and 10.18 also confirm that the constrained-DDPTPILC outperforms the constrained-DDILC for multiple points tracking tasks with a faster convergence and reduced control effort.

## 10.7 Summary

Taking a class of nonlinear nonaffine discrete-time systems into consideration, this chapter proposes a constrained-DDILC for a trajectory tracking problem and a constrained DDPTPILC for a point-to-point tracking problem, respectively. The key-point before proceeding to the controller design and analysis is the iterative dynamic linearization method introduced. Then the following derivations depend on the I/O measurements without the need of explicit model knowledge. The proposed constrained methods attain a good performance without violations of constraints on system output and control input by incorporating the I/O constraints into the controller design. The proposed constrained methods do not depend on the linear model

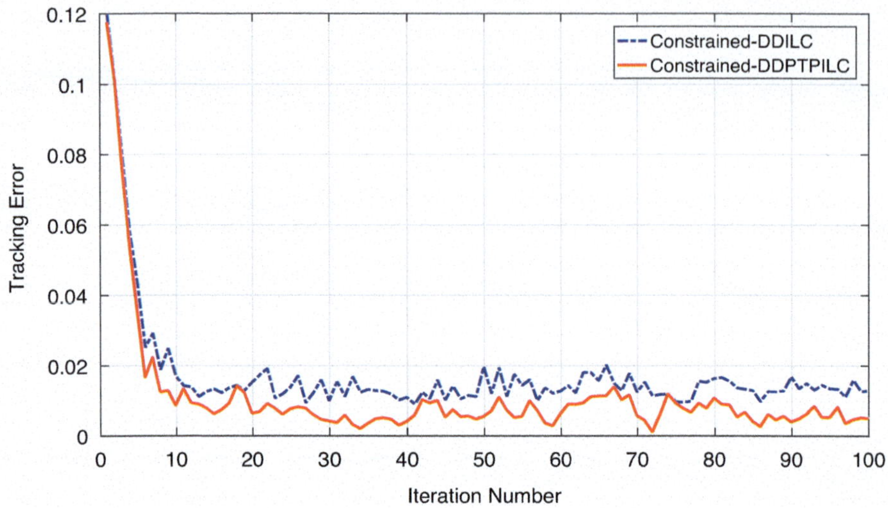


**Fig. 10.17** Error convergence of constrained DDILC and constrained DDPTPILC in case 2.3

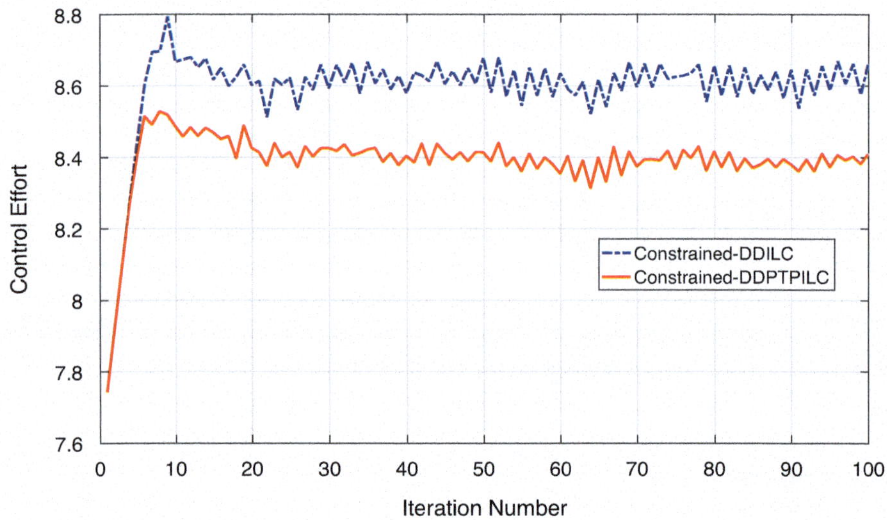


**Fig. 10.18** Control effort of constrained DDILC and constrained DDPTPILC in case 2.3

and thus can achieve a satisfactory performance in the presence of large disturbances to the system model. The robustness of the proposed constrained methods to exogenous uncertainties has been illustrated by taking a realistic nonlinear batch reactor under different levels of stochastic disturbances as an example.

## References


Amann N, Owens DH, Rogers E (1996) Iterative learning control using optimal feedback and feedforward actions. Int J Control 65(2):29–277

Amann N, Owens DH, Rogers E (1996) Iterative learning control for discrete-time systems with exponential rate of convergence. IEE Proc Control Theory Appl 143(2):217–224

Amann N, Owens DH, Rogers E (1996c) Robustness of norm-optimal iterative learning control. In: Proceedings of the UKACC international conference on control. IEEE. https://ieeexplore.ieee.org/document/656192

Amann N, Owens DH, Rogers E (1998) Predictive optimal iterative learning control. Int J Control 69(2):203–226

Chi R, Hou Z, Xu J (2008) Adaptive ILC for a class of discrete-time systems with iteration-varying trajectory and random initial condition. Automatica 44(8):2207–2213

Chi R, Hou Z, Huang B, Jin S (2015) A unified data-driven design framework of optimality-based generalized iterative learning control. Comput Chem Eng 77:10–23

Chin I, Qin SJ, Lee KS, Cho M (2004) A two-stage iterative learning control technique combined with real-time feedback for independent disturbance rejection. Automatica 40(11):1913–1922

Chu B, Owens DH (2010) Iterative learning control for constrained linear systems. Int J Control 83(7):1397–1413

Doyle FJ III (1999) An anti-windup input-output linearization scheme for SISO systems. J Process Control 9(3):213–220

Driessen BJ, Sadegh N (2002) Multi-input square iterative learning control with input rate limits and bounds. IEEE Trans Syst Man Cybern Part B Cybern 32(4):545–550

Huang D, Xu J, Li X, Xu C, Yu M (2013) D-type anticipatory iterative learning control for a class of inhomogeneous heat equations. Automatica 49(8):2397–2408

Janssens P, Pipeleers G, Swevers J (2012) A data-driven constrained norm-optimal iterative learning control framework for LTI systems. IEEE Trans Control Syst Technol 21(2):546–551

Jin X, Xu J (2013) Iterative learning control for output-constrained systems with both parametric and nonparametric uncertainties. Automatica 49(8):2508–2516

Lee KS, Chin IS, Lee HJ, Lee JH (1999) Model predictive control technique combined with iterative learning for batch processes. AIChE J 45(10):2175–2187

Lee JH, Lee KS, Kim WC (2000) Model-based iterative learning control with a quadratic criterion for time-varying linear systems. Automatica 36(5):641–657

Li X, Xiao T, Zheng H (2011) Adaptive discrete-time iterative learning control for non-linear multiple input multiple output systems with iteration-varying initial error and reference trajectory. IET Control Theory Appl 5(9):1131–1139

Liu T, Wang Y (2012) A synthetic approach for robust constrained iterative learning control of piecewise affine batch processes. Automatica 48(11):2762–2775

Logsdon JS, Biegler LT (1992) Decomposition strategies for large-scale dynamic optimization problems. Chem Eng Sci 47(4):851–864

Mishra S, Topcu U, Tomizuka M (2010) Optimization-based constrained iterative learning control. IEEE Trans Control Syst Technol 19(6):1613–1621

Moore KL (1999) Iterative learning control: An expository overview. In: Applied and computational control, signals, and circuits. Birkhäuser, Boston, MA, pp 151–214

Raó W (1981) Advanced process control. McGraw-Hill, New York

Sun M, Wang D (2001) Anticipatory iterative learning control for nonlinear systems with arbitrary relative degree. IEEE Trans Autom Control 46(5):783–788

Tan Y, Xu J, Norrlöf M, Freeman C (2011) On reference governor in iterative learning control for dynamic systems with input saturation. Automatica 47(11):2412–2419

Tan Y, Dai H, Huang D, Xu J (2012) Unified iterative learning control schemes for nonlinear dynamic systems with nonlinear input uncertainties. Automatica 48(12):3173–3182

Wang L, Freeman CT, Chai S, Rogers E (2013) Predictive-repetitive control with constraints: from design to implementation. J Process Control 23(7):956–967

Xu J, Jin X (2012) State-constrained iterative learning control for a class of MIMO systems. IEEE Trans Autom Control 58(5):1322–1327

Xu J, Tan Y, Lee TH (2004) Iterative learning control design based on composite energy function with input saturation. Automatica 40(8):1371–1377

Yin C, Xu J, Hou Z (2010) A high-order internal model based iterative learning control scheme for nonlinear systems with time-iteration-varying parameters. IEEE Trans Autom Control 55(11):2665–2670

Yu M, Wang J, Qi D (2013) Discrete-time adaptive iterative learning control for high-order nonlinear systems with unknown control directions. Int J Control 86(2):299–308

# Chapter 11
# ESO-based Data-Driven Iterative Learning Control

## 11.1 Introduction

Many variables in practical processes are unmeasurable or inaccurate due to the presence of disturbances, which significantly affect the control performance. Generally, an observer is often used to estimate such variables intervened by the noises. Extended state observer (ESO) has been widely used in designing control systems for its ability in rejecting uncertainties.

In Zhao and Guo (2017), an ESO-based feedback control is proposed for MIMO systems to compensate the uncertainties. An event-triggered active disturbance reject control (ADRC) is proposed in Sun et al. (2017) to reduce the communication frequency with a satisfactory control performance. In Xue et al. (2015), an adaptive ESO-based ADRC is proposed for the control task of gasoline engines. A sliding mode control combined with ESO (Li et al. 2015) is proposed for the control of an electronic throttle valve. In order to deal with the parameter falling problem in motor system, an ESO-based dynamic surface control is proposed in Das and Talole (2016). Similarly, several composite controllers are designed in Kobayashi et al. (2007), Wei and Guo (2010), and Guo and Wen (2011) using disturbance observer to estimate multiple disturbances. In Wei et al. (2016), the stochastic disturbance observer design is proposed for a class of stochastic systems.

The combination of ESO with ILC is investigated in Sun et al. (2014), but no information from the previous iterations is utilized in the ESO. More recently, the ESO method has been extended to iteration domain for a repetitive system (Li et al., 2016), where the estimation of the extended states is conducted along the iteration direction. Most of the ESO-based control methods, however, have mainly focused on continuous-time systems and require the prior model structure affine to control input.

This chapter presents an ESO-based DDILC for a nonlinear nonaffine discrete-time system with exogenous disturbances. No modeling step is required for the



R. Chi et al., *Data-Driven Iterative Learning Control for Discrete-Time Systems*, Intelligent Control and Learning Systems 2,
https://doi.org/10.1007/978-981-19-5950-9_11

proposed method. At first, an IDL is introduced to gain the LDM equivalent to the original nonlinear nonaffine system, where the randomly iteration-varying initial states and disturbances as well as the corresponding partial derivatives are retained and regarded as new states. Then, the ESO is extended to iteration domain to iteratively estimate the new states. In the sequel, the ESO-based DDILC is presented by minimizing two objective functions, respectively, where the observed values from the ESO are incorporated. The ESO estimation is conducted along iteration direction instead of the time direction, and thus the iterative ESO is capable of reducing the estimation error from iteration to iteration by using its learning ability from repetitions. Both the theoretical analysis and the simulation study are provided.

This chapter is organized as follows. Section 11.2 shows the problem formulation, and the ESO-based DDILC is designed and analyzed in Sects. 11.3 and 11.4, respectively. Simulation study is shown in Sect. 11.5. Section 11.6 presents some conclusions.

## 11.2 Problem Formulation

Consider a class of nonlinear nonaffine discrete-time systems

$$\begin{aligned} y_k(t+1) = f\left(y_k(t), y_k(t-1), \dots y_k\left(t-n_y\right), u_k(t), u_k(t-1), \dots u_k\left(t-n_u\right)\right. \\ \left. d_k(t), d_k(t-1), \dots, d_k\left(t-n_d\right)\right) \end{aligned} \tag{11.1}$$

where $y_k(t) \in R$, $u_k(t) \in R$, and $d_k(t) \in R$ are the output, input, and disturbance, respectively; $y_k(t) = 0$, $u_k(t) = 0$, and $d_k(t) = 0$, $\forall t < 0$; $n_y$, $n_u$, and $n_d$ are unknown orders; $t \in \{0, 1, \dots, N\}$ is the time instant and $N \in Z^+$ represents the endpoint; $k \in \{0, 1, \dots\}$ indicates the iteration number; $f(\cdots) : R^{n_u+n_y+n_d+3} \mapsto R$ is an unknown nonlinear scalar function, and $f(0, 0, 0) = 0$.

Similar to (4.4), one has

$$y_k(t+1) = g_t\left(y_k(0), \boldsymbol{u}_k(t), \boldsymbol{d}_k(t)\right) \tag{11.2}$$

where $g_t(\cdot)$ is a compound function with respect to $f(\cdot)$ and has the same properties as $f(\cdot)$; $\boldsymbol{u}_k(t) = [u_k(0), u_k(1), \dots, u_k(t)]^{\mathrm{T}} \in R^{t+1}$, $\boldsymbol{d}_k(t) = [d_k(0), d_k(1), \dots, d_k(t)]^{\mathrm{T}} \in R^{t+1}$.

The nonlinear system (11.1) satisfies Assumptions 4.2 and 4.3 and the following two assumptions.

**Assumption 11.1** The initial value is bounded, i.e., $|y_k(0)| < b_{y0}$, $\forall k$, where $b_{y0} > 0$ is a positive constant.

**Assumption 11.2** The disturbance is bounded, that is, $\|\boldsymbol{d}_k(t)\| < b_d$, $\forall t \in \{0, 1, \dots, N\}$, $\forall k$, where $b_d > 0$ is a positive constant.

The new dynamic linear data model containing nonlinear uncertainties is given as follows.

**Theorem 11.1** *Consider nonlinear system (11.1) satisfying Assumptions 4.2, 4.3, 11.1, and 11.2. There exists an time-iteration-varying parameter* $\boldsymbol{\varphi}_k(t) \in R^{1\times(t+1)}$, *such that system (11.1) can be transformed into the following dynamic linear data model,*

$$\Delta y_k(k+1) = \boldsymbol{\varphi}_k(t)\Delta \boldsymbol{u}_k(t) + \xi_k(t) \tag{11.3}$$

*where* $\Delta y_k(t+1) = y_k(t+1) - y_{k-1}(t+1)$ *and* $\Delta \boldsymbol{u}_k(t) = \boldsymbol{u}_k(t) - \boldsymbol{u}_{k-1}(t)$; $\boldsymbol{\varphi}_k(t) = [\partial g_t(\cdot)/\partial \boldsymbol{u}_k(t)]^{\mathrm{T}} \in R^{1\times(t+1)}$; $\xi_k(t)$ *is an time-iteration-varying uncertainty;* $\|\boldsymbol{\varphi}_k(t)\| \le b_\varphi$ *and* $|\xi_k(t)| \le b_\xi$ *are bounded, in which* $b_\varphi > 0$ *and* $b_\xi > 0$ *are constants.*

***Proof*** Using differential mean value theorem, one can obtain from (11.2) that,

$$\begin{aligned}\Delta y_k(t+1) &= g_t\left(y_k(0), \boldsymbol{u}_k(t), \boldsymbol{d}_k(t)\right) - g_t\left(y_{k-1}(0), \boldsymbol{u}_{k-1}(t), \boldsymbol{d}_{k-1}(t)\right)\\ &= \frac{\partial g_t(\cdot)}{\partial y_k(0)}\left(y_k(0) - y_{k-1}(0)\right) + \left(\boldsymbol{u}_k(t) - \boldsymbol{u}_{k-1}(t)\right)^{\mathrm{T}}\frac{\partial g_t(\cdot)}{\partial \boldsymbol{u}_k(t)}\\ &\quad + \left(\boldsymbol{d}_k(t) - \boldsymbol{d}_{k-1}(t)\right)^{\mathrm{T}}\frac{\partial g_t(\cdot)}{\partial \boldsymbol{d}_k(t)}\end{aligned} \tag{11.4}$$

Let $\boldsymbol{\varphi}_k(t) = [\partial g_t(\cdot)/\partial \boldsymbol{u}_k(t)]^{\mathrm{T}} \in R^{1\times(t+1)}$, $\xi_k(t) = \partial g_t(\cdot)/\partial y_k(0)\left(y_k(0) - y_{k-1}(0)\right) + \left(\boldsymbol{d}_k(t) - \boldsymbol{d}_{k-1}(t)\right)^{\mathrm{T}}\left(\partial g_t(\cdot)/\partial \boldsymbol{d}_k(t)\right)$, one has,

$$\Delta y_k(t+1) = \boldsymbol{\varphi}_k(t)\Delta \boldsymbol{u}_k(t) + \xi_k(t)$$

According to Assumption 4.2, one has $\|\boldsymbol{\varphi}_k(t)\| \le b_\varphi$. In addition, we have $\partial g_t(\cdot)/\partial y_k(0)$ and $\partial g_t(\cdot)/\partial \boldsymbol{d}_k(t)$ are bounded. Since the initial state and disturbances are bounded, the boundedness of $\xi_k(t)$ can be obtained, i.e., $|\xi_k(t)| \le b_\xi$. □

**Remark 11.1** Different from Theorem 4.1, the STIDL-based LDM proposed in this chapter contains the nonlinear uncertainties such as the random initial states, exogenous disturbances, etc., which are regarded as an extended state to be estimated in the subsequent analysis. As a result, the LDM proposed in Theorem 11.1 does not need the identical initial condition which is a significant condition relaxation of the traditional ILC methods.

## 11.3 Controller Design

Consider a desired output $y_d(t), t \in \{0, 1, \ldots, N\}$. The control objective is to find an optimal control input $u_k(t)$, such that system output $y_k(t+1)$ precisely tracks $y_d(t+1)$ for all time instants, i.e., the output tracking error $e_k(t+1) = y_d(t+1) - y_k(t+1)$ converges iteratively.

The objective function with respect to control input is designed as follows,

$$J(u_k(t)) = |e_k(t+1)|^2 + \lambda |u_k(t) - u_{k-1}(t)|^2 \tag{11.5}$$

where $\lambda > 0$ is a weighting factor.

In terms of (11.3), one has

$$\begin{aligned} J(u_k(t)) = & \left| e_{k-1}(t+1) - \sum_{l=0}^{t-1} \phi_k(l) \Delta u_k(l) - \phi_k(t) \Delta u_k(t) - \xi_k(t) \right|^2 \\ & + \lambda |u_k(t) - u_{k-1}(t)|^2 \end{aligned} \tag{11.6}$$

where $\phi_k(l)$ and $u_k(l)$ are the $(l+1)$th elements of $\boldsymbol{\varphi}_k(t)$ and $\boldsymbol{u}_k(t)$, respectively; $\Delta u_k(l) = u_k(l) - u_{k-1}(l), l \in \{0, 1, \ldots, t\}$.

Differentiating (11.6) with respect to $u_k(t)$ and setting it zero, gives

$$u_k(t) = u_{k-1}(t) + \frac{\rho \phi_k(t) \left(e_{k-1}(t+1) - \xi_k(t)\right)}{\lambda + |\phi_k(t)|^2} - \frac{\rho \phi_k(t) \sum_{l=0}^{t-1} \phi_k(l) \Delta u_k(l)}{\lambda + |\phi_k(t)|^2} \tag{11.7}$$

where $\rho > 0$ is a step-size factor.

Since the value of $\boldsymbol{\varphi}_k(t)$ in the learning control law (11.7) is unknown, its estimation law is developed along iteration direction by using the following objective function of $\boldsymbol{\varphi}_k(t)$,

$$\begin{aligned} J\left(\hat{\boldsymbol{\varphi}}_k(t)\right) = & \left\| \Delta y_{k-1}(t+1) - \hat{\boldsymbol{\varphi}}_k(t) \Delta \boldsymbol{u}_{k-1}(t) - \xi_{k-1}(t) \right\|^2 \\ & + \mu \left\| \hat{\boldsymbol{\varphi}}_k(t) - \hat{\boldsymbol{\varphi}}_{k-1}(t) \right\|^2 \end{aligned} \tag{11.8}$$

where $\mu > 0$ denotes a weighting factor and $\hat{\boldsymbol{\varphi}}_{k-1}(t)$ is the estimation of $\boldsymbol{\varphi}_{k-1}(t)$.

By minimizing (11.8) with respect to $\hat{\boldsymbol{\varphi}}_k(t)$, one can derive that

$$\begin{aligned} \hat{\boldsymbol{\varphi}}_k(t) = & \hat{\boldsymbol{\varphi}}_{k-1}(t) \\ & + \frac{\eta \left( \Delta y_{k-1}(t+1) - \hat{\boldsymbol{\varphi}}_{k-1}(t) \Delta \boldsymbol{u}_{k-1}(t) - \xi_{k-1}(t) \right) \Delta \boldsymbol{u}_{k-1}^{\mathrm{T}}(t)}{\mu + \|\Delta \boldsymbol{u}_{k-1}(t)\|^2} \end{aligned} \tag{11.9}$$

where $0 < \eta < 2$ is a step-size factor.

Also, the nonlinear uncertainty, $\xi_k(t)$, is unknown in (11.7) and (11.9), so an iterative ESO is designed as follows to estimate it using the output information from the previous iterations,

$$\hat{\xi}_k(t) = \hat{\xi}_{k-1}(t) + \alpha\left(y_{k-1}(t+1) - y_d(t+1)\right) \tag{11.10}$$

$$\hat{\xi}_k(t) = \begin{cases} b_{\hat{\xi}} & \text{if } \hat{\xi}_k(t) > b_{\hat{\xi}} \\ \hat{\xi}_k(t) & \text{if } \left|\hat{\xi}_k(t)\right| \leq b_{\hat{\xi}} \\ -b_{\hat{\xi}} & \text{if } \hat{\xi}_k(t) < -b_{\hat{\xi}} \end{cases} \tag{11.11}$$

where $\alpha$ is an adjustable observer gain; $b_{\hat{\xi}} > 0$ is a saturation constant; and $\hat{\xi}_k(t)$ is the estimation of $\xi_k(t)$.

The complete ESO-based DDILC proposed in this chapter is as follows,

$$\begin{aligned} u_k(t) = u_{k-1}(t) &+ \frac{\rho\hat{\phi}_k(t)\left(e_{k-1}(t+1) - \hat{\xi}_k(t)\right)}{\lambda + \left|\hat{\phi}_k(t)\right|^2} \\ &- \frac{\rho\hat{\phi}_k(t)\sum_{l=0}^{t-1}\hat{\phi}_k(l)\Delta u_k(l)}{\lambda + \left|\hat{\phi}_k(t)\right|^2} \end{aligned} \tag{11.12}$$

$$\begin{aligned} \hat{\boldsymbol{\varphi}}_k(t) &= \hat{\boldsymbol{\varphi}}_{k-1}(t) \\ &+ \frac{\eta\left(\Delta y_{k-1}(t+1) - \hat{\boldsymbol{\varphi}}_{k-1}(t)\Delta \boldsymbol{u}_{k-1}(t) - \hat{\xi}_{k-1}(t)\right)\Delta \boldsymbol{u}_{k-1}^{\mathrm{T}}(t)}{\mu + \|\Delta \boldsymbol{u}_{k-1}(t)\|^2} \end{aligned} \tag{11.13}$$

$$\begin{aligned} \hat{\boldsymbol{\varphi}}_k(t) &= \hat{\boldsymbol{\varphi}}_0(t), \text{ if } \operatorname{sgn}\left(\hat{\phi}_k(l)\right) \neq \operatorname{sgn}\left(\hat{\phi}_0(l)\right) \text{ or} \\ &\left\|\hat{\boldsymbol{\varphi}}_k(t)\right\| \leq \varepsilon, l \in \{0, 1, \ldots, t\} \end{aligned} \tag{11.14}$$

$$\begin{aligned} \hat{\xi}_k(t) &= \hat{\xi}_{k-1}(t) + \alpha\left(y_{k-1}(t+1) - y_d(t+1)\right) \\ \hat{\xi}_k(t) &= \begin{cases} b_{\hat{\xi}}, & \text{if } \hat{\xi}_k(t) > b_{\hat{\xi}} \\ \hat{\xi}_k(t), & \text{if } \left|\hat{\xi}_k(t)\right| \leq b_{\hat{\xi}} \\ -b_{\hat{\xi}}, & \text{if } \hat{\xi}_k(t) < -b_{\hat{\xi}} \end{cases} \end{aligned} \tag{11.15}$$

where $\varepsilon$ is a small positive scalar; $\hat{\varphi}_0(t)$ is the initial value of $\hat{\varphi}_k(t)$; (11.14) is a reset algorithm, which is added to enhance the tracking performance of the estimation algorithm (11.13) for the time-iteration-varying $\boldsymbol{\varphi}_k(t)$.

**Remark 11.2** Different from the ESO in Zhao and Guo (2017), Sun et al. (2017), Xue et al. (2015), Li et al. (2015), Das and Talole (2016) and the DOB in Kobayashi et al. (2007), Wei and Guo (2010), Guo and Wen (2011), Wei et al. (2016), the proposed ESO estimation law (11.15) is conducted iteratively and utilizes additional operation data from the previous iterations such that it is more efficient for repetitive processes.

**Remark 11.3** Compared with Sect. 4.4, the proposed ESO-based DDILC can compensate the uncertainties by containing the estimations of ESO, and thus can deal with the issues of random varying initial conditions and the nonrepetitive exogenous disturbances.

**Remark 11.4** When both the initial states and the exogenous disturbances are strictly iteration-invariant, the ESO is no longer needed. Therefore, one can set $b_{\xi} = 0$ in (11.15) such that $\hat{\xi}_k(t) \equiv 0$ which means that the ESO is removed from (11.12) and (11.13) and the ESO-based DDILC becomes the DDILC in Sect. 4.4.

**Remark 11.5** The proposed ESO-based DDILC (11.12)–(11.15) is a data-driven approach and can be directly applied to the plant formulated by a general nonlinear system mathematically where no mechanistic modeling step is required anymore.

## 11.4 Convergence Analysis

The convergence of the ESO-based DDILC is guaranteed by the following theorem.

**Theorem 11.2** *For a class of nonlinear nonaffine discrete-time systems (11.1), if Assumptions 4.2, 4.3, 11.1, and 11.2 hold, by selecting the controller parameters properly, the proposed ESO-based DDILC method (11.12)–(11.15) guarantees that:*

*i) the parameter estimation $\hat{\varphi}_k(t)$ is bounded over the finite time interval $t \in \{0, 1, \ldots, N\}, \forall k$;*
*ii) the tracking error $e_k(t)$ converges to a small bound even though both the initial states and the disturbances are iteration-varying.*

***Proof*** *Part 1. The boundedness of $\hat{\varphi}_k(t)$*

Subtracting $\boldsymbol{\varphi}_k(t)$ from both sides of (11.13) and utilizing (11.3), we have

$$
\begin{aligned}
\tilde{\boldsymbol{\varphi}}_k(t) &= \tilde{\boldsymbol{\varphi}}_{k-1}(t) \\
&\quad - \frac{\eta\left(\Delta y_{k-1}(t+1) - \hat{\boldsymbol{\varphi}}_{k-1}(t)\Delta \boldsymbol{u}_{k-1}(t) - \hat{\xi}_{k-1}(t)\right)\Delta \boldsymbol{u}_{k-1}^{\mathrm{T}}(t)}{\mu + \|\Delta \boldsymbol{u}_{k-1}(t)\|^2} \\
&\quad + \boldsymbol{\varphi}_k(t) - \boldsymbol{\varphi}_{k-1}(t) \\
&= \tilde{\boldsymbol{\varphi}}_{k-1}(t)\left(I - \frac{\eta \Delta \boldsymbol{u}_{k-1}(t)\Delta \boldsymbol{u}_{k-1}^{\mathrm{T}}(t)}{\mu + \|\Delta \boldsymbol{u}_{k-1}(t)\|^2}\right) - \frac{\eta \xi_{k-1}(t)\Delta \boldsymbol{u}_{k-1}^{\mathrm{T}}(t)}{\mu + \|\Delta \boldsymbol{u}_{k-1}(t)\|^2} \\
&\quad + \frac{\eta \hat{\xi}_{k-1}(t)\Delta \boldsymbol{u}_{k-1}^{\mathrm{T}}(t)}{\mu + \|\Delta \boldsymbol{u}_{k-1}(t)\|^2} + \boldsymbol{\varphi}_k(t) - \boldsymbol{\varphi}_{k-1}(t)
\end{aligned}
\tag{11.16}
$$

where $\tilde{\boldsymbol{\varphi}}_k(t) = \boldsymbol{\varphi}_k(t) - \hat{\boldsymbol{\varphi}}_k(t)$ denotes the parameter estimation error.

Then, taking the norm of (11.16), we have

$$
\begin{aligned}
\|\tilde{\boldsymbol{\varphi}}_k(t)\| &\le \left\| \tilde{\boldsymbol{\varphi}}_{k-1}(t)\left(I - \frac{\eta \Delta \boldsymbol{u}_{k-1}(t)\Delta \boldsymbol{u}_{k-1}^{\mathrm{T}}(t)}{\mu + \|\Delta \boldsymbol{u}_{k-1}(t)\|^2}\right)\right\| \\
&\quad + \frac{\eta\,|\xi_{k-1}(t)|\,\|\Delta \boldsymbol{u}_{k-1}(t)\|}{\mu + \|\Delta \boldsymbol{u}_{k-1}(t)\|^2} + \frac{\eta\left|\hat{\xi}_{k-1}(t)\right|\|\Delta \boldsymbol{u}_{k-1}(t)\|}{\mu + \|\Delta \boldsymbol{u}_{k-1}(t)\|^2} \\
&\quad + \left\|\boldsymbol{\varphi}_k(t) - \boldsymbol{\varphi}_{k-1}(t)\right\|
\end{aligned}
\tag{11.17}
$$

From (11.17), it is easy to get the following equation,

$$
\begin{aligned}
\left\| \tilde{\boldsymbol{\varphi}}_{k-1}(t)\left(I - \frac{\eta \Delta \boldsymbol{u}_{k-1}(t)\Delta \boldsymbol{u}_{k-1}^{\mathrm{T}}(t)}{\mu + \|\Delta \boldsymbol{u}_{k-1}(t)\|^2}\right)\right\|^2 &= \left\|\tilde{\boldsymbol{\varphi}}_{k-1}(t)\right\|^2 \\
+ \left(-2 + \frac{\eta\,\|\Delta \boldsymbol{u}_{k-1}(t)\|^2}{\mu + \|\Delta \boldsymbol{u}_{k-1}(t)\|^2}\right) \frac{\eta\left\|\Delta \boldsymbol{u}_{k-1}(t)\tilde{\boldsymbol{\varphi}}_{k-1}(t)\right\|^2}{\mu + \|\Delta \boldsymbol{u}_{k-1}(t)\|^2}
\end{aligned}
\tag{11.18}
$$

Since $0 < \eta < 2$ and $\mu > 0$, then

$$
\left(-2 + \frac{\eta\,\|\Delta \boldsymbol{u}_{k-1}(t)\|^2}{\mu + \|\Delta \boldsymbol{u}_{k-1}(t)\|^2}\right) < 0 \tag{11.19}
$$

So, (11.18) and (11.19) together imply that

$$
\left\| \tilde{\boldsymbol{\varphi}}_{k-1}(t)\left(I - \frac{\eta \Delta \boldsymbol{u}_{k-1}(t)\Delta \boldsymbol{u}_{k-1}^{\mathrm{T}}(t)}{\mu + \|\Delta \boldsymbol{u}_{k-1}(t)\|^2}\right)\right\| \le d_1 \left\|\tilde{\boldsymbol{\varphi}}_{k-1}(t)\right\| \tag{11.20}
$$

where $0 < d_1 < 1$ is a positive constant.

Evidently, the following inequality holds,

$$\frac{\eta\,|\xi_{k-1}(t)|\,\|\Delta \boldsymbol{u}_{k-1}(t)\|}{\mu+\|\Delta \boldsymbol{u}_{k-1}(t)\|^2} \le \frac{\eta b_{\xi}\,\|\Delta \boldsymbol{u}_{k-1}(t)\|}{2\sqrt{\mu}\,\|\Delta \boldsymbol{u}_{k-1}(t)\|} = \frac{\eta b_{\xi}}{2\sqrt{\mu}} \tag{11.21}$$

Similarly, one has

$$\frac{\eta\left|\hat{\xi}_{k-1}(t)\right|\,\|\Delta \boldsymbol{u}_{k-1}(t)\|}{\mu+\|\Delta \boldsymbol{u}_{k-1}(t)\|^2} \le \frac{\eta b_{\hat{\xi}}}{2\sqrt{\mu}} \tag{11.22}$$

According to Theorem 11.1, we have

$$\left\|\boldsymbol{\varphi}_k(t)-\boldsymbol{\varphi}_{k-1}(t)\right\| \le 2b_{\varphi} \tag{11.23}$$

Therefore, in terms of (11.20)–(11.23), inequality (11.17) leads to

$$\begin{aligned}\left\|\tilde{\boldsymbol{\varphi}}_k(t)\right\| &\le d_1\left\|\tilde{\boldsymbol{\varphi}}_{k-1}(t)\right\|+\varsigma_1 \le d_1^2\left\|\tilde{\boldsymbol{\varphi}}_{k-2}(t)\right\|+d_1\varsigma_1+\varsigma_1\\ &\le \cdots \le d_1^k\left\|\tilde{\boldsymbol{\varphi}}_0(t)\right\|+d_1^{k-1}\varsigma_1+\cdots+\varsigma_1\\ &\le d_1^k\left\|\tilde{\boldsymbol{\varphi}}_0(t)\right\|+\frac{\varsigma_1}{1-d_1}\end{aligned} \tag{11.24}$$

where $\varsigma_1=[\eta/(2\sqrt{\mu})]\left(b_{\xi}+b_{\hat{\xi}}\right)+2b_{\varphi}$.

Inequality (11.24) implies $\tilde{\boldsymbol{\varphi}}_k(t)$ is bounded, because $\hat{\boldsymbol{\varphi}}_0(t)$ and $\boldsymbol{\varphi}_k(t)$ are bounded. Therefore, the boundedness of $\hat{\boldsymbol{\varphi}}_k(t)$ is evident.

*Part 2. The convergence of tracking error*

Since $\boldsymbol{\varphi}_k(t)$ is bounded, and the boundedness of $\hat{\boldsymbol{\varphi}}_k(t)$ has been proved, by selecting $\rho$ properly, there exist constants $M_1$, $M_2$, $M_3$, and $\lambda$, such that

$$\left|\frac{\hat{\phi}_k(t)}{\lambda+\left|\hat{\phi}_k(t)\right|^2}\right| \le \left|\frac{\hat{\phi}_k(t)}{2\sqrt{\lambda}\left|\hat{\phi}_k(t)\right|}\right| = \frac{1}{2\sqrt{\lambda}} = M_1 < \frac{1}{\rho} \tag{11.25}$$

$$\left|\frac{\phi_k(t)\hat{\phi}_k(t)}{\lambda+\left|\hat{\phi}_k(t)\right|^2}\right| \le b_{\varphi}\left|\frac{\hat{\phi}_k(t)}{2\sqrt{\lambda}\left|\hat{\phi}_k(t)\right|}\right| = \frac{b_{\varphi}}{2\sqrt{\lambda}} = M_2 < \frac{1}{\rho} \tag{11.26}$$

$$\sum_{l=0}^{N-1}\left|\frac{\hat{\phi}_k(t)\hat{\phi}_k(l)}{\lambda+\left|\hat{\phi}_k(t)\right|^2}\right| = M_3 < \frac{0.5}{\rho} \tag{11.27}$$

Let $\boldsymbol{A}_k(t) = \begin{bmatrix} 0 & 1 & 0 & \cdots & 0 \\ 0 & 0 & 1 & \cdots & 0 \\ \vdots & \vdots & \vdots & \ddots & \vdots \\ 0 & 0 & 0 & \cdots & 1 \\ 0 & -\frac{\rho\hat{\phi}_k(t)\hat{\phi}_k(0)}{\lambda+\left|\hat{\phi}_k(t)\right|^2} & -\frac{\rho\hat{\phi}_k(t)\hat{\phi}_k(1)}{\lambda+\left|\hat{\phi}_k(t)\right|^2} & \cdots & -\frac{\rho\hat{\phi}_k(t)\hat{\phi}_k(t-1)}{\lambda+\left|\hat{\phi}_k(t)\right|^2} \end{bmatrix} \in R^{(t+1)\times(t+1)}$ and $\boldsymbol{C} = [0, \ldots, 0, 1]^{\mathrm{T}} \in R^{t+1}$, then according to (11.12), one has

$$\Delta \boldsymbol{u}_k(t) = \boldsymbol{A}_k(t)\Delta \boldsymbol{u}_k(t-1) + \frac{\rho\hat{\phi}_k(t)}{\lambda + \left|\hat{\phi}_k(t)\right|^2}\boldsymbol{C}\left(e_{k-1}(t+1) - \hat{\xi}_k(t)\right) \tag{11.28}$$

From (11.27), and by selecting $\rho$ properly, we have

$$\rho\sum_{l=0}^{t-1}\left|\frac{\hat{\phi}_k(t)\hat{\phi}_k(l)}{\lambda + \left|\hat{\phi}_k(t)\right|^2}\right| \le \rho\sum_{l=0}^{N-1}\left|\frac{\hat{\phi}_k(t)\hat{\phi}_k(l)}{\lambda + \left|\hat{\phi}_k(t)\right|^2}\right| = \rho M_3 < 0.5 \tag{11.29}$$

Based on Lemma 4.2, (11.29) means that $s\left(\boldsymbol{A}_k(t)\right) < 0.5$.
Then, according to Lemma 3.2, we have

$$\left\|\boldsymbol{A}_k(t)\right\|_v \le s\left(\boldsymbol{A}_k(t)\right) + \delta = d_2 < 0.5 \tag{11.30}$$

where $\left\|\boldsymbol{A}_k(t)\right\|_v$ is a proper matrix norm of $\boldsymbol{A}_k(t)$; $\delta$ is a small enough positive constant; $0 < d_2 < 0.5$ is a positive constant.

Taking norms on both sides of (11.28), leads to

$$\begin{aligned}
\left\|\Delta \boldsymbol{u}_k(t)\right\|_v \le\ & \left\|\boldsymbol{A}_k(t)\right\|_v \left\|\Delta \boldsymbol{u}_k(t-1)\right\|_v + \left|\frac{\rho\hat{\phi}_k(t)}{\lambda + \left|\hat{\phi}_k(t)\right|^2}\right| \\
& \times \left|e_{k-1}(t+1)\right| + \left|\frac{\rho\hat{\phi}_k(t)}{\lambda + \left|\hat{\phi}_k(t)\right|^2}\right|\left|\hat{\xi}_k(t)\right| \\
\le\ & d_2\left\|\Delta \boldsymbol{u}_k(t-1)\right\|_v + \rho M_1\left|e_{k-1}(t+1)\right| + \rho M_1 b_{\hat{\xi}} \\
& \vdots \\
=\ & \rho M_1\sum_{l=0}^{t} d_2^l\left|e_{k-1}(t+1-l)\right| + \frac{\rho M_1 b_{\hat{\xi}}}{1-d_2}
\end{aligned} \tag{11.31}$$

where the input $\boldsymbol{u}_k(t)$ is set as zero for $t < 0$.

By virtue of (11.3) and (11.28), the error dynamics can be derived as follows,

$$\begin{aligned}
e_k(t+1) &= e_{k-1}(t+1) - \boldsymbol{\varphi}_k(t)\Delta \boldsymbol{u}_k(t) - \xi_k(t) \\
&= e_{k-1}(t+1) - \xi_k(t) - \boldsymbol{\varphi}_k(t) \\
&\quad \times \left( \boldsymbol{A}_k(t)\Delta \boldsymbol{u}_k(t-1) + \frac{\rho\hat{\phi}_k(t)}{\lambda + \left|\hat{\phi}_k(t)\right|^2}\boldsymbol{C}\left(e_{k-1}(t+1) - \hat{\xi}_k(t)\right)\right) \\
&= \left(1 - \frac{\rho\phi_k(t)\hat{\phi}_k(t)}{\lambda + \left|\hat{\phi}_k(t)\right|^2}\right) e_{k-1}(t+1) - \boldsymbol{\varphi}_k(t)\boldsymbol{A}_k(t) \\
&\quad \times \Delta \boldsymbol{u}_k(t-1) + \frac{\rho\phi_k(t)\hat{\phi}_k(t)}{\lambda + \left|\hat{\phi}_k(t)\right|^2}\hat{\xi}_k(t) - \xi_k(t).
\end{aligned} \tag{11.32}$$

According to Assumption 4.3, the values of $\hat{\boldsymbol{\varphi}}_0(t)$ can be selected such that its elements' signs are the same as that of $\boldsymbol{\varphi}_k(t)$. Then, according to the reset algorithm (11.14), one can guarantee that $\hat{\phi}_k(t)\phi_k(t) > 0$, $\forall t$ and $\forall k$.

Based on (11.26), one can select $\rho$ properly, such that

$$0 < \frac{\rho\phi_k(t)\hat{\phi}_k(t)}{\lambda + \left|\hat{\phi}_k(t)\right|^2} \le \rho M_2 = d_3 < 1 \tag{11.33}$$

where $0 < d_3 < 1$ is a positive constant.

Taking norms on both sides of (11.32), yields

$$\begin{aligned}
|e_k(t+1)| &\le \left|\left(1 - \frac{\rho\phi_k(t)\hat{\phi}_k(t)}{\lambda + \left|\hat{\phi}_k(t)\right|^2}\right)\right| |e_{k-1}(t+1)| + |\xi_k(t)| \\
&\quad + \left\|\boldsymbol{\varphi}_k(t)\right\|_v \|\boldsymbol{A}_k(t)\|_v \|\Delta \boldsymbol{u}_k(t-1)\|_v \\
&\quad + \left|\frac{\rho\phi_k(t)\hat{\phi}_k(t)}{\lambda + \left|\hat{\phi}_k(t)\right|^2}\right| \left|\hat{\xi}_k(t)\right| \\
&\le d_4 |e_{k-1}(t+1)| + d_2 \left\|\boldsymbol{\varphi}_k(t)\right\|_v \|\Delta \boldsymbol{u}_k(t-1)\|_v \\
&\quad + d_3 b_{\hat{\xi}} + b_\xi
\end{aligned} \tag{11.34}$$

where $0 < d_4 = 1 - d_3 < 1$.

Substituting (11.31) into (11.34), it follows

$$
\begin{aligned}
|e_k(t+1)| \le {} & d_4 |e_{k-1}(t+1)| \\
& + d_2 \left\|\boldsymbol{\varphi}_k(t)\right\|_v \rho M_1 \sum_{l=0}^{t-1} d_2^l |e_{k-1}(t-l)| \\
& + d_2 \left\|\boldsymbol{\varphi}_k(t)\right\|_v \frac{\rho M_1 b_{\xi}}{1-d_2} + d_3 b_{\xi} + b_{\xi}
\end{aligned} \tag{11.35}
$$

Since $\left\|\boldsymbol{\varphi}_k(t)\right\| \le b_{\varphi}$, letting $d_5 = \rho M_1$, we have

$$
|e_k(t+1)| \le d_4 |e_{k-1}(t+1)| + b_{\varphi} d_2 d_5 \sum_{l=0}^{t-1} d_2^l |e_{k-1}(t-l)| + \varsigma_2 \tag{11.36}
$$

where $\varsigma_2 = \left[\left(d_2 d_5 b_{\xi} b_{\varphi}\right) / (1-d_2)\right] + d_3 b_{\xi} + b_{\xi}$.

Define $e_{\max}(k) = \max_{t\in\{1,\ldots,N\}} \{|e_k(t)|\}$. Assume $e_{\max}(k) = |e_k(\tau)|$, where $\tau$ is some time instant within $\{1, \ldots, N\}$. Then, according to (11.36), one has

$$
\begin{aligned}
e_{\max}(k) = |e_k(\tau)| \le {} & d_4 |e_{k-1}(\tau)| \\
& + b_{\varphi} d_2 d_5 \sum_{l=0}^{\tau-2} d_2^l |e_{k-1}(\tau-1-l)| + \varsigma_2 \\
\le {} & d_4 e_{\max}(k-1) + b_{\varphi} d_5 e_{\max}(k-1) \sum_{l=0}^{\tau-2} d_2^{l+1} + \varsigma_2 \\
\le {} & \left(d_4 + \frac{d_2 d_5 b_{\varphi}}{1-d_2}\right) e_{\max}(k-1) + \varsigma_2
\end{aligned} \tag{11.37}
$$

Since $d_3 = b_{\varphi} d_5$, we have

$$
0 < d_4 + \frac{d_2 d_5 b_{\varphi}}{1-d_2} = 1 - d_3 + \frac{d_2}{1-d_2} d_3 = d_6 < 1 \tag{11.38}
$$

where $0 < d_6 < 1$ is a positive constant

Then, from (11.37), we have

$$
\begin{aligned}
e_{\max}(k) \le d_6 e_{\max}(k-1) + \varsigma_2 \le d_6^2 e_{\max}(k-2) + d_6 \varsigma_2 + \varsigma_2 \\
\le \cdots \le d_6^k e_{\max}(0) + \frac{\varsigma_2}{1-d_6}
\end{aligned} \tag{11.39}
$$

Because the initial maximum tracking error $e_{\max}(0)$ is bounded, the inequality (11.39) leads to a bounded convergence of the tracking error directly, i.e., $\lim_{k\to\infty} e_k(t) \le [\varsigma_2/(1-d_6)]$. Thus, the conclusion ii) of Theorem 11.2 is obtained.

□

## 11.5 Simulation Study

Consider a permanent magnet linear motor system,

$$\begin{cases} \dot{v}_k(t) = \frac{u_k(t) - f_{\text{ripple},k}(t) - f_{\text{friction},k}(t) - f_{l,k}(t)}{M} \\ f_{\text{friction},k}(t) = \left[ f_c + (f_s - f_c)\, e^{-(v_k(t)/v_s)^2} \right] \text{sgn}\,(v_k(t)) + B_v v_k(t) \\ f_{\text{ripple},k}(t) = A_r \sin\,(\omega p_k(t) + j) \end{cases} \quad (11.40)$$

where $v_k(t)$ denotes the motor velocity (m/s); $u_k(t)$ denotes the time-iteration-varying motor terminal voltage (V); $p_k(t)$ denotes the motor position (m); $f_{l,k}(t)$ denotes load disturbance; and $j$ is the position dependent phase angle measured in radians, which will determine the spread of the ripple with respect to position (rad). Other parameters can be queried in Tan et al. (2000) and Lee et al. (2000).

The parameters of PMLM (11.40) are given in Table 11.1. The target velocity of PMLM is

$$v_d(t) = 2\sin(\pi t/500) \quad (11.41)$$

The nonrepetitive initial velocity of PMLM is $v_k(0) = 0.02\cos(k/300)$; the nonrepetitive initial position is $p_k(0) = 0.02\sin(k\pi/2)$; the nonrepetitive load disturbance is $f_{l,k}(t) = 2 + 2\sin(t + 2k) + d_k(t)$, where $d_k(t)$ is randomly varying within $[0, 0.1]$ with times and iterations.

The simulation time is $t_f = 1$ s, and the sampling time is $h = 0.002$ s. The initial terminal voltage is $u_0(t) = 0$; the initial parameter is $\hat{\varphi}_0(t) = 0.001, t \in \{0, 1, \ldots, 500\}$; the initial value $\hat{\xi}_0(t) = 0$. The controller parameters are set as $\lambda = 0.000006$, $\rho = 0.0048$, $\mu = 1$, $\eta = 1.6$, $\alpha = 0.000003$ and $b_{\hat{\xi}} = 1$. Applying the proposed ESO-based DDILC (11.12)–(11.15), the simulation results are shown in Figs. 11.1 and 11.2.

Figure 11.1 is the velocity tracking performance of the PMLM, where X-axis denotes time and Y-axis denotes motor speed. The solid line in the figure represents the desired velocity, and the dashed line represents the actual output velocity of the PMLM. It can be seen that the proposed ESO-based DDILC can realize the bounded tracking performance of the motor velocity even though all the initial speed, initial position, and load disturbances are nonrepetitive.

Figure 11.2 shows the profile of velocity tracking error of the PMLM, where the horizontal axis shows the number of iterations, and the vertical axis represents the maximum absolute error during each iteration, $e_{\max}(k) = \max_{t\in\{1,\ldots,500\}} \{|e_k(t)|\}$. It can be seen that the proposed ESO-based DDILC can realize the bounded convergence of velocity tracking error in the presence of nonrepetitive initial speed, initial position, and load disturbances in PMLM.

For the comparison, the STIDL-DDILC (4.42)–(4.44) proposed in Sect. 4.4 is also applied to the PMLM. Under the same simulation condition, i.e., $v_k(0) = 0.02\cos(k/300)$, $p_k(0) = 0.02\sin(k\pi/2)$, $f_{l,k}(t) = 2 + 2\sin(t + 2k) + d_k(t)$, and with the same controller parameters, i.e., $\lambda = 0.000006$, $\rho = 0.0048$, $\mu = 1$, $\eta =$

**Table 11.1** PARAMETERS OF PMLM

| Symbol | Quantity |
|---|---|
| $B_v/\left(\mathrm{N}\cdot\mathrm{s}\cdot\mathrm{m}^{-1}\right)$ | 10 |
| $f_c/\mathrm{N}$ | 10 |
| $f_s/\mathrm{N}$ | 20 |
| $A_r/\mathrm{N}$ | 8.5 |
| $\omega/\left(\mathrm{rad}\cdot\mathrm{m}^{-1}\right)$ | 314 |
| $j/\mathrm{rad}$ | $0.05\pi$ |
| $\mathrm{v}_s/\left(\mathrm{m}\cdot\mathrm{s}^{-1}\right)$ | 0.1 |
| $M/\mathrm{kg}$ | 0.59 |

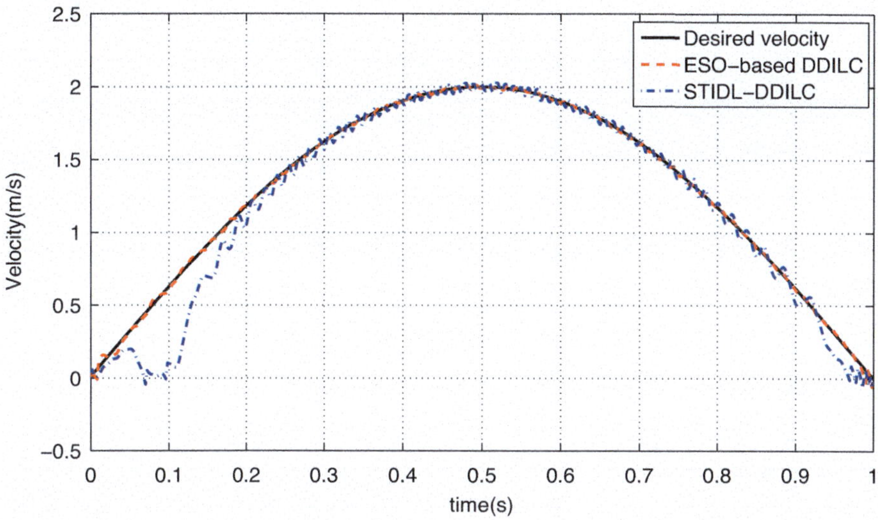


**Fig. 11.1** Profile of velocity of PMLM

1.6, $u_0(t) = 0$, $\hat{\varphi}_0(t) = 0.001$, the control performance by using STIDL-DDILC is illustrated by the dash-dotted line in Figs. 11.1 and 11.2. It can be observed that STIDL-DDILC is also robust to the nonrepetitive initial states and disturbances of the PMLM because the uncertainties of system can be reflected and estimated by $\hat{\varphi}_k(t)$ to some extent. However, the proposed ESO-based DDILC achieves a better control performance than the STIDL-DDILC in Sect. 4.4.

Aiming to compare the capability between STIDL-DDILC and ESO-based DDILC more clearly, we define an error index function $J(e) = \sum_{l=0}^{K} \max_{t\in\{1,\ldots,500\}} \{|e_l(t+1)|\}$, in which $K$ denotes the total iterations. The error index function value of ESO-based DDILC is calculated as 223.1153; the error index function value of STIDL-DDILC is calculated as 251.6231. Apparently, the proposed ESO-based DDILC can deal with the uncertainties more effectively than the STIDL-DDILC

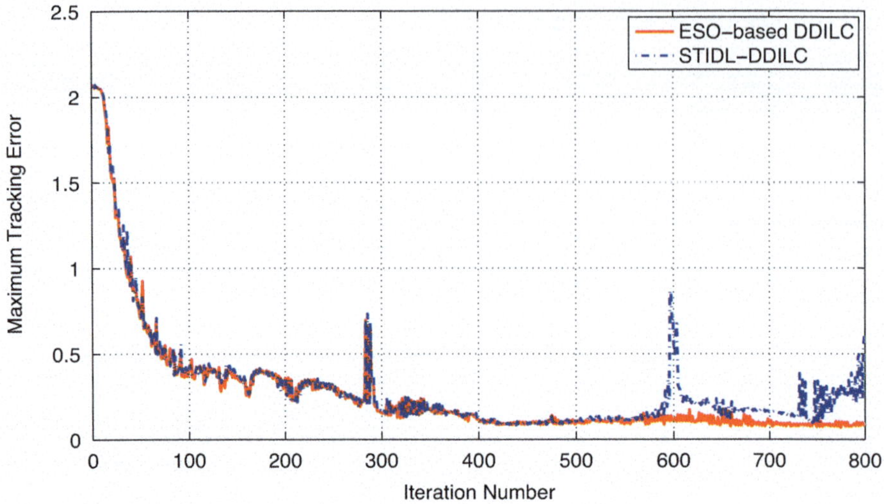


**Fig. 11.2** Profile of velocity tracking error of PMLM

proposed in Sect. 4.4 owing to the ESO utilized to estimate and to compensate the uncertainties.

## 11.6 Summary

This chapter proposes an ESO-based DDILC for a nonlinear nonaffine discrete-time system. The identical conditions on initial states and disturbances are not required any more but a robust convergence can be warranted. By introducing an ESO, which is conducted along iteration direction instead of the time direction, the nonrepetitive uncertainties can be estimated effectively. Moreover, the ESO-based DDILC is data-driven, i.e., only I/O data of the system is utilized rather than an explicit model.

## References

Chi R, Hou Z, Jin S, Huang B (2018) Computationally-light non-lifted data-driven norm-optimal iterative learning control. Asian J Control 20(1):115–124

Das S, Talole SE (2016) Robust steering autopilot design for marine surface vessels. IEEE J Oceanic Eng 41(4):913–922

Guo L, Wen X (2011) Hierarchical anti-disturbance adaptive control for non-linear systems with composite disturbances and applications to missile systems. T I Meas Control 33(8):942–956

Kobayashi H, Katsura S, Ohnishi K (2007) An analysis of parameter variations of disturbance observer for motion control. IEEE Trans Ind Electron 54(6):3413–3421

Lee TH, Tan KK, Lim SY, Dou H (2000) Iterative learning control of permanent magnet linear motor with relay automatic tuning. Mechatronics 10(1–2):169–190

Li Y, Yang B, Zheng T, Li Y, Cui M, Peeta S (2015) Extended-state-observer-based double-loop integral sliding-mode control of electronic throttle valve. IEEE trans Intell Transp Syst 16(5):2501–2510

Li X, Tian S, Ai W (2016) Active disturbance rejection based iterative learning control. In: 2016 Chinese control and decision conference. IEEE. https://ieeexplore.ieee.org/document/7532193

Sun J, Yang J, Li S, Zheng W (2017) Sampled-data-based event-triggered active disturbance rejection control for disturbed systems in networked environment. IEEE Trans Cybern 49(2):556–566

Sun J, Li S, Yang J (2014) Iterative learning control with extended state observer for iteration-varying disturbance rejection. In: The 11th world congress on intelligent control and automation. IEEE. https://ieeexplore.ieee.org/document/7052880

Tan KK, Lim SY, Lee TH, Dou H (2000) High-precision control of linear actuators incorporating acceleration sensing. Robot Comput Integr Manuf 16(5):295–305

Wei X, Guo L (2010) Composite disturbance-observer-based control and $H_\infty$ control for complex continuous models. Int J Robust Nonlinear Control 20(1):106–118

Wei XJ, Wu ZJ, Karimi HR (2016) Disturbance observer-based disturbance attenuation control for a class of stochastic systems. Automatica 63:21–25

Xue W, Bai W, Yang S, Song K, Huang Y, Xie H (2015) ADRC with adaptive extended state observer and its application to air-fuel ratio control in gasoline engines. IEEE Trans Ind Electron 62(9):5847–5857

Zhao Z, Guo B (2017) A novel extended state observer for output tracking of MIMO systems with mismatched uncertainty. IEEE Trans Automat Control 63(1):211–218

# Chapter 12
# Event-Triggered Data-Driven Iterative Learning Control

## 12.1 Introduction

Event-triggered control (ETC) means to execute the control task after the event occurs, rather than after a certain fixed period of time, where the event is driven by well-designed event-triggering conditions (Peng and Li 2018; Ding et al. 2017). Since the control is only executed when needed, the utilization of resources is greatly reduced, while the control performance is guaranteed (Kishida 2018; Liu and Yang 2018; Lin et al. 2019). Therefore, ETC is an effective alternative way to solve the limited network resources in networked control systems (NCSs).

Recently, there have been a lot of works (Shen et al. 2017; Bu et al. 2013; Pan et al. 2009; Bu et al. 2017; Meng and Moore 2017) reporting the applications of ILC in NCSs. For example, several stochastic ILC laws (Shen et al. 2017; Bu et al. 2013) have been presented to address the problem of data dropout. The robust ILC methods (Pan et al. 2009; Bu et al. 2017; Meng and Moore 2017) have been investigated to address time-varying transmission delays, data packet dropouts, and external disturbances. Besides the problems of data dropouts, transmission delay, and data noise, the limited network channel is also a critical issue to be solved in the network control systems. Therefore, a new research topic of combining ETC and ILC has come to light with the objective toward improving the efficiency of a NCS.

In Xiong et al. (2016), two event-triggering schemes with and without quantization by utilizing ILC laws are proposed for linear discrete-time systems to iteratively reduce controller updating numbers in each batch. An event-triggered robust learning control with quantization is proposed for linear multiagent systems (Zhang and Li 2018). Further, the problem of iterative learning fault-tolerant control with event-triggered transmission strategy is studied in Tang and Sheng (2018) for networked batch processes. Apparently, compared with the ETC methods for one-dimensional (1D) time dynamic systems, the study of ETC for the two-dimensional (2D) repetitive systems is still in its infancy and many problems are yet to explore, including the research for nonlinear systems.

 
R. Chi et al., *Data-Driven Iterative Learning Control for Discrete-Time Systems*, Intelligent Control and Learning Systems 2,
https://doi.org/10.1007/978-981-19-5950-9_12

This chapter proposes an event-triggered DDILC (ET-DDILC) method for nonlinear nonaffine systems. The 2D dynamic behavior along both time and iteration directions is considered when designing event-triggering conditions for the ET-DDILC. A virtual linear data model is used to describe the input and output relationship between the consecutive batches by means of the IDL technique where the uncertainties of the original nonlinear system are lumped into an unknown parameter. Consequently, the ET-DDILC method is proposed by designing an event-triggering condition based on the Lyapunov stability analysis in the iteration domain. The control input of ET-DDILC is updated along the iterative direction only if the event-triggering condition is satisfied, which reduces the action time of control devices, computation burden, and the required network resource. The event-triggering condition can be verified by offline computation using tracking errors, event-triggering errors, and estimated parameters since the ET-DDILC is feedforward with respect to the time domain. The convergence of the proposed ET-DDILC is proved mathematically. Furthermore, the results are also extended to a MIMO nonlinear nonaffine system, where the convergence is proved by using the property of input-to-state stability as the basic mathematical tool.

This chapter is organized as follows. Section 12.2 shows the problem formulation. Sections 12.3 and 12.4 present the ET-DDILC with detail analysis. Section 12.5 extends the results to MIMO nonlinear nonaffine systems. Section 12.6 provides some simulation examples. Section 12.7 gives some conclusions.

## 12.2 Problem Formulation

Consider a class of nonlinear nonaffine discrete-time systems

$$y_k(t+1) = f\left(y_k(t), \ldots, y_k(t-n_y), u_k(t), \ldots, u_k(t-n_u)\right) \tag{12.1}$$

where $y_k(t) \in R$ and $u_k(t) \in R$ are system output and input, respectively, and $y_k(t) = 0$ and $u_k(t) = 0$, $\forall t < 0$; $n_y$ and $n_u$ are two unknown orders; $t \in \{0, 1, \ldots, N\}$ is time instant and $N \in Z^+$ represents the endpoint; $k \in \{0, 1, \ldots\}$ indicates the iteration number; $f(\cdots) : R^{n_u+n_y+2} \mapsto R$ is an unknown nonlinear scalar function, and $f(0, 0) = 0$.

Denote $y_d(t)$ as the desired trajectory. The control goal of this chapter is to design an event-triggered ILC to make the tracking error $\tilde{y}_k(t) = y_d(t) - y_k(t)$ converge within a satisfied bound while the required control input updating number is reduced according to the event-triggering condition.

Define an event-triggered error $e_{k-1}(t+1) = \tilde{y}_{k_{l-1}}(t+1) - \tilde{y}_{k-1}(t+1)$, $(k-1) \in \left[k_{l-1}, k_l\right)$ where $k_l, l \in \{0, 1, \ldots\}$, denotes the event-triggered iterative sequence generated by an event- triggering function $\chi\left(e_{k-1}(t+1)\right)$ which is to be designed in the following analysis. The sequence of the event-triggered iteration is determined by

$$k_l = \inf\left\{k \,|\, k > k_{l-1}, \chi\left(e_{k-1}(t+1)\right) \geq 0\right\}$$

At a specified operation time, the control input is updated only at the event-triggered iterations; otherwise, the control input remains the same as that in the last iteration.

Consider nonlinear system (12.1) satisfying Assumptions 2.1, 2.2, and 2.3.

According to Theorem 2.1, the CFIDL model can be obtained as follows:

$$\Delta y_k(t+1) = \theta_k(t)\Delta u_k(t) \tag{12.2}$$

where $\Delta y_k(t+1) = y_k(t+1) - y_{k-1}(t+1)$, $\Delta u_k(t) = u_k(t) - u_{k-1}(t)$, $\theta_k(t)$ is the IPPD parameter.

According to Assumption 2.2, $|\theta_k(t)| \leq b_\theta$, where $b_\theta$ is a positive constant.

## 12.3 Controller Design

The following is the design procedure of the event-triggered data-driven ILC.

When $k = k_l$, define an objective function as follows:

$$J\left(u_{k_l}(t)\right) = \left|y_d(t+1) - y_{k_l}(t+1)\right|^2 + \lambda_s\left|u_{k_l}(t) - u_{k_{l-1}}(t)\right|^2 \tag{12.3}$$

where $\lambda_s > 0$.

Differentiating (12.3) with respect to $u_{k_l}(t)$, combining it with (12.2), and setting it to be 0, one can get

$$u_{k_l}(t) = u_{k_{l-1}}(t) + P_{k_l}(t)\tilde{y}_{k_{l-1}}(t+1) \tag{12.4}$$

where $P_{k_l}(t) = \frac{\rho_s\theta_{k_l}(t)}{\lambda_s+\theta_{k_l}(t)^2}$, and $\rho_s$ is added additionally as a step-size factor to make the control algorithm (12.4) more general.

When $k \in (k_{l-1}, k_l)$, the actuator is not triggered, so we have $u_k(t) = u_{k_{l-1}}(t)$.

Therefore, the event-triggered ILC law is proposed as follows:

$$u_k(t) = \begin{cases} u_{k_{l-1}}(t) + \hat{P}_k(t)\tilde{y}_{k_{l-1}}(t+1), k = k_l \\ u_{k_{l-1}}(t), \ \ k \in (k_{l-1}, k_l) \end{cases} \tag{12.5}$$

where $\hat{P}_k(t) = \frac{\rho_s\hat{\theta}_k(t)}{\lambda_s+\hat{\theta}_k(t)^2}$; $\hat{\theta}_k(t)$ is the estimation of $\theta_k(t)$.

The parameter estimation law of $\hat{\theta}_k(t)$ is presented by designing an objective function as follows:

$$\begin{aligned} J\left(\hat{\theta}_k(t)\right) = & \left(y_{k-1}(t+1) - y_{k-2}(t+1) - \hat{\theta}_k(t)\Delta u_{k-1}(t)\right)^2 \\ & + \mu_s\left(\hat{\theta}_k(t) - \hat{\theta}_{k-1}(t)\right)^2 \end{aligned} \tag{12.6}$$

where $\mu_s > 0$.

So, we can get the iterative learning parameter estimation law by optimizing (12.6), that is

$$\hat{\theta}_k(t) = \hat{\theta}_{k-1}(t) + \frac{\eta_s \Delta u_{k-1}(t)\left(\Delta y_{k-1}(t+1) - \hat{\theta}_{k-1}(t)\Delta u_{k-1}(t)\right)}{\mu_s + \Delta u_{k-1}(t)^2} \tag{12.7}$$

where $0 < \eta_s < 2$ is a step-size factor, which is added additionally to make the estimation algorithm (12.7) more general.

According to the definition of tracking error and (12.2), we have

$$\begin{aligned}\tilde{y}_k(t+1) &= y_d(t+1) - y_{k-1}(t+1) - \theta_k(t)\Delta u_k(t)\\ &= \tilde{y}_{k-1}(t+1) - \theta_k(t)\Delta u_k(t).\end{aligned} \tag{12.8}$$

In the case of the triggered iterations, i.e., $k = k_l$, considering the event-triggered law (12.5), Eq. (12.8) becomes

$$\begin{aligned}\tilde{y}_k(t+1) &= \tilde{y}_{k_{l-1}}(t+1) - \theta_k(t)\hat{P}_k(t)\tilde{y}_{k_{l-1}}(t+1)\\ &= \left(1 - \theta_k(t)\hat{P}_k(t)\right)\tilde{y}_{k_{l-1}}(t+1)\end{aligned} \tag{12.9}$$

According to the definition of event-triggered error, we have $\tilde{y}_{k_{l-1}}(t+1) = \tilde{y}_{k-1}(t+1) + e_{k-1}(t+1)$, $(k-1) \in [k_{l-1}, k_l)$. For $k = k_l$, it is clear that $(k-1) \in [k_{l-1}, k_l)$. Thus, we can further get from (12.9) that

$$\tilde{y}_k(t+1) = \left(1 - \theta_k(t)\hat{P}_k(t)\right)\left(\tilde{y}_{k-1}(t+1) + e_{k-1}(t+1)\right). \tag{12.10}$$

Define a Lyapunov function $V_k(t+1) = \tilde{y}_k^2(t+1)$. If $k \in (k_{l-1}, k_l)$, it is obvious that $(k-1) \in [k_{l-1}, k_l)$, $u_k(t) = u_{k_{l-1}}(t)$ and $u_{k-1}(t) = u_{k_{l-1}}(t)$. Then we can get $y_k(t+1) = y_{k-1}(t+1)$ when neither noise nor disturbance is considered. As a result, for $k \in (k_{l-1}, k_l)$, we have

$$\begin{aligned}\Delta V_k(t+1) &= V_k(t+1) - V_{k-1}(t+1)\\ &= \tilde{y}_k^2(t+1) - \tilde{y}_{k-1}^2(t+1)\\ &= \left(y_d(t+1) - y_k(t+1)\right)^2 - \left(y_d(t+1) - y_{k-1}(t+1)\right)^2\\ &= 0.\end{aligned} \tag{12.11}$$

For the case of $k = k_l$, according to (12.10), we have

$$\begin{aligned}\Delta V_k(t+1) = &\left(1 - \theta_k(t)\hat{P}_k(t)\right)^2\left(\tilde{y}_{k-1}(t+1) + e_{k-1}(t+1)\right)^2\\ &- \tilde{y}_{k-1}^2(t+1).\end{aligned} \tag{12.12}$$

Using Cauchy-Schwartz inequality $(a+b)^2 \le 2\left(a^2+b^2\right)$, from (12.12), we have

$$\Delta V_k(t+1) \le 2\Big(1-\theta_k(t)\hat{P}_k(t)\Big)^2 \tilde{y}_{k-1}^2(t+1) + 2\Big(1-\theta_k(t)\hat{P}_k(t)\Big)^2 e_{k-1}^2(t+1) - \tilde{y}_{k-1}^2(t+1). \quad (12.13)$$

Letting $\Delta V_k(t+1) < 0$, it follows that

$$e_{k-1}^2(t+1) < \frac{\tilde{y}_{k-1}^2(t+1) - 2\Big(1-\theta_k(t)\hat{P}_k(t)\Big)^2 \tilde{y}_{k-1}^2(t+1)}{2\Big(1-\theta_k(t)\hat{P}_k(t)\Big)^2} \quad (12.14)$$

Since $\theta_k(t)$ in (12.14) is unknown, the parameter updating law (12.7) is used to estimate it. So, we can get the following inequality:

$$e_{k-1}^2(t+1) < \frac{\tilde{y}_{k-1}^2(t+1) - 2\Big(1-\hat{\theta}_k(t)\hat{P}_k(t)\Big)^2 \tilde{y}_{k-1}^2(t+1)}{2\Big(1-\hat{\theta}_k(t)\hat{P}_k(t)\Big)^2}. \quad (12.15)$$

Combining the definition of event-triggered iterative sequence, the event-triggering condition is that the following inequality holds

$$\chi\left(e_{k-1}(t+1)\right) = e_{k-1}^2(t+1) - \frac{H_{k-1}(t)}{2\Big(1-\hat{\theta}_k(t)\hat{P}_k(t)\Big)^2} \ge 0 \quad (12.16)$$

where $H_{k-1}(t) = \tilde{y}_{k-1}^2(t+1) - 2\Big(1-\hat{\theta}_k(t)\hat{P}_k(t)\Big)^2 \tilde{y}_{k-1}^2(t+1)$.

By nature, combining (12.13)–(12.16), if the condition $\Delta V_k(t+1) < 0$ holds, the controller is not triggered; otherwise, the controller is triggered.

Furthermore, define a tracking error threshold as $\sup_k \max_{t\in\{0,1,\ldots N\}} |\tilde{y}_k(t)| \le \vartheta$, where $\vartheta > 0$ is a constant. If the obtained tracking error is larger than the threshold, the control input is also updated even though the event-triggering condition (12.16) is satisfied.

As stated in Tallapragada and Chopra (2013), the tracking error threshold is necessary because the absence of it may result in an accumulation of control updates and the role of threshold limits is to ensure that the error is within a small range. In other words, the actual triggering action of the ILC mechanism is up to both the event-triggering condition (12.16) and the tracking error threshold.

In summary, the proposed event-triggered DDILC (ET-DDILC) method is shown as follows:

$$k_l = k,\ l \in \{0, 1, \ldots\}, \text{if } |\tilde{y}_k(t)| > \vartheta \text{ or } \chi\left(e_{k-1}(t+1)\right) \geq 0 \tag{12.17}$$

$$u_k(t) = \begin{cases} u_{k_{l-1}}(t) + \hat{P}_k(t)\tilde{y}_{k_{l-1}}(t+1)\,, k = k_l \\ u_{k_{l-1}}(t),\ k \in (k_{l-1}, k_l) \end{cases} \tag{12.18}$$

$$\hat{\theta}_k(t) = \hat{\theta}_{k-1}(t) + \frac{\eta_s \Delta u_{k-1}(t)\left(\Delta y_{k-1}(t+1) - \hat{\theta}_{k-1}(t)\Delta u_{k-1}(t)\right)}{\mu_s + \Delta u_{k-1}(t)^2} \tag{12.19}$$

$$\hat{\theta}_k(t) = \hat{\theta}_0(t), \text{ if } \left|\hat{\theta}_k(t)\right| \leq \varepsilon \text{ or sgn}\left(\hat{\theta}_k(t)\right) \neq \text{sgn}\left(\hat{\theta}_0(t)\right) \tag{12.20}$$

where the reset algorithm (12.20) is introduced to make the parameter estimation law (12.19) have stronger tracking ability for time-iteration-varying parameters and to ensure $\hat{\theta}_k(t)$ be nonzero, which is similar to a persistent excitation condition (Hou and Xiong 2019); $\varepsilon$ is a sufficiently small positive constant.

**Remark 12.1** Note that $\hat{P}_k(t)$ is adjustable through the time-iteration-varying $\hat{\theta}_k(t)$, which is estimated merely using the real-time I/O data. Further, the computation burden of $\hat{P}_k(t)$ is light because no complex matrix operation is involved.

**Remark 12.2** Different from Shen et al. (2017), Bu et al. (2013), Pan et al. (2009), Bu et al. (2017), and Meng and Moore (2017), the ILC law proposed in this chapter is event-triggered such that the action times of control devices and the transmission data required can be greatly reduced. On the other hand, it is seen from (12.18) that the learning control law is feedforward and the input signal is updated by using the control knowledge in the previous iterations. The controller update is from one point to the same point along the iteration direction, regardless of the time-dynamics of the systems.

**Remark 12.3** There exists a switching mechanism in control law (12.18) such that the control law is discontinuous. Note that such a switching mechanism is conducted under the convergence conditions. So, the convergence and stability of the proposed method can still be guaranteed even though chattering may occur. Up to now, it is still an open problem to reduce the chattering in event-triggered control methods.

**Remark 12.4** The threshold-triggering mechanism in (12.17) is similar to a dead-zone strategy (Ibrir and Su 2010; Lai et al. 2019) to make the controller be triggered only when the tracking error does not enter the given range; otherwise, the controller may hold without being updated.

## 12.4 Convergence Analysis

**Theorem 12.1** *For a class of nonlinear nonaffine discrete-time systems (12.1), if assumptions 2.1, 2.2, and 2.3 hold, by selecting controller parameters properly, the proposed ET-DDILC method (12.17)–(12.20) guarantees that:*

*(i) the parameter estimation $\hat{\theta}_k(t)$ is bounded over the finite time interval $t \in \{0, 1, \ldots, N\}$, $\forall k$;*
*(ii) the tracking error $\tilde{y}_k(t)$ converges over the finite time interval $t \in \{0, 1, \ldots, N\}$ as $k$ approaches to infinity.*

***Proof*** The boundedness of $\hat{\theta}_k(t)$ can be obtained by following the similar steps in Theorem 2.2.

In the triggered iteration, i.e., $k = k_l$, according to (12.9), we have

$$\tilde{y}_{k_l}(t+1) = \left(1 - \theta_{k_l}(t)\hat{P}_{k_l}(t)\right)\tilde{y}_{k_{l-1}}(t+1). \tag{12.21}$$

Taking norms on both sides of (12.21), it results

$$\left|\tilde{y}_{k_l}(t+1)\right| \leq \left|1 - \theta_{k_l}(t)\hat{P}_{k_l}(t)\right| \left|\tilde{y}_{k_{l-1}}(t+1)\right| \tag{12.22}$$

By selecting the parameters properly, such that $0 < \frac{\rho_s b_\theta}{\sqrt{\lambda_s}} < 4$, one can derive according to Assumption 2.3 and reset algorithm (12.20) that

$$0 < \frac{\rho_s \theta_{k_l}(t)\hat{\theta}_{k_l}(t)}{\lambda_s + \hat{\theta}^2_{k_l}(t)} < \frac{\rho_s \left|\theta_{k_l}(t)\right| \left|\hat{\theta}_{k_l}(t)\right|}{2\sqrt{\lambda_s}\left|\hat{\theta}_{k_l}(t)\right|} < \frac{\rho_s b_\theta}{2\sqrt{\lambda_s}} < 2 \tag{12.23}$$

which implies that

$$\left|1 - \theta_{k_l}(t)\hat{P}_{k_l}(t)\right| = \left|1 - \frac{\rho_s \theta_{k_l}(t)\hat{\theta}_{k_l}(t)}{\lambda_s + \hat{\theta}^2_{k_l}(t)}\right| \leq d_1 < 1 \tag{12.24}$$

where $0 < d_1 < 1$ is a constant.

According to (12.22) and (12.24), one gets

$$\left|\tilde{y}_{k_l}(t+1)\right| \leq d_1 \left|\tilde{y}_{k_{l-1}}(t+1)\right| \leq \cdots \leq d_1^l \left|\tilde{y}_{k_0}(t+1)\right| \tag{12.25}$$

from which, it is clear that $\lim_{k_l \to \infty} \left|\tilde{y}_{k_l}(t+1)\right| = 0$. In other words, the tracking error converges to zero along with the triggered iterations.

For the inter-event iterations, i.e., $k \in (k_{l-1}, k_l)$, the control input remains the same as that in the previous triggered iteration. Since the tracking error at the triggering iteration has been proved to be convergent, it is evident that the tracking error at the non-triggering iterations remains unchanged.

Therefore, in general, the tracking error of the system decreases with the increasing iterations. □

## 12.5 Extension to MIMO Nonlinear Nonaffine Systems

### 12.5.1 Problem Formulation

Consider a class of MIMO nonlinear nonaffine discrete-time systems

$$\boldsymbol{y}_k(t+1) = \boldsymbol{f}\left(\boldsymbol{y}_k(t), \ldots, \boldsymbol{y}_k(t-n_y), \boldsymbol{u}_k(t), \ldots, \boldsymbol{u}_k(t-n_u)\right) \tag{12.26}$$

where $\boldsymbol{y}_k(t) \in R^m$ represents system output; $\boldsymbol{u}_k(t) \in R^m$ represents control input; $\boldsymbol{f}(\cdots) = (f_1(\cdot), \ldots, f_m(\cdots))^{\mathrm{T}} \in \prod_{n_y+n_u+2} R^m \mapsto R^m$ is an unknown nonlinear function, and $\boldsymbol{f}(\mathbf{0}, \mathbf{0}) = \mathbf{0}$.

Some assumptions are made as follows:

**Assumption 12.1** The partial derivative of $\boldsymbol{f}(\cdot)$ with respect to control input $\boldsymbol{u}_k(t)$ is continuous.

**Assumption 12.2** System (12.26) satisfies the generalized *Lipschitz* condition for all $k$, that is

$$\left\|\boldsymbol{y}_{k_1}(t+1) - \boldsymbol{y}_{k_2}(t+1)\right\| \leq \bar{b}\left\|\boldsymbol{u}_{k_1}(t) - \boldsymbol{u}_{k_2}(t)\right\| \tag{12.27}$$

for $\boldsymbol{u}_k(t) \neq \boldsymbol{u}_{k-1}(t)$ and any $k_1 \neq k_2, k_1, k_2 \geq 0$, where $\boldsymbol{y}_{k_i}(t+1) = \boldsymbol{f}(\boldsymbol{y}_{k_i}(t), \boldsymbol{y}_{k_i}(t-1), \ldots, \boldsymbol{y}_{k_i}(t-n_y), \boldsymbol{u}_{k_i}(t), \boldsymbol{u}_{k_i}(t-1), \ldots, \boldsymbol{u}_{k_i}(t-n_u)), i = 1, 2$, and $\bar{b}$ is a positive constant.

**Theorem 12.2** *For MIMO nonlinear system (12.26), satisfying Assumptions 12.1 and 12.2 with* $\|\Delta\boldsymbol{u}_k(t)\| \neq 0$, *there must exist an iterative pseudo Jacobi Matrix (IPJM),* $\boldsymbol{\Theta}_k(t)$, *such that*

$$\Delta\boldsymbol{y}_k(t+1) = \boldsymbol{\Theta}_k(t)\Delta\boldsymbol{u}_k(t) \tag{12.28}$$

*where* $\boldsymbol{\Theta}_k(t) = \left(\theta_{ji,k}(t)\right) \in R^{m\times m}$, $j = 1, 2, \ldots, m$, *and* $i = 1, 2, \ldots, m$, *and* $\|\boldsymbol{\Theta}_k(t)\| \leq \bar{b}$, $t \in \{0, 1, \ldots, N\}$.

***Proof*** From system (12.26), we have

$$
\begin{aligned}
\Delta \boldsymbol{y}_k(t+1) = \ & \boldsymbol{f}(\boldsymbol{y}_k(t), \boldsymbol{y}_k(t-1), \ldots, \boldsymbol{y}_k\left(t-n_y\right), \\
& \quad \boldsymbol{u}_k(t), \boldsymbol{u}_k(t-1), \ldots, \boldsymbol{u}_k\left(t-n_u\right)) \\
& -\boldsymbol{f}(\boldsymbol{y}_{k-1}(t), \boldsymbol{y}_{k-1}(t-1), \ldots, \boldsymbol{y}_{k-1}\left(t-n_y\right), \\
& \quad \boldsymbol{u}_{k-1}(t), \boldsymbol{u}_{k-1}(t-1), \ldots, \boldsymbol{u}_{k-1}\left(t-n_u\right)) \\
= \ & \boldsymbol{f}(\boldsymbol{y}_k(t), \boldsymbol{y}_k(t-1), \ldots, \boldsymbol{y}_k\left(t-n_y\right), \\
& \quad \boldsymbol{u}_k(t), \boldsymbol{u}_k(t-1), \ldots, \boldsymbol{u}_k\left(t-n_u\right)) \\
& -\boldsymbol{f}(\boldsymbol{y}_k(t), \boldsymbol{y}_k(t-1), \ldots, \boldsymbol{y}_k\left(t-n_y\right), \\
& \quad \boldsymbol{u}_{k-1}(t), \boldsymbol{u}_k(t-1), \ldots, \boldsymbol{u}_k\left(t-n_u\right)) \\
& +\boldsymbol{f}(\boldsymbol{y}_k(t), \boldsymbol{y}_k(t-1), \ldots, \boldsymbol{y}_k\left(t-n_y\right), \\
& \quad \boldsymbol{u}_{k-1}(t), \boldsymbol{u}_k(t-1), \ldots, \boldsymbol{u}_k\left(t-n_u\right)) \\
& -\boldsymbol{f}(\boldsymbol{y}_{k-1}(t), \boldsymbol{y}_{k-1}(t-1), \ldots, \boldsymbol{y}_{k-1}\left(t-n_y\right), \\
& \quad \boldsymbol{u}_{k-1}(t), \boldsymbol{u}_{k-1}(t-1), \ldots, \boldsymbol{u}_{k-1}\left(t-n_u\right)).
\end{aligned} \tag{12.29}
$$

By virtue of Assumption 12.1 and the differential mean value theorem, Eq. (12.29) can be rewritten as

$$
\Delta \boldsymbol{y}_k(t+1) = \frac{\partial \boldsymbol{f}}{\partial \boldsymbol{u}_k^{\mathrm{T}}(t)} \Delta \boldsymbol{u}_k(t) + \boldsymbol{\Xi}_k(t) \tag{12.30}
$$

where $\partial \boldsymbol{f}/\partial \boldsymbol{u}_k^{\mathrm{T}}(t) = \begin{bmatrix} \frac{\partial f_1}{\partial u_{1,k}(t)} & \frac{\partial f_1}{\partial u_{2,k}(t)} & \cdots & \frac{\partial f_1}{\partial u_{m,k}(t)} \\ \frac{\partial f_2}{\partial u_{1,k}(t)} & \frac{\partial f_2}{\partial u_{2,k}(t)} & \cdots & \frac{\partial f_2}{\partial u_{m,k}(t)} \\ \vdots & \vdots & \ddots & \vdots \\ \frac{\partial f_m}{\partial u_{1,k}(t)} & \frac{\partial f_m}{\partial u_{2,k}(t)} & \cdots & \frac{\partial f_m}{\partial u_{m,k}(t)} \end{bmatrix} \in R^{m \times m}$, $\partial f_i/\partial u_{j,k}(t)$ represents the partial derivative value of $f_i$ with respect to input $u_{j,k}(t)$ at some point in the interval $\left[u_{j,k}(t), u_{j,k-1}(t)\right]$, $i, j \in \{1, \ldots, m\}$, $t \in \{0, \ldots, N\}$, and

$$
\begin{aligned}
\boldsymbol{\Xi}_k(t) = \ & \boldsymbol{f}(\boldsymbol{y}_k(t), \boldsymbol{y}_k(t-1), \ldots, \boldsymbol{y}_k\left(t-n_y\right), \\
& \quad \boldsymbol{u}_{k-1}(t), \boldsymbol{u}_k(t-1), \ldots, \boldsymbol{u}_k\left(t-n_u\right)) \\
& -\boldsymbol{f}(\boldsymbol{y}_{k-1}(t), \boldsymbol{y}_{k-1}(t-1), \ldots, \boldsymbol{y}_{k-1}\left(t-n_y\right), \\
& \quad \boldsymbol{u}_{k-1}(t), \boldsymbol{u}_{k-1}(t-1), \ldots, \boldsymbol{u}_{k-1}\left(t-n_u\right)).
\end{aligned} \tag{12.31}
$$

For each fixed $k$, we consider the following equation with numerical matrix $\boldsymbol{\varsigma}_k(t)$ with $m$ rows and $m$ columns

$$
\boldsymbol{\Xi}_k(t) = \boldsymbol{\varsigma}_k(t) \Delta \boldsymbol{u}_k(t), \quad t \in \{0, \ldots, N\}. \tag{12.32}
$$

It is clear that (12.32) must have a solution $\boldsymbol{\varsigma}_k(t)$ since $\|\Delta \boldsymbol{u}_k(t)\| \neq 0$ for all $t \in \{0, 1, \ldots, N\}$ and $k \in \{0, 1, \ldots\}$.

Let $\boldsymbol{\Theta}_k(t) = \boldsymbol{\varsigma}_k(t) + \left(\partial \boldsymbol{f}/\partial \boldsymbol{u}_k^{\mathrm{T}}(t)\right)$, then we have $\Delta \boldsymbol{y}_k(t+1) = \boldsymbol{\Theta}_k(t) \Delta \boldsymbol{u}_k(t)$. The result of $\|\boldsymbol{\Theta}_k(t)\| \leq \bar{b}$ is the direct conclusion of Assumption 12.2. □

### 12.5.2 Controller Design

When $k = k_l$, define an objective function as follows:

$$J\left(\boldsymbol{u}_{k_l}(t)\right) = \left\|\boldsymbol{y}_d(t+1) - \boldsymbol{y}_{k_l}(t+1)\right\|^2 + \lambda_m \left\|\boldsymbol{u}_{k_l}(t) - \boldsymbol{u}_{k_{l-1}}(t)\right\|^2$$

where $\lambda_m > 0$.

Then, the event-triggered ILC law for MIMO nonlinear systems is proposed as follows:

$$\boldsymbol{u}_k(t) = \begin{cases} \boldsymbol{u}_{k_{l-1}}(t) + \hat{\boldsymbol{P}}_k(t)\tilde{\boldsymbol{y}}_{k_{l-1}}(t+1), k = k_l \\ \boldsymbol{u}_{k_{l-1}}(t), k \in (k_{l-1}, k_l) \end{cases} \tag{12.33}$$

where $\hat{\boldsymbol{P}}_k(t) = \frac{\rho_m \hat{\boldsymbol{\Theta}}_k^{\mathrm{T}}(t)}{\lambda_m + \left\|\hat{\boldsymbol{\Theta}}_k(t)\right\|^2}$, and $\rho_m$ is an adjustable positive parameter. $\hat{\boldsymbol{\Theta}}_k(t)$ is the estimation of $\boldsymbol{\Theta}_k(t)$.

Given an objective function

$$J\left(\hat{\boldsymbol{\Theta}}_k(t)\right) = \left\|\boldsymbol{y}_{k-1}(t+1) - \boldsymbol{y}_{k-2}(t+1) - \hat{\boldsymbol{\Theta}}_k(t)\Delta\boldsymbol{u}_{k-1}(t)\right\|^2 + \mu_m \left\|\hat{\boldsymbol{\Theta}}_k(t) - \hat{\boldsymbol{\Theta}}_{k-1}(t)\right\|^2.$$

Taking the derivative of this objective function with respect to $\hat{\boldsymbol{\Theta}}_k(t)$, and letting it be zero, we can derive the parameter estimate law as

$$\hat{\boldsymbol{\Theta}}_k(t) = \hat{\boldsymbol{\Theta}}_{k-1}(t) + \frac{\eta_m \left(\Delta\boldsymbol{y}_{k-1}(t+1) - \hat{\boldsymbol{\Theta}}_{k-1}(t)\Delta\boldsymbol{u}_{k-1}(t)\right)\Delta\boldsymbol{u}_{k-1}^{\mathrm{T}}(t)}{\mu_m + \left\|\Delta\boldsymbol{u}_{k-1}(t)\right\|^2} \tag{12.34}$$

where $0 < \eta_m < 2$ and $\mu_m > 0$ are the two constants.

In the case of the triggered iterations, i.e., $k = k_l$, similar to the steps (12.8)–(12.9), it obtains

$$\tilde{\boldsymbol{y}}_k(t+1) = \left(\boldsymbol{I} - \boldsymbol{\Theta}_k(t)\hat{\boldsymbol{P}}_k(t)\right)\tilde{\boldsymbol{y}}_{k_{l-1}}(t+1). \tag{12.35}$$

Define an event-triggered error $\boldsymbol{e}_{k-1}(t+1) = \tilde{\boldsymbol{y}}_{k_{l-1}}(t+1) - \tilde{\boldsymbol{y}}_{k-1}(t+1)$, $(k-1) \in \left[k_{l-1}, k_l\right)$, (12.35) becomes

$$\tilde{\boldsymbol{y}}_k(t+1) = \left(\boldsymbol{I} - \boldsymbol{\Theta}_k(t)\hat{\boldsymbol{P}}_k(t)\right)\left(\tilde{\boldsymbol{y}}_{k-1}(t+1) + \boldsymbol{e}_{k-1}(t+1)\right). \tag{12.36}$$

Define a Lyapunov function as $V_k(t+1) = \left\|\tilde{\boldsymbol{y}}_k(t+1)\right\|^2$. Similarly, for the case of $k \in (k_{l-1}, k_l)$, we have

$$\Delta V_k(t+1) = \left\|\tilde{\boldsymbol{y}}_k(t+1)\right\|^2 - \left\|\tilde{\boldsymbol{y}}_{k-1}(t+1)\right\|^2 = 0. \tag{12.37}$$

For the case of $k = k_l$, we have

$$\begin{aligned}\Delta V_k(t+1) &= \left\| \left(\boldsymbol{I} - \boldsymbol{\Theta}_{k_l}(t)\hat{\boldsymbol{P}}_k(t)\right)\left(\tilde{\boldsymbol{y}}_{k-1}(t+1) + \boldsymbol{e}_{k-1}(t+1)\right)\right\|^2 \\ &-\left\|\tilde{\boldsymbol{y}}_{k-1}(t+1)\right\|^2 \\ &\le 2\left\|\boldsymbol{I} - \boldsymbol{\Theta}_k(t)\hat{\boldsymbol{P}}_k(t)\right\|^2 \left\|\tilde{\boldsymbol{y}}_{k-1}(t+1)\right\|^2 \\ &+2\left\|\boldsymbol{I} - \boldsymbol{\Theta}_k(t)\hat{\boldsymbol{P}}_k(t)\right\|^2 \left\|\boldsymbol{e}_{k-1}(t+1)\right\|^2 - \left\|\tilde{\boldsymbol{y}}_{k-1}(t+1)\right\|^2 \\ &= \left(2\left\|\boldsymbol{I} - \boldsymbol{\Theta}_k(t)\hat{\boldsymbol{P}}_k(t)\right\|^2 - 1\right)\left\|\tilde{\boldsymbol{y}}_{k-1}(t+1)\right\|^2 \\ &+2\left\|\boldsymbol{I} - \boldsymbol{\Theta}_k(t)\hat{\boldsymbol{P}}_k(t)\right\|^2 \left\|\boldsymbol{e}_{k-1}(t+1)\right\|^2.\end{aligned} \tag{12.38}$$

Letting $\Delta V_k(t+1) < 0$, it obtains

$$\left\|\boldsymbol{e}_{k-1}(t+1)\right\|^2 < \frac{\left(1 - 2\left\|\boldsymbol{I} - \boldsymbol{\Theta}_k(t)\hat{\boldsymbol{P}}_k(t)\right\|^2\right)\left\|\tilde{\boldsymbol{y}}_{k-1}(t+1)\right\|^2}{2\left\|\boldsymbol{I} - \boldsymbol{\Theta}_k(t)\hat{\boldsymbol{P}}_k(t)\right\|^2}. \tag{12.39}$$

Then, the event-triggering condition for MIMO systems can be written as

$$\chi\left(\boldsymbol{e}_{k-1}(t+1)\right) = \left\|\boldsymbol{e}_{k-1}(t+1)\right\|^2 - \frac{G_{k-1}(t)}{2\left\|\boldsymbol{I} - \hat{\boldsymbol{\Theta}}_k(t)\hat{\boldsymbol{P}}_k(t)\right\|^2} \ge 0 \tag{12.40}$$

where $G_{k-1}(t) = \left(1 - 2\left\|\boldsymbol{I} - \hat{\boldsymbol{\Theta}}_k(t)\hat{\boldsymbol{P}}_k(t)\right\|^2\right)\left\|\tilde{\boldsymbol{y}}_{k-1}(t+1)\right\|^2$.

Similarly, the tracking error threshold is still considered, that is $\sup_k \max_{t\in\{0,1,\ldots N\}} \left|\tilde{y}_{i,k}(t)\right| \le \vartheta$, $i = 1, 2, \ldots, m$.

In summary, the proposed ET-DDILC method for MIMO systems is given as follows:

$$k_l = k,\ l \in \{0, 1, \ldots\},\ \text{if } \left|\tilde{y}_{i,k}(t)\right| > \vartheta \text{ or } \chi\left(\boldsymbol{e}_{k-1}(t+1)\right) \ge 0 \tag{12.41}$$

$$\boldsymbol{u}_k(t) = \begin{cases}\boldsymbol{u}_{k_{l-1}}(t) + \hat{\boldsymbol{P}}_k(t)\tilde{\boldsymbol{y}}_{k_{l-1}}(t+1), & k = k_l \\ \boldsymbol{u}_{k_{l-1}}(t), & k \in (k_{l-1}, k_l)\end{cases} \tag{12.42}$$

$$\hat{\boldsymbol{\Theta}}_k(t) = \hat{\boldsymbol{\Theta}}_{k-1}(t) + \frac{\eta_m\left(\Delta\boldsymbol{y}_{k-1}(t+1) - \hat{\boldsymbol{\Theta}}_{k-1}(t)\Delta\boldsymbol{u}_{k-1}(t)\right)\Delta\boldsymbol{u}_{k-1}^{\mathrm{T}}(t)}{\mu_m + \left\|\Delta\boldsymbol{u}_{k-1}(t)\right\|^2} \tag{12.43}$$

$$\hat{\theta}_{ji,k}(t) = \hat{\theta}_{ji,0}(t),\ \text{if } \operatorname{sgn}\left(\hat{\theta}_{ji,k}(t)\right) \ne \operatorname{sgn}\left(\hat{\theta}_{ji,0}(t)\right),\ j, i = 1, 2, \ldots, m \tag{12.44}$$

where (12.44) is a reset algorithm added to make the estimate algorithm (12.43) more flexible.

### 12.5.3 Convergence Analysis

**Lemma 12.1** (Meng and Moore 2016) *Given the system*

$$\boldsymbol{x}_{k+1}(t) = \boldsymbol{A}_k(t)\boldsymbol{x}_k(t) + \boldsymbol{\sigma}_k(t) \tag{12.45}$$

*where* $\boldsymbol{x}_k(t) \in R^P$ *is the state,* $\boldsymbol{\sigma}_k(t) \in R^p$ *is the external input, and* $\boldsymbol{A}_k(t) \in R^{p\times p}$ *is a bounded mapping matrix, i.e.,* $\sup_{t\in\{0,1,\ldots,N\},k\in Z^+} \|\boldsymbol{A}_k(t)\| \leq \beta_{\boldsymbol{A}}$ *with* $\beta_{\boldsymbol{A}} > 0$ *being a positive constant. If for any given integer* $t \in \{0, 1, \ldots, N\}$*, there exists some iteration sequence* $\{\xi_q(t) : q \in Z^+\}$*, where* $\{\xi_0(t) = 0\}$ *and* $0 < \xi_{q+1}(t) - \xi_q(t) \leq \delta(t)$ *for some finite positive integer* $1 \leq \delta(t) < \infty$*, such that*

$$\left\| \prod_{j=\xi_q(t)}^{\xi_{q+1}(t)-1} \boldsymbol{A}_j(t) \right\| \leq \alpha < 1 \ for\ all\ q \in Z^+ \tag{12.46}$$

*then for any initial state* $\boldsymbol{x}_{k_0}(t)$ $(k_0 \in Z^+)$ *and any bounded input* $\boldsymbol{\sigma}_k(t)$*, the solution to (12.45) exists for all* $k \geq k_0$ *and satisfies*

$$\begin{aligned}\|\boldsymbol{x}_k(t)\| &\leq \alpha^{\bar{q}_k(t)-q_{k_0}(t)}\beta_{\boldsymbol{A}}^{2(\delta(t)-1)}\|\boldsymbol{x}_{k_0}(t)\| \\ &+ \left[\alpha^{\bar{q}_k(t)-q_{k_0}(t)}\beta_{\boldsymbol{A}}^{\delta(t)-1}\frac{1-\beta_{\boldsymbol{A}}^{\delta(t)-1}}{1-\beta_{\boldsymbol{A}}} + \frac{1-\beta_{\boldsymbol{A}}^{\delta(t)-1}}{1-\beta_{\boldsymbol{A}}}\right. \\ &\left. + \frac{1-\alpha^{\bar{q}_k(t)-k_0(t)}}{1-\alpha}\beta_{\boldsymbol{A}}^{\delta(t)-1}\frac{1-\beta_{\boldsymbol{A}}^{\delta(t)}}{1-\beta_{\boldsymbol{A}}}\right]\max_{k_0\leq i\leq k-1}\|\boldsymbol{\sigma}_i(t)\|\end{aligned} \tag{12.47}$$

*where* $\bar{q}_k(t) \in Z^+$ *and* $q_k(t) \in Z^+$*,* $k \in Z^+$*, denote the greatest and least integers such that* $\xi_{\bar{q}_k(t)} \leq k < \xi_{\bar{q}_k(t)+1}(t)$ *and* $\xi_{q_k(t)-1}(t) < k \leq \xi_{q_k(t)}(t)$ *hold, respectively.*

The following is the convergence theorem of the proposed ET-DDILC for MIMO systems.

**Theorem 12.3** *For a class of MIMO nonlinear nonaffine discrete-time systems (12.26), if Assumptions 12.1 and 12.2 hold, and*

$$\left\| \prod_{k_l=\xi_q(t)}^{\xi_{q+1}(t)-1} \boldsymbol{Q}_{k_l}(t) \right\| \leq \gamma < 1,\ \forall q \in Z^+ \tag{12.48}$$

*hold for some iteration* $\{\xi_q(t) : q \in Z^+\}$, *where* $0 < \xi_{q+1}(t) - \xi_q(t) \le \delta(t)$, $1 \le \delta(t) < \infty$, $\{\xi_0(t) = 0\}$, $\boldsymbol{Q}_{k_l}(t) = \boldsymbol{I} - \boldsymbol{\Theta}_{k_l}(t)\hat{\boldsymbol{P}}_{k_l}(t)$, *then the proposed ET-DDILC approach (12.41)–(12.44) guarantees:*

*(i) the parameter estimation* $\hat{\boldsymbol{\Theta}}_k(t)$ *is bounded over the finite time interval* $t \in \{0, 1, \ldots, N\}$, $\forall k$;

*(ii) the tracking error* $\tilde{\boldsymbol{y}}_k(t)$ *is convergent over the finite time interval* $t \in \{0, 1, \ldots, N\}$ *as k approaches to infinity.*

***Proof*** *Part 1. The boundedness of* $\hat{\boldsymbol{\Theta}}_k(t)$

Define a parameter estimation error $\tilde{\boldsymbol{\Theta}}_k(t) = \boldsymbol{\Theta}_k(t) - \hat{\boldsymbol{\Theta}}_k(t)$. Following the similar step of (2.23), we can get

$$\left\|\tilde{\boldsymbol{\Theta}}_k(t)\right\| \le \left\|\tilde{\boldsymbol{\Theta}}_{k-1}(t)\left(\boldsymbol{I} - \frac{\eta_m \Delta \boldsymbol{u}_{k-1}(t)\Delta \boldsymbol{u}_{k-1}^T(t)}{\mu_m + \left\|\Delta \boldsymbol{u}_{k-1}(t)\right\|^2}\right)\right\| + 2b_\Theta. \tag{12.49}$$

Squaring the first term on the right-hand side of (12.49), one obtains

$$\begin{aligned}\left\|\tilde{\boldsymbol{\Theta}}_{k-1}(t)\left(\boldsymbol{I} - \frac{\eta_m \Delta \boldsymbol{u}_{k_{l-1}}(t)\Delta \boldsymbol{u}_{k-1}^{\mathrm{T}}(t)}{\mu_m + \left\|\Delta \boldsymbol{u}_{k-1}(t)\right\|^2}\right)\right\|^2 &= \left\|\tilde{\boldsymbol{\Theta}}_{k-1}(t)\right\|^2 \\ &+ \left(-2 + \frac{\eta_m \|\Delta \boldsymbol{u}_{k-1}(t)\|^2}{\mu_m + \|\Delta \boldsymbol{u}_{k-1}(t)\|^2}\right)\frac{\eta_m \left\|\Delta \boldsymbol{u}_{k-1}(t)\tilde{\boldsymbol{\Theta}}_{k-1}(t)\right\|^2}{\mu_m + \|\Delta \boldsymbol{u}_{k-1}(t)\|^2}.\end{aligned} \tag{12.50}$$

Since $0 < \eta_m < 2$, and $\mu_m > 0$, then

$$-2 + \frac{\eta_m \|\Delta \boldsymbol{u}_{k-1}(t)\|^2}{\mu_m + \|\Delta \boldsymbol{u}_{k-1}(t)\|^2} < 0. \tag{12.51}$$

From (12.50) and (12.51), it obtains

$$\left\|\tilde{\boldsymbol{\Theta}}_{k-1}(t)\left(\boldsymbol{I} - \frac{\eta_m \Delta \boldsymbol{u}_{k-1}(t)\Delta \boldsymbol{u}_{k-1}^{\mathrm{T}}(t)}{\mu_m + \left\|\Delta \boldsymbol{u}_{k-1}(t)\right\|^2}\right)\right\|^2 < \left\|\tilde{\boldsymbol{\Theta}}_{k-1}(t)\right\|^2. \tag{12.52}$$

So, there must exist a constant $0 < d_2 < 1$, such that

$$\left\|\tilde{\boldsymbol{\Theta}}_k(t)\right\| \le d_2\left\|\tilde{\boldsymbol{\Theta}}_{k-1}(t)\right\| + 2b_\Theta \le \cdots \le d_2^k\left\|\tilde{\boldsymbol{\Theta}}_0(t)\right\| + \frac{2b_\Theta}{1 - d_2}. \tag{12.53}$$

Hence, $\tilde{\boldsymbol{\Theta}}_k(t)$ is bounded because $\hat{\boldsymbol{\Theta}}_0(t)$ is chosen bounded and $\boldsymbol{\Theta}_k(t)$ is also bounded according to Theorem 12.2. Therefore, the boundedness of $\hat{\boldsymbol{\Theta}}_k(t)$ can be guaranteed $\forall t \in \{0, \ldots, N\}$, and $\forall k$.

*Part 2. The convergence of tracking error* $\tilde{\boldsymbol{y}}_k(t)$

We have known $\hat{\boldsymbol{\Theta}}_k(t)$ and $\boldsymbol{\Theta}_k(t)$ are bounded, the boundedness of $\boldsymbol{I} - \boldsymbol{\Theta}_{k_l}(t)\hat{\boldsymbol{P}}_{k_l}(t)$ can be easy to obtain.

Let $\boldsymbol{Q}_{k_l}(t) = \boldsymbol{I} - \boldsymbol{\Theta}_{k_l}(t)\hat{\boldsymbol{P}}_{k_l}(t)$. For $k = k_l$, one can rewrite Eq. (12.35) as

$$\tilde{\boldsymbol{y}}_{k_l}(t+1) = \boldsymbol{Q}_{k_l}(t)\tilde{\boldsymbol{y}}_{k_{l-1}}(t+1). \tag{12.54}$$

By comparing equations (12.54) and (12.45), we can find that $\boldsymbol{x}_{k+1}(t)$ and $\boldsymbol{A}_k(t)$ in Eq. (12.45) correspond to $\tilde{\boldsymbol{y}}_{k_l}(t+1)$ and $\boldsymbol{Q}_{k_l}(t)$ in Eq. (12.54), respectively, and $\boldsymbol{\sigma}_k(t)$ in Eq. (12.45) corresponds to the term $\boldsymbol{0}$ in Eq. (12.54). According to Lemma 12.1 and Lemma 7.1, it is easy to get $\lim_{k_l\to\infty} \tilde{\boldsymbol{y}}_{k_l}(t+1) = 0$ if (12.48) is satisfied, i.e., the tracking error in the triggered iteration converges to zero with the increase of iteration under the condition of (12.48).

For the inter-event interval, i.e. $k \in (k_{l-1}, k_l)$, the tracking error will be maintained because the control input remains the same as the input at the latest triggered iteration.

Therefore, the tracking error $\tilde{\boldsymbol{y}}_k(t+1)$ converges gradually with the increase of iteration. □

## 12.6 Simulation Study

**Example 12.1** *Consider a two-input and two-output nonlinear discrete-time system*

$$\begin{cases} y_{1,k}(t+1) = \frac{2.5y_{1,k}(t)}{1+y_{1,k}^2(t)+y_{2,k}^2(t)} + 1.2u_{1,k}(t) + 0.5u_{2,k}(t) \\ y_{2,k}(t+1) = \frac{5y_{2,k}(t)}{1+y_{1,k}^2(t)+y_{2,k}^2(t)} + 0.5u_{1,k}(t) + u_{2,k}(t). \end{cases}$$

*The desired tracking trajectory is*

$$\begin{cases} y_{d,1}(t) = \cos(t/20) + \sin(t/50) \\ y_{d,2}(t) = \sin(t/50) + 5\cos(t/20) \end{cases}, t \in \{0, 1, \ldots, 500\}.$$

In this simulation, the tracking error threshold is selected as $\vartheta = 0.1$. Select parameters as $\hat{\boldsymbol{\Theta}}_0(t) = \begin{bmatrix} 2 & 1 \\ 1 & 2 \end{bmatrix}$, $\eta_m = 1.8$, $\mu_m = 2$, $\rho_m = 1.6$, and $\lambda_m = 0.1$. Applying the proposed ET-DDILC method (12.41)–(12.44), the simulation results are shown in Figs. 12.1, 12.2, 12.3, 12.4, 12.5.

Figures 12.1 and 12.2 demonstrate the output tracking performance, where the vertical axis denotes the system output and the horizontal axis denotes the time instant. Figure 12.3 shows the total triggered instances of every iteration, where the

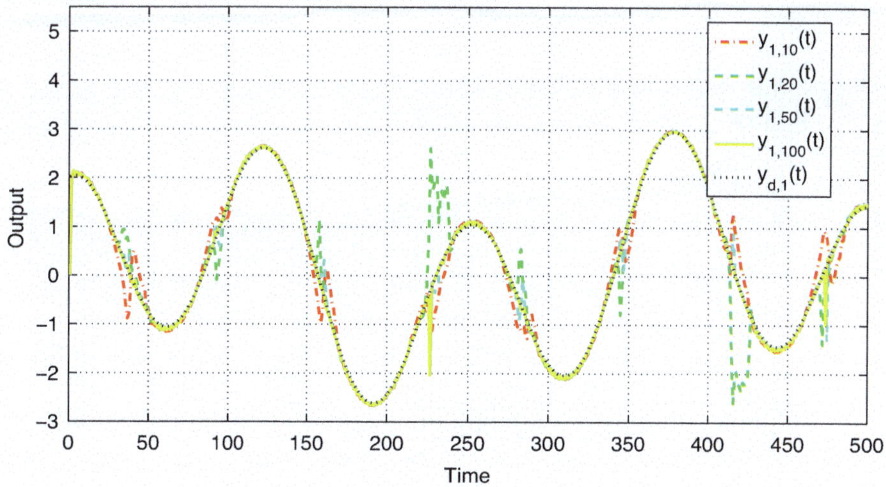


**Fig. 12.1** Output tracking performance of $y_{1,k}(t)$

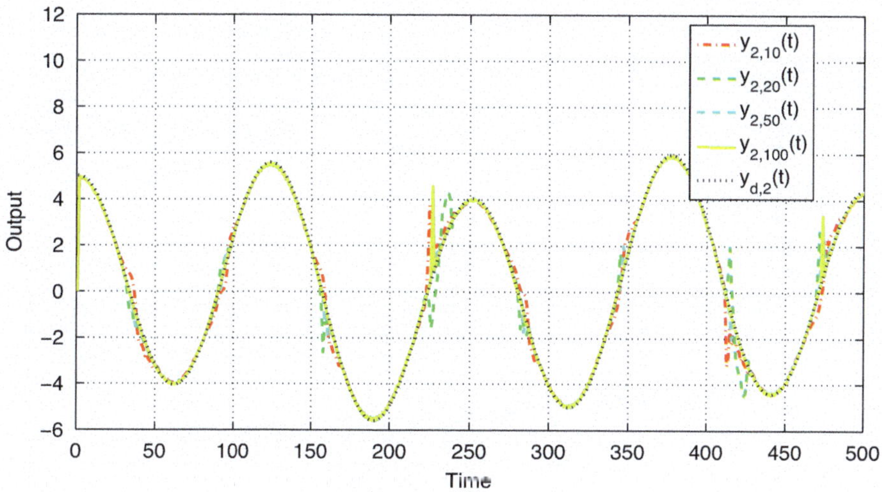


**Fig. 12.2** Output tracking performance of $y_{2,k}(t)$

Y-axis denotes the event-triggered number of the output tracking error and the X-axis denotes the iteration numbers. Figures 12.4 and 12.5 illustrate the convergence of tracking error where the Y-axis is the average tracking error, defined as $y_{i,k}^{e} = \sum_{j=1}^{N} \tilde{y}_{i,k}(j) \Big/ N$, $i = 1, 2$, and the X-axis denotes the iteration numbers.

It is seen that, as the iteration number increases, the total triggered instances in an iteration decrease, which verifies that the proposed ET-DDILC method (12.41)–

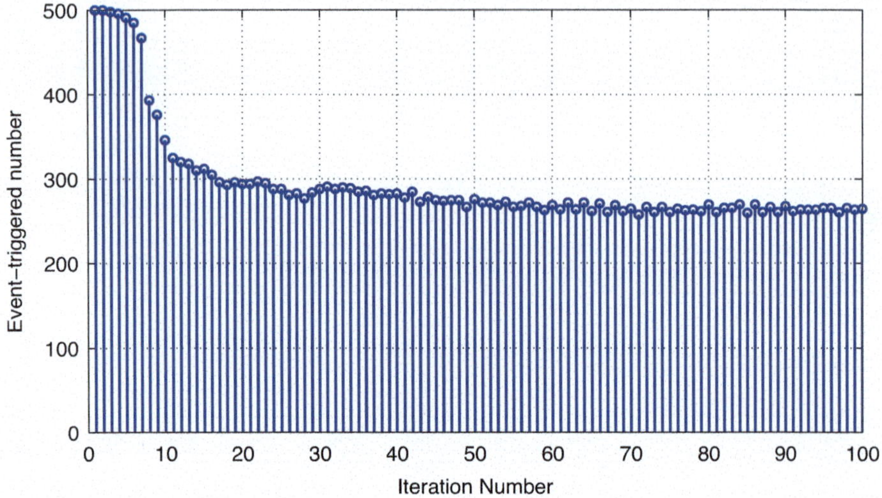


**Fig. 12.3** Total number of event-triggered moments per iteration

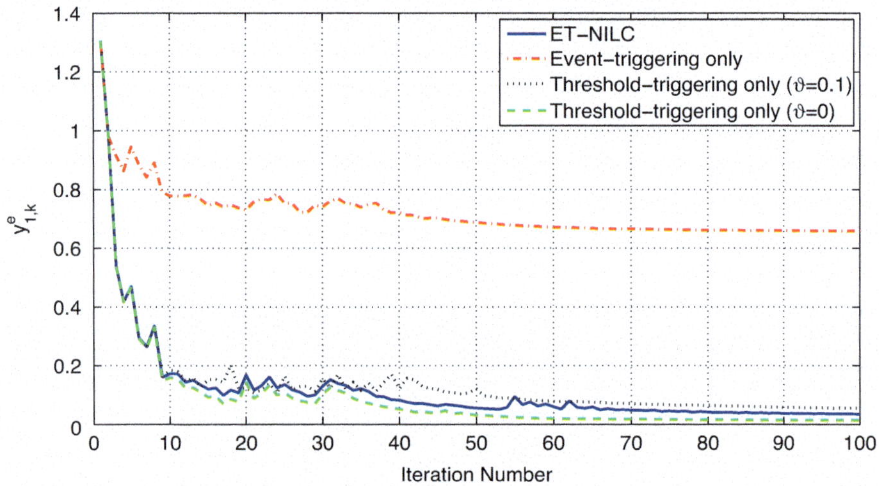


**Fig. 12.4** Absolute mean tracking error $y^e_{1,k}$

(12.44) can decrease the number of the control activation while ensuring a good tracking performance.

By nature, the proposed ET-DDILC includes both an event-triggering condition derived according to Lyapunov stability and a threshold condition together. The former is used to reduce controller triggering times and ensure system stability. The latter is used to reduce the controller triggering times while making the tracking error converge to within a small bound.

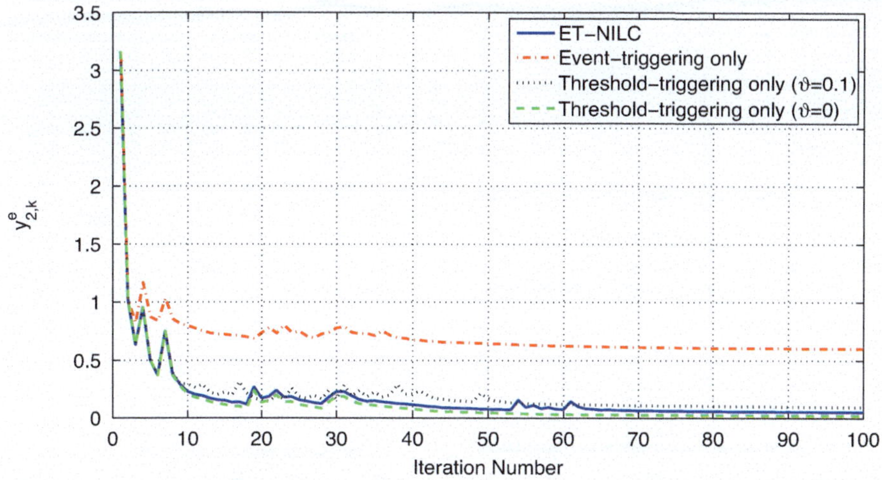


**Fig. 12.5** Absolute mean tracking error $y^e_{2,k}$

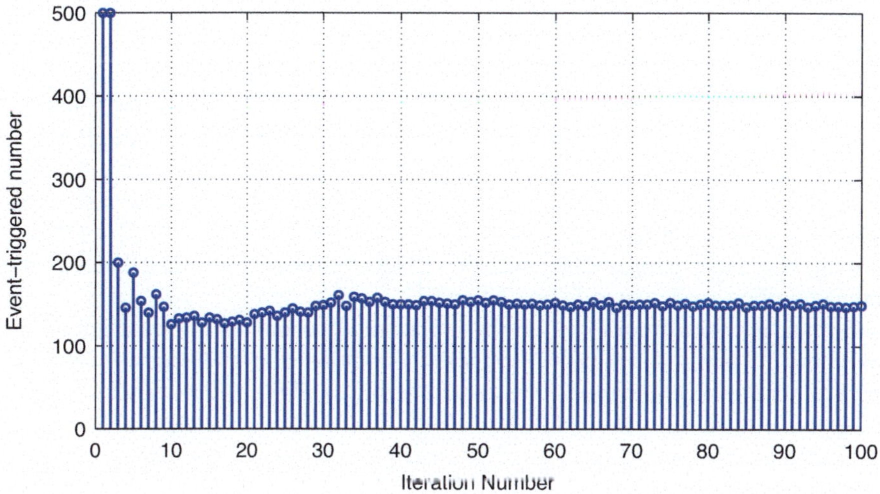


**Fig. 12.6** Total number of event-triggered moments per iteration with the event-triggering condition (12.40) only

In order to clearly show the roles of the two conditions, applying the event-triggering condition (12.40) solely under the same simulation background, the results are shown in Figs. 12.4, 12.5, 12.6, and 12.7. The dash-dotted lines in Figs. 12.4 and 12.5 show the convergence of the tracking error. Figure 12.6 shows the total triggered time instances of every iteration. Figure 12.7 shows the event-triggering cases along iteration at the time points of 49, 93, 352, and 484 by using condition (12.40) only

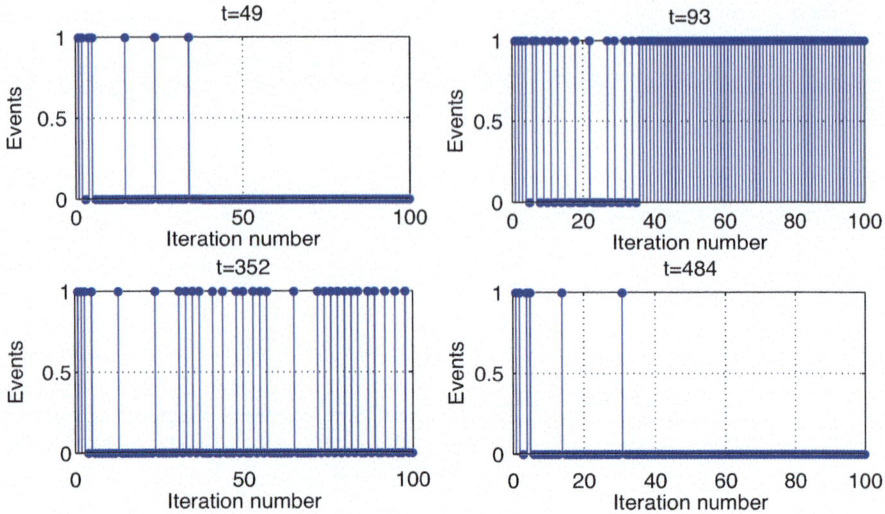


**Fig. 12.7** Event-triggering cases at the time points of 49, 93, 352, and 484 by using condition (12.40) only

where the vertical axis represents whether the controller is triggered or not, and "1" means the controller is triggered and "0" indicates no update for the controller.

Compared with Figs. 12.3 and 12.6, one can see that the total triggered times of the ET-DDILC with the single event-triggering condition (12.40) are less than that with both conditions together. While the convergence performance of the ET-DDILC with the single event-triggering condition (12.40) becomes worse than that with the both conditions, as shown in Figs. 12.4 and 12.5.

Further, one sees from Fig. 12.6 that a flat trend occurs after several iterations, but it does not imply that the event-triggering condition is invalid. One can see from Fig. 12.7 that for a certain fixed time, the triggering condition (12.40) always works.

Applying the threshold condition solely under the same simulation background, the results are shown in Figs. 12.4, 12.5, and 12.8. Similarly, one can see that the total triggered times of the ET-DDILC with the single threshold condition are less than that with the both conditions together. But the convergence of the ET-DDILC with the single threshold condition becomes worse than that with the both conditions. Furthermore, if we set the threshold value as zero, the ET-DDILC becomes all-time-iteration-triggered one, and a good control performance can be achieved, but the cost is that no controller actions are reduced, as shown in Figs. 12.4, 12.5, and 12.9 under the same simulation background.

In summary, the proposed ET-DDILC with both conditions cannot only reduce the controller triggering times, but also ensure the tracking error converges to an arbitrarily predesigned small bound.

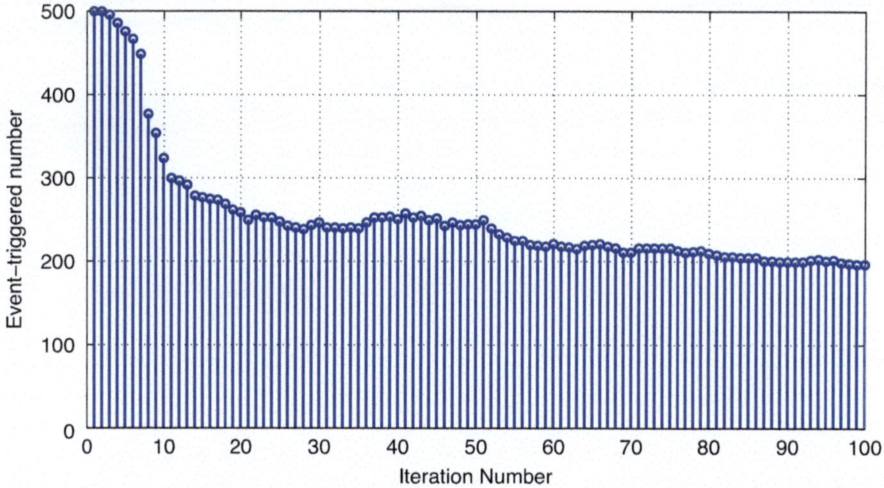


**Fig. 12.8** Total number of event-triggered moments per iteration with threshold-triggering condition only ($\vartheta = 0.1$)

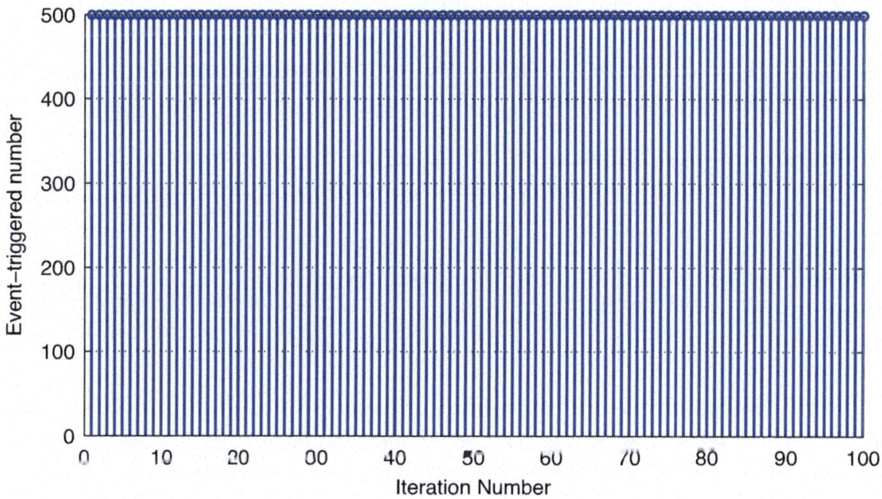


**Fig. 12.9** Total number of event-triggered moments per iteration with the threshold-triggering condition only ($\vartheta = 0$)

**Fig. 12.10** Heat exchanger

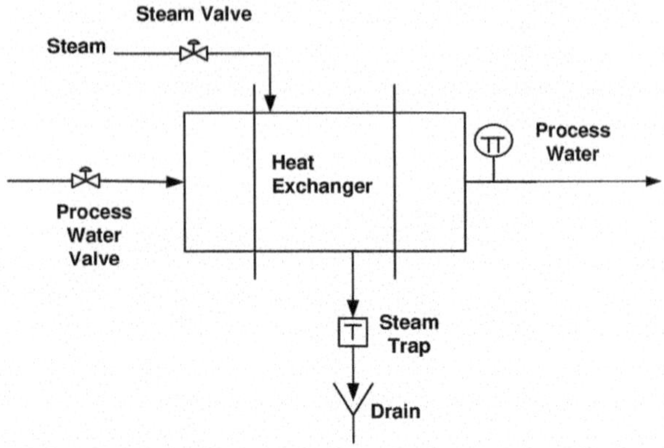


**Example 12.2** *Consider a steam-water heat exchanger (Eskinat et al. 1991; Yamamoto et al. 2008),*

$$
\begin{aligned}
&x_k(t) = 1.5u_k(t) - 1.5u_k(t)^2 + 0.5u_k(t)^3 \\
&y_k(t+1) = 0.6y_k(t) - 0.1y_k(t-1) + 1.2x_k(t) - 0.1x_k(t-1) + w_k(t)
\end{aligned}
$$

*where* $u_k(t)$ *represents the process water flow rate;* $y_k(t)$ *represents the process water exit temperature;* $w_k(t) = \frac{0.01\text{randn}}{k}$ *is a disturbance varying with time and iteration.*

The experimental equipment of the steam-water heat exchanger is shown in Fig. 12.10. The control objective is to track the following desired output:

$$
y_d(t) = \begin{cases} 0.5, \ t \in [0, 50) \\ 1, \ t \in [50, 100) \\ 2, \ t \in [100, 150) \\ 1.5, \ t \in [150, 200]. \end{cases}
$$

In this simulation, the tracking error threshold is selected as $\vartheta = 0.001$. The parameters are selected as $\hat{\theta}_0(t) = 1.5$, $\eta_s = 0.01$, $\mu_s = 1$, $\rho_s = 0.6$, and $\lambda_s = 0.6$. Applying the proposed ET-DDILC method (12.17)–(12.20), the simulation results are shown in Figs. 12.11 and 12.12. It is seen that the tracking performance under disturbance can also be guaranteed although not all time the control is triggered. Further, the total number of triggers per iteration has a decreasing trend with the increase of iterations.

As a summary, the simulation results illustrate that the presented ET-DDILC method can reduce the control activations while ensuring tracking performance, so as to achieve the purpose of reducing control cost.

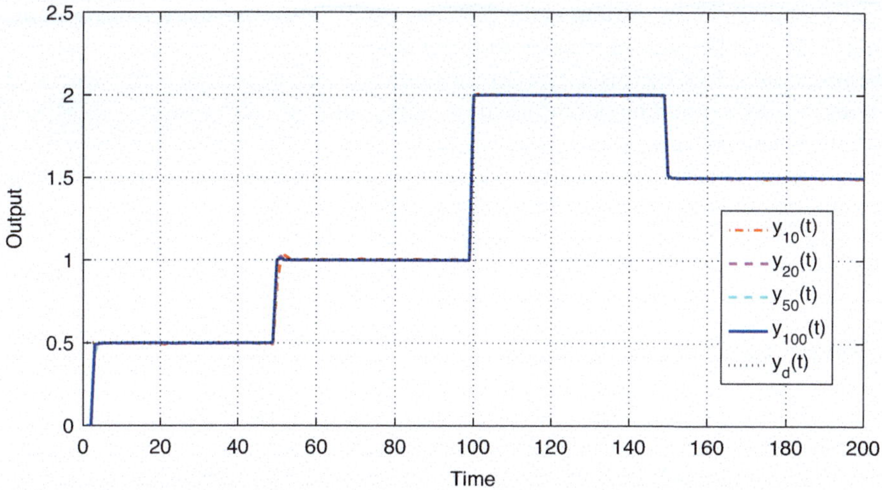


**Fig. 12.11** Output tracking performance

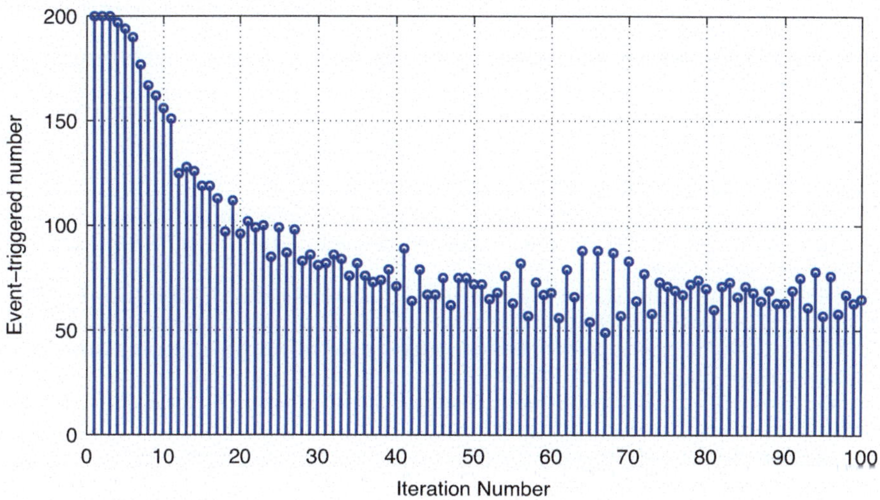


**Fig. 12.12** Total number of event-triggered moments per iteration

## 12.7 Summary

An ET-DDILC method is presented in this chapter for nonlinear nonaffine systems via a virtual linear data model. The 2D dynamic behavior along both time and iteration directions is considered when designing event-triggering conditions. The ET-DDILC is updated iteratively when the triggering condition is satisfied; otherwise, the control input remains the same as that in the previous batch, where the learning gain of the control law can be calculated using I/O data in real time to enhance the robustness. The proposed ET-DDILC is a data-driven method using the I/O data only instead of relying on any explicit model information. The results are also extended to MIMO systems where the convergence analysis is provided by introducing the property of input-to-state stability for discrete-time systems. The effectiveness of the presented methods is verified by two simulation examples.

## References

Bu X, Yu F, Hou Z, Wang F (2013) Iterative learning control for a class of nonlinear systems with random packet losses. Nonlinear Anal Real World Appl 14(1):567–580

Bu X, Zhang H, Hou Z, Yang J (2017) Robust iterative learning control design for linear systems with time-varying delays and packet dropouts. Adv Differ Eqs 1:1–17

Ding L, Han Q, Ge X, Zhang X (2017) An overview of recent advances in event-triggered consensus of multiagent systems. IEEE Trans Cybern 48(4):1110–1123

Eskinat E, Johnson SH, Luyben WL (1991) Use of Hammerstein models in identification of non-linear systems. AIChE J 37(2):255–268

Hou Z, Xiong S (2019) On model-free adaptive control and its stability analysis. IEEE Trans Autom Control 64(11):4555–4569

Ibrir S, Su CY (2010) Simultaneous state and dead-zone parameter estimation for a class of bounded-state nonlinear systems. IEEE Trans Control Syst Technol 19(4):911–919

Kishida M (2018) Event-triggered control with self-triggered sampling for discrete-time uncertain systems. IEEE Trans Autom Control 64(3):1273–1279

Lai G, Tao G, Zhang Y, Liu Z (2019) Adaptive control of noncanonical neural-network nonlinear systems with unknown input dead-zone characteristics. IEEE Trans Neural Netw Learn Syst 31(9):3346–3360

Lin N, Chi R, Huang B (2019) Event-triggered model-free adaptive control. IEEE Trans Syst Man Cybern: Syst 51(6):3358–3369

Liu D, Yang G (2018) Neural network-based event-triggered MFAC for nonlinear discrete-time processes. Neurocomputing 272:356–364

Meng D, Moore KL (2016) Robust iterative learning control for nonrepetitive uncertain systems. IEEE Trans Autom Control 62(2):907–913

Meng D, Moore KL (2017) Robust cooperative learning control for directed networks with nonlinear dynamics. Automatica 75:172–181

Pan Y, Marquez HJ, Chen T, Sheng L (2009) Effects of network communications on a class of learning controlled non-linear systems. Int J Syst Sci 40(7):757–767

Peng C, Li F (2018) A survey on recent advances in event-triggered communication and control. Inf Sci 457:113–125

Shen D, Zhang C, Xu Y (2017) Two updating schemes of iterative learning control for networked control systems with random data dropouts. Inf Sci 381:352–370

Tallapragada P, Chopra N (2013) On event triggered tracking for nonlinear systems. IEEE Trans Autom Control 58(9):2343–2348

Tang J, Sheng L (2018) Iterative learning fault-tolerant control for networked batch processes with event-triggered transmission strategy and data dropouts. Syst Sci Control Eng 6(3):44–53
Xiong W, Yu X, Patel R, Yu W (2016) Iterative learning control for discrete-time systems with event-triggered transmission strategy and quantization. Automatica 72:84–91
Yamamoto T, Takao K, Yamada T (2008) Design of a data-driven PID controller. IEEE Trans Control Syst Technol 17(1):29–39
Zhang T, Li J (2018) Event-triggered iterative learning control for multi-agent systems with quantization. Asian J Control 20(3):1088–1101

The manufacturer's authorised representative in the EU is Springer Nature Customer Service Centre GmbH, Europaplatz 3, 69115 Heidelberg, Germany. If you have any concerns regarding our products, please contact ProductSafety@springernature.com

Printed and bound by CPI Group (UK) Ltd, Croydon, CR0 4YY
30/09/2026
02225315-0004